M000311414

Conservation of Wildlife Populations

Dedication

For my father, who instilled in me a profound respect and appreciation for the natural world.

Conservation of Wildlife Populations

Demography, Genetics, and Management

L. Scott Mills

Blackwell
Publishing

© 2007 by L. Scott Mills

BLACKWELL PUBLISHING
350 Main Street, Malden, MA 02148-5020, USA
9600 Garsington Road, Oxford OX4 2DQ, UK
550 Swanston Street, Carlton, Victoria 3053, Australia

The right of L. Scott Mills to be identified as the Author of this Work has been asserted in accordance with the
UK Copyright, Designs, and Patents Act 1988.

All rights reserved. No part of this publication may be reproduced, stored in a retrieval system, or transmitted,
in any form or by any means, electronic, mechanical, photocopying, recording or otherwise, except as permitted by the UK Copyright, Designs, and Patents Act 1988, without the prior permission of the publisher.

First published 2007 by Blackwell Publishing Ltd

5   2009

Library of Congress Cataloging-in-Publication Data

Mills, L. Scott.
  Conservation of wildlife populations : demography, genetics, and management / by L. Scott Mills.
    p. cm.
  Includes bibliographical references (p.) and index.
  ISBN 978-1-4051-2146-0 (pbk. : alk. paper)
  1. Wildlife management. 2. Animal populations. I. Title.

SK355.M55 2006
639.9–dc22
           2006009953

A catalogue record for this title is available from the British Library.

Set in 10.5 on 13pt Minion
by SNP Best-set Typesetter Ltd, Hong Kong
Printed and bound in Singapore
by C.O.S. Printers Pte Ltd

The publisher's policy is to use permanent paper from mills that operate a sustainable forestry policy, and
which has been manufactured from pulp processed using acid-free and elementary chlorine-free practices.
Furthermore, the publisher ensures that the text paper and cover board used have met acceptable
environmental accreditation standards.

For further information on
Blackwell Publishing, visit our website:
www.blackwellpublishing.com

Contents

List of boxes

List of symbols

List of symbols commonly used in text, and the chapter where they are primarily discussed.

Symbol	Meaning	Chapter
α	Probability of Type I error (falsely concluding a difference) in a frequentist statistical test	2
β	Probability of Type II error (falsely concluding no difference) in a frequentist statistical test	2
dN/dt	Instantaneous change in population size over tiny time interval; also yield or recruitment in harvest models.	5, 6, 14
ϕ	Apparent survival (does not separate death from probability of permanent emigration)	4, 10
F_{ST}	Fixation index, or reduction in heterozygosity due to genetic drift (typically equivalent to "inbreeding coefficient" in wild populations)	9, 10
K	Carrying capacity	6, 14
λ	Geometric growth rate or population multiplication rate	5
$\mathbf{M}$	Projection matrix of survival and reproductive rates	7
$\hat{N}$	Estimated abundance.	4
$\mathbf{n}$	Vector of number of individuals by age or stage class	7
N_e	Genetic effective population size	9, 14
N_t	Abundance at some time t	5
$\hat{p}$	Estimated probability of detection (or capture) of an animal	4
Ψ	Probability of moving from one population to another	10
r	Exponential growth rate *or* instantaneous growth rate per capita or intrinsic growth rate	5, 6
$\hat{r}$ or $\hat{\mu}$	Estimate, from real data, of r	5
S	Survival (if estimated from data: $\hat{S}$)	4, 10
SSD	Stable stage distribution	7
Var (or σ^2)	Variance	2, 4, 5

Preface

Population biology spans the wide and fascinating world of population ecology, demography, and population genetics. This book extracts from these fields the most relevant concepts and principles for solving real-world management problems in wildlife and conservation biology. It will build on your training in basic ecology and genetics, then move deeper into areas where ecological and genetic concepts and theory are applied. Because this is an applied book on population biology, it will not derive or prove theoretical premises, nor will it dwell on theory not directly applicable to management problems. There are already excellent books that cover basic ecology, theoretical and mathematical population biology, and conservation genetics. There are also excellent books describing case studies of wildlife management and conservation practice. With this book, I hope to fill in the space between these texts, providing a conceptual framework spanning from fieldwork to demographic models and genetic analysis as they inform applied decision-making.

The book is organized into three sections. The first provides a background to the science of applied wildlife population biology. Here I cover the context of historical and current extinction rates, the dynamics of human population growth, an overview of study design and ethics, essential background on genetics necessary for understanding the interface between genetic and demographic approaches, and the estimation of within-population vital rates. The second section covers population processes that form the basis for applied management. Beginning with exponential and then density-dependent population growth, I will next cover stage-structured population dynamics, predation (a necessary background for understanding the impacts of harvest by humans), effects of genetic variation on population dynamics, and animal spacing within and among populations. The final section brings together concepts and principles from the first two sections. The emphasis here is less on introducing new conceptual material and more on synthesizing the previous chapters by applying the ideas to specific problems of declining, small, or harvestable populations. Chapters deal with deterministic factors leading to population decline, specific issues related to small and declining populations, the use of focal species to bridge population biology and ecosystem approaches, and harvest theory and practice.

Although I do review fundamental concepts and relevant theory, I assume that readers will have taken the equivalent of a basic ecology class and (hopefully) a basic genetics course. Ideally, you will have also taken at least one semester of statistics and calculus, although I will review the math and keep it honed to that which is critical to application.

Throughout I will present rules of thumb. These guidelines represent simple answers to complex questions, always a dangerous undertaking. I hope they will be useful for distilling subtle and complicated topics into the bare essence that can inform management as a starting point. Of course there will always be complexities, and in many specific cases the rules of thumb will, ultimately, be wrong. I will explain the primary caveats that accompany the rules of thumb, and in all cases will give references that explore the intricacies in more detail.

Finally, you'll notice two major themes running throughout this book: in class, I call these bumper stickers. One of them is embrace uncertainty. Do not be intimidated by the fact that ecological processes are complex. Do not feel that wide variance in estimates of parameters come from weak science. Do not freeze up from a lack of knowledge about all the pieces that are necessary for understanding a problem. Rather, recognize and illuminate the complexity of ecological systems as you deal with applied questions. The lack of full scientific certainty should never be used as a reason for inaction in the face of a wildlife population problem. Embrace uncertainty.

My other bumper sticker is that ecological processes are not democratic. All vital rates are not equal in their effect on population growth, all age classes are not equal in their importance to population dynamics, all individuals do not contribute equally to genetic composition of a population, all species are not equal in their effects on community structure and stability, and so on. Because ecological processes are not democratic, we can rank and act on both research priorities and threats to wildlife populations.

L.S. Mills

Acknowledgments

A book like this can only be written on the backs of hundreds of researchers who did the hard work that I touch on very briefly. Writing the book was also nurtured by a wonderful work environment, and for that I thank the Wildlife Biology Program at the University of Montana. I began this book while on sabbatical at Virginia Tech University, and I appreciate the support provided by the Department of Fisheries and Wildlife Sciences during that critical first year.

I am deeply indebted to the many colleagues who commented on one or more chapters in this book, including Fred Allendorf, Larissa Bailey, Joe Ball, Roman Biek, Juan Bouzat, Ellen Cheng, John Citta, Kevin Crooks, Patrick Devers, Dan Doak, Dave Freddy, Eric Hallerman, Mark Hebblewhite, Steve Hoekman, Jean-Yves Humbert, Heather Johnson, Matt Kauffman, Bill Laurance, Jay McGee, Peter Laver, Winsor Lowe, Gordon Luikart, Shawn Meagher, Tammy Mildenstein, Barry Noon, Dan Pletscher, Charles Romesburg, Jon Runge, Mike Schwartz, Bob Steidl, Dalit Ucitel, Rebecca Wahl, and Mike Wisdom. Several heroic people read all or most of the book, sometimes more than once and usually coordinating reviews by their students; for their detailed and insightful feedback I am particularly grateful: Barry Brook, Elizabeth Crone, Curt Griffin, Rich Harris, Karen Hodges, Marcella Kelly, David Tallmon, and John Vucetich. Although they are too many to name, I thank the cohorts of undergraduate and graduate students at Virginia Tech University, University of Montana, and University of British Columbia Okanagan who reviewed various drafts: their detailed, constructive, frank, and extraordinarily helpful comments demonstrated that the often-maligned "students of today" burn with commitment, enthusiasm, integrity, and intelligence.

I also thank Blackwell Publishing – especially Rosie Hayden – for helping me throughout this project. Copy-editor Nik Prowse read the manuscript with sharp eyes that caught several embarrassing typos, and Janey Fisher expertly managed the project. My student helpers – Anna Semple, Aira Kidder, and Mike McDonald – helped keep things organized, as did Administrative Assistant Jeanne Franz.

Finally, I thank my family, both here in Montana and elsewhere, for helping in so many ways to bring this book to fruition. I stand in special gratitude before my wife Lisa and children Nicholas and Linnea, whose happiness and awe of nature serve as constant reminders that science has special meaning when it is used for good in this world.

I

Background to applied population biology

The big picture: human population dynamics meets applied population biology

The metabolic rate of history is too fast for us to observe it. It's as if, attending to the day-long life cycle of a single mayfly, we lose sight of the species and its fate. At the same time, the metabolic rate of geology is too slow for us to perceive it, so that, from birth to death, it seems to us who are caught in the beat of our own individual human hearts that everything happening on this planet is what happens to us, personally, privately, secretly. We can stand at night on a high, cold plain and look out toward the scrabbled, snow-covered mountains in the west, the same in a suburb of Denver as outside a village in Baluchistan in Pakistan, and even though beneath our feet continent-sized chunks of earth grind inexorably against one another, go on driving one or the other continent down so as to rise up and over it, as if desiring to replace it on the map, we poke with our tongue for a piece of meat caught between two back teeth and think of sarcastic remarks we should have made to our brother-in-law at dinner.

Russell Banks (1985:36–7), *Continental Drift*

Experience with game has shown, however, that a determination to conserve, even when supported by public sentiment, protective legislation, and a few public reservations or parks, is an insufficient conservation program. Notwithstanding these safeguards, non-game wild life is year by year being decimated in numbers and restricted in distribution by the identical economic trends – such as clean farming, close grazing, and drainage – which are decimating and restricting game. The fact that game is legally shot while other wild life is only illegally shot in no [way] alters the deadly truth of the principle that it cannot nest in a cornstalk.

Aldo Leopold (1933:404), *Game Management*

Introduction

Should Texas panthers be brought in to breed with Florida panthers? What factors are most likely to explain global amphibian declines, and what is the most efficient path to reverse the decline? Do wolves reduce the numbers of caribou available for hunters? What factors affect harvest regulations for waterfowl? Was the introduction of foxes to

Australia likely to have driven native prey species to extinction, and how best to decrease the numbers of the exotic predator? These are just a few samples of the sort of real-world questions that can be informed by knowledge of applied population biology.

To set the stage for this book, consider some of the key words in the title. **Population** has many meanings, but for the purposes of this book the term will be used in a broad sense, referring to a collection of individuals of a species in a defined area; the individuals in a population may or may not breed with other groups of that species in other places. A similar definition was presented by Caughley (1977), who quoted Cole (1957:2) to define a population as "a biological unit at the level of ecological integration where it is meaningful to speak of a birth rate, a death rate, a sex ratio, and an age structure in describing the properties of the unit." The advantage of such vague yet practical definitions is that they allow discussion of both single and multiple populations, with and without gene flow and demographic influence from other populations.

Next, some thoughts on the term **wildlife**. Although about 1.5 million species have been described on Earth, vertebrates comprise only about 3% of the total and terrestrial vertebrates less than 2%. Yet policy, public opinion, and even ecological research still revolve around vertebrates, particularly birds and mammals (Leader-Williams & Dublin 2000, Clark & May 2002). Certainly, harvest management outside of fisheries and forestry deals primarily with terrestrial vertebrates.

Currently the term wildlife means considerably more than merely terrestrial game (harvested) species. Even Aldo Leopold's classic book *Game Management* (Leopold 1933) made clear that harvested species should be considered a narrow segment of "wild life" (two words). Recognition of "The little things that run the world" (Wilson 1987) has emphasized the importance of small creatures – especially insects – to ecosystem structure and function. And of course, Leopold (1953) reminded us more than 50 years ago that "To keep every cog and wheel is the first precaution of intelligent tinkering," an admonition that our focus should be on all the parts. Happily, it seems that now, more than ever, people value the conservation of all species (Czech et al. 1998). Reflecting these philosophies, US federal wildlife law in its broadest sense recognizes all nonhuman and nondomesticated animals (plants occupy a different conceptual status in law; Bean & Rowland 1997). Recent texts with wildlife in the title have considered all free-ranging undomesticated animals, and in some cases plants (e.g. Moulton & Sanderson 1997, Krausman 2002, Bolen & Robinson 2003). This perspective has historical precedent: the first issue of the *Journal of Wildlife Management* (1937) stated that wildlife management actions ". . . along sound biological lines are also part of the greater movement for conservation of our entire native fauna and flora."

This book will embrace a broad view of wildlife, because most concepts in population biology can be applied to all taxa. However, several core ecological, genetic, and life-history phenomena are idiosyncratic to plants, insects, or fish (e.g. seed banks, larval instars, anadromous breeding, self-fertilization, etc.), and so would require detailed treatment to understand population biology in detail for those taxa. For one book to effectively convey applications for species that are – at this point in human civilization – most prominent in the public eye[1], the majority of examples and case studies in this book will focus on the subset of wildlife consisting of amphibians, reptiles, birds, and mammals.

Finally, some thoughts on the word **management**. This term is a pejorative in the minds of some, conjuring up images of manipulation and arrogance. And it is certainly true that, in most cases, humans and human actions are ultimately what is managed, not the animals themselves. For others, the inclusion of management in the same book title with **conservation** is repetitive. Nevertheless, I have included management in the title because it is convenient shorthand for applied outcomes of population biology, ranging from measuring and interpreting trends, to setting harvest limits, to evaluating viability of endangered species, to determining the effects of predation on prey populations.

The overall influence of a species – any species – on its community and ecosystem is a function of its local density, its geographic range, and the per-capita impact of each member of the population. Virtually every problem related to wildlife conservation can be traced at least in part to human population growth – in terms of absolute numbers and distribution – as well as the per-capita impact of humans as strong interactors on the global stage (Channell & Lomolino 2000, Pletscher & Schwartz 2000). In the spirit of acknowledging that managing wildlife populations really comes back to managing anthropogenic factors, the following section considers human population ecology, both emphasizing the role that humans play in affecting other species and conveying several principles we will use throughout the book.

Population ecology of humans

Human population growth

Humans are remarkable because they are one of the few species to show nearly exponential growth (see Chapter 5) for thousands of years, and this growth has resulted in enormous abundances. However, the population growth that humans have shown has not been constant. Consider human population growth beginning about 12,000 years ago, some 30,000 years after the evolution of indisputably modern humans. The last major ice age had just ended, humans were beginning village life in some parts of the world, and people had recently spread into and through the Americas. Plant and animal domestication by humans would begin in one or two thousand years (Diamond 1999). At this point, somewhere between 1 and 10 million humans existed worldwide. It took about 10,000 years – until roughly 1 AD – to increase to about a quarter of a billion humans (Fig. 1.1). Thus, our population growth has historically been low, with

---

[1]Fish and fisheries have equal worldwide attention in applied biology. But literature on fish and fisheries is extensive enough, and the details of both biology and management different enough from that of terrestrial vertebrates, that this book will address fisheries-related issues only occasionally.

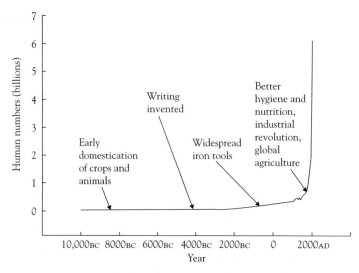

Fig. 1.1 Human population growth from 10,000 BC to present day. Data are from sources compiled on the US Census Bureau website (www.census.gov/ipc/www/worldhis.html). The dip in the 14th century represents deaths due to bubonic plague.

increases of a tiny fraction of a percent per year[2]. This relatively low growth rate continued over the next 1600 years, with some noticeable setbacks such as the outbreaks of Black Death (bubonic plague) that killed one-quarter of the people in Europe between 1346 and 1352.

Between 1650 and 1850, growth of human numbers began to rocket (Fig. 1.1), following development of global agriculture, the initiation of the Industrial Revolution in western Europe, and improved nutrition and hygiene across much of the world. Death rates fell dramatically, leading to longer life expectancies, improved infant survival, and larger human numbers.

By the late 1960s, the Earth held about 3.6 billion humans. At that point the rate of increase of our species had just passed its peak of just over 2% per year (Fig. 1.2). Think about it: it took 10,000 years to increase by a quarter of a billion, but by 1968 our numbers were increasing by that much every 4 years.

What about now? The global population growth rate has declined since the late 1960s (Fig. 1.2). But current growth is still positive, and multiplying by the ever larger numbers of our current population size results in enormous increases in abundance. Human numbers passed the 6 billion benchmark in 1999, and as of 2005 a total of nearly 6.5 billion humans live on Earth (Box 1.1). At current rates of growth and population size we are adding about 70 million people per year to the planet. That's nearly 200,000 per day, or about 8000 additional people during a 1-hour lecture. And these are net additions, births minus deaths.

---

[2]See Diamond 1999 for a fascinating treatise on how local biological diversity and environment have historically caused great variation in the growth of human populations around the world.

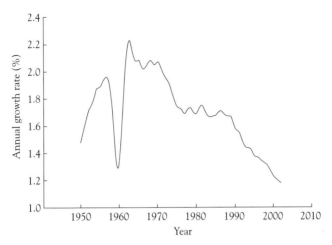

Fig. 1.2 Global human population growth rate (presented as the percentage change per year) since 1950. Data from the US Census Bureau, International Database (www.census.gov/ipc/www/worldpop.html). The dip in global population growth rate 1959–61 was due to the Great Leap Forward in China which resulted in more than 20 million premature deaths from famine in a 2-year period (Becker 1996).

Box 1.1 Grasping the meaning of billions of people

A billion is a hard number to fathom. First, count the zeros. A billion is 1000 million, otherwise written as 1,000,000,000, or in scientific notation as 10^9. (Here I acknowledge my American perspective; in some European countries this number is the **milliard**, with **billion** referring not to a thousand million but rather to a million million, adding three more zeros; Cohen 1995.)

So how much is a billion (or 10^9)? Line up a billion soccer balls on the equator and they would go around the world 5.5 times. Space a billion people 38 cm apart in a straight line and the line of humans would go from the Earth to the moon. Put a billion people in a square field, with each person getting 1 m² to call their own, and the field would measure nearly 32 km along each side. Live a billion seconds on Earth and you would be 31.7 years old. That's 1 billion. Times six is 6 billion: a lot of humans.

What next? Humans cannot escape the factors that constrain the numbers of all species. Resources (physical, chemical, biological, technical, institutional) cannot be without limit on the planet, so no species can increase indefinitely (see Chapter 6). Pinpointing where this population limit is – or when or how we will reach it – is highly uncertain, but credible scientists estimate that humans will stabilize at numbers around 10 billion by the end of the 21st century (Smil 1999, Lutz et al. 2001).

Population biology can help elucidate how changes in certain human vital rates, such as number of offspring per mother, age at first reproduction, or survival at different ages, will influence population change. For humans as for wildlife, these effects are not necessarily intuitive. The historical increase in the human growth rate, for example, occurred as much or more from increased survival as from increases in the number of children per female. Less obviously, population growth can also be strongly affected by the age when reproduction begins. Bongaarts (1994) provides a striking example for humans: the world population could be decreased by 0.6 billion over 100 years if the mean age of childbearing increased by 2.5 years and by 1.2 billion if the mean age increased by 5 years[3]. An increase in average childbearing age could occur as a by-product of education, because when girls and women stay in school longer they tend to get married later and delay childbearing. Overall, women with a high level of education have their first child about 5 years later than those with low education (Bongaarts 1994, Beets 1999).

The distribution of individuals across different ages also interacts with vital rates and population growth. In developed regions of the world, increased survival and declines in fertility prior to 1950 led to an increase in the percentage of the population older than 65, from 7.6% in 1950 to 12.1% in 1990 (Cohen 1995:98). Because women generally outlive men (for reasons discussed in Chapter 4), a shift to older age classes also shifted to a higher proportion of women.

Age structure also creates what is known as **population momentum**, which can cause population growth to be very different than expected from current birth and death rates alone. For humans, population momentum would lead to our numbers continuing to increase for several decades even if female reproduction dropped immediately to the replacement level of two surviving children per female (Fig. 1.3). Why? Because the large cohort of youngsters characteristic of the increasing population will be reproductive for decades, inflating population growth until the age distribution adjusts to the smaller proportion of young, child-bearing individuals expected at stable age distribution (Chapter 7). Conversely, population momentum could cause declining endangered species to continue to decline for some time even after management has increased birth or survival rates to replacement levels.

Human impacts on wildlife through effects other than population size

Obviously, human numbers affect other species. But the overall influence of any species on its community and ecosystem is a function not just of numbers, but also of per-capita (or per-individual) interactions. Human use of inanimate energy (wood, oil, etc.) grew from 0.9 MWh per year per person in 1860 to nearly 19 MWh in 1990; this is the energy equivalent of each person in the world keeping two light bulbs (40-watt) burning continuously through the year in 1860 compared to each person burning 52 bulbs continuously in 1990 (Cohen 1995). Extraction and use of this energy has profoundly affected the global environment and other species (Chapter 11).

---

[3]The current mean age of first reproduction varies a lot by country and region, but tends to be in the mid-20s, with the Netherlands having the oldest mothers at first birth, at 29 years (Beets 1999).

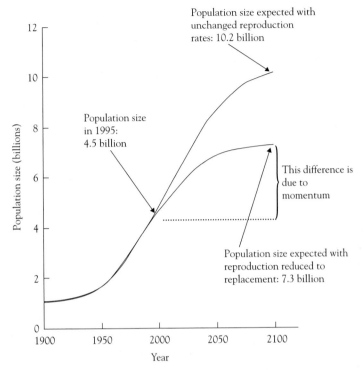

Fig. 1.3 Population momentum due to age structure in humans. The top line shows a projection of human population size from 1995 through 2100 with unchanged reproduction (projection made by World Bank in the mid 1990s). The bottom line shows the population projection with the hypothetical assumption that in 1995, at 4.5 billion people, reproduction was immediately reduced to replacement level of two surviving offspring per female. Even though births are reduced to replacement levels, the number of humans would continue to grow to 7.3 billion because the growing 1995 population has a high proportion of younger females with a greater chance of survival and reproduction; population momentum means that it takes a while to shift to the older age structure and slower growth-rate characteristic of the reduced reproduction. Figure modified from Bongaarts 1994. Copyright (1994) AAAS.

The distribution and social grouping of humans has also changed. In 1880 only about 2% of people lived in cities with more than 20,000 people; by 1995 about 45% lived in cities and more than 17% of all people lived in cities larger than 750,000 (Cohen 1995:13). Humans have also shifted the number of people per household, going from 3.2 to 2.5 per household in more developed nations and from 5.1 to 4.4 in less developed countries over the last 30 years (Keilman 2003). Fewer people per household means more houses, and more resource use per person. For example, as China goes from 3.5 people/house in 2000 to a projected 2.7 by the year 2015, 126 million new households will be added, even if China's population size remains constant (that's more new households in China over 15 years than the total current number of US households; Diamond 2005). Thus, not only population size drives extinction rates of other species, but also household occupancy and per-capita resource use (Liu et al.

2003). And of course, the **footprint** humans have on other species also involves complex interactions with cultural norms, wealth distribution, and per-capita consumption rates. In short, humans are strong interactors (Chapter 13) whose impact on other species comes from not only numbers and distribution but also large per-capita influences of each human on other species.

Extinction rates of other species

As the human population has climbed, the extinction rate of other species has also gone up; indeed, humans began causing extinctions about 45,000 years ago (Caughley & Gunn 1996, Brook & Bowman 2004). To place current human impacts on other species into a broad context of applied population biology, I next consider how many species are on Earth and how current extinction rates compare to background rates over geologic time.

Number of species on Earth: described and not yet described

My discussion of number of species on Earth will focus on eukaryotic species, excluding bacteria and viruses. Part of the decision to ignore the domain of bacteria – those wondrous organisms that have been on Earth for more than 3.5 billion years, metabolizing sulfur in deep-sea trenches to make carbohydrates without sunlight, thriving in boiling mud, in steam vents, and 4 km deep in the Earth at temperatures exceeding 100°C, filling our mouths to the tune of 4 billion bacteria per mouthful (between brushings!), and comprising nearly 90 trillion of the 100 trillion cells that make up an individual human[4] – is because this book focuses primarily on terrestrial vertebrates. But another, perhaps more legitimate, reason is that bacterial life histories make a classic species definition dubious for these organisms (and even more so for viruses, those important pseudo-life forms that can persist but not replicate outside the cells they infect).

About 1.5 million species of nonbacterial eukaryotes have been described (Fig. 1.4). Although there are more animal species described than any of the other three eukaryotic kingdoms (protoctista, plants, fungi), vertebrates make up only a small slice (3%) of the life-on-Earth pie. Mammals, birds, amphibians, and reptiles collectively amount to less than 2% (approximately 30,000 species) of described species. By contrast, beetles rule the world in terms of numbers of species, with 300,000 described species, so one out of five species on Earth is a beetle. More than 85% of all recorded species are terrestrial, although aquatic systems have a greater variety. For example, 32 of the 33 multicellular animal phyla are found in the sea (21 are exclusively marine), with only 12 on land (only one exclusively; May 1994).

Without a doubt, only a fraction of the species on Earth has been described (Box 1.2). So if 1.5 million species have been described, how many are there really? The plau-

---

[4]For good overviews of these and other astonishing feats of bacteria see Stevens (1996), Nisbet and Sleep (2001), Rothschild and Mancinelli (2001), and Buckman (2003).

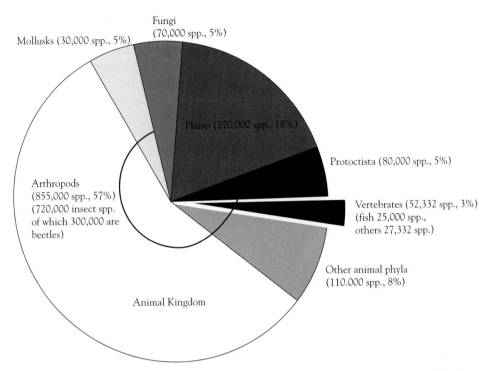

Fig. 1.4 Composition of the 1.5 million described eukaryotic species on Earth. Three of the four kingdoms (protoctista, plants, fungi) are each represented by one pie slice. The fourth kingdom – animals – is shown and further divided to highlight three of the 37 phyla that make up the animal kingdom (mollusks, arthropods, vertebrates). The exploded slice is the vertebrates. Although the total count of eukaryotic species comes to slightly more than 1.5 million, uncertainties warrant rounding to this figure. I used the following data sources. For vertebrates I used Groombridge and Jenkins (2002) because they use updated figures for this phylum, which is the most studied – and therefore the most reliably counted – group of taxa. For all other taxa I used May (1997) because this study gives particular care to minimizing the overcounting of species due to single species being recorded multiple times under different names.

sible range for numbers of eukaryotic species on Earth is 5–50 million species, with the best guess somewhere around 14 million (see May 1997, Stork 1999, Groombridge & Jenkins 2002). We are left with the disconcerting realization that only 10% or so of the species that exist on Earth have been described.

Historic versus current rates of extinction

Just as every human dies, every species goes extinct. Clearly, neither deaths nor extinction per se cause concern so much as whether changes have occurred in the rate of human death (say due to epidemic disease) or the rate of species extinctions. Although the trend over geologic time has been one of increasing **species richness** (Fig. 1.5), the species currently on Earth represent less than 2% of all species that have ever lived over the past 3 billion years. (The parallel with human life and death holds here as well;

Box 1.2 Erwin's estimate of the number of tropical arthropod species on Earth

Although there have been many different approaches to estimating the number of existing species on Earth, one of the most creative and high profile was by Terry Erwin (1982), who used an insecticide fogger to kill beetles in 19 trees of *Luehea seemannii* in tropical rainforests of Panama. Erwin estimated that the dead beetles comprised 1200 species, with 163 species depending strongly on this one tree species (high host specificity). Extending the 163 host-specific beetle species to an estimated 50,000 other tropical tree species led to an estimated 8.1 million tropical-canopy beetle species. Assuming that 40% of arthropod species are beetles leads to an extrapolated 20 million canopy arthropod species. Finally, assuming next that the canopy fauna is twice as rich as the fauna of the forest floor adds another 10 million non-canopy arthropods, leading to a richness estimate for tropical arthropods of 30 million species.

Erwin emphasized that his figure of 30 million tropical arthropods should be taken with the proper grain of salt: "I would hope someone will challenge these figures with more data" (Erwin 1982:75; see also Stork 1999).

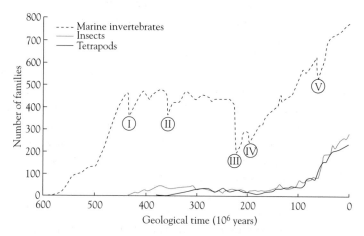

Fig. 1.5 Historical extinctions and increasing diversity. The increase in biodiversity (animal family diversity) over time has been punctuated by occasional mass extinctions. The top line represents marine invertebrates, the second line insects, and the bottom line tetrapods (amphibians, reptiles, birds, and mammals); to maintain clarity fish are not shown, but their line tracks almost directly atop the tetrapods. Families are used instead of species because the fossil record is more complete at the family level. The five classic mass-extinction events are indicated with roman numerals: (I) late Ordovician (440 million years ago); (II) late Devonian (365 million years ago); (III) late Permian (250 million years ago); (IV) late Triassic (205 million years ago); (V) late Cretaceous (66 million years ago). Modified from Groombridge and Jenkins (2002). Reproduced by permission, The Regents of the University of California.

although far more humans exist on Earth than ever before, the humans currently on Earth represent less than about 6% of all those who ever lived; Haub 2002). Given that species have always "died" over geologic time, we can use the fossil record to ask, what is the background rate of extinction? In other words, what is the average species lifespan, from origination to extinction, over the last 600 million years since the explosive diversification of multicellular animals? Assembling estimates from many sources indicates that the average species has had a lifespan of about 1–10 million years (May et al. 1995). Superimposed on the background extinctions during geologic time are five major extinction periods (Fig. 1.5). Although uncertainty in the time span of extinction spasms makes it difficult to express mass extinction events as a rate (Jablonski 1991), the losses were massive, with some 75–95% of all species on Earth becoming extinct during each mass extinction period[5] (Fig. 1.5).

So the key question is how current rates of extinction compare to background extinction rates. To state it more provocatively: are humans now causing the sixth **mass extinction**, with species lifespans shortened to lengths similar to those during the five geologic mass extinctions? To answer this question, we need to estimate current extinction rates, a difficult task since we know so little about how many species exist! One approach estimates current losses of both described and undescribed species by coupling rates of deforestation in the tropics with species-area relationships, leading to a predicted 27,000 species lost per year (Wilson 1992).

A more direct and conservative approach estimates current extinction rates using only described species and documented extinctions. During the past 400 years extinctions have been recorded for about 60 mammal, 120 bird, 26 reptile and amphibian, 81 fish, 375 invertebrate, and 380 plant species (Groombridge & Jenkins 2002). Focusing on birds and mammals (because written records of extinctions and skeletal remains are most reliable for these taxa) leads to 180 observed extinctions over 400 years, an average of about 0.5 extinctions per year. How does this current rate compare to what would be expected based on historical species lifespan? If an historical 10-million-year lifespan per species were applied to the currently existing 15,000 described bird and mammal species, we would expect (15,000/10 million =) 0.0015 extinctions per year; using a 1-million-year lifespan gives an annual expectation of 0.015 extinctions per year.

Thus the current rate of extinction for birds and mammals of 0.5/year is 33–333 times greater than expected by the background extinction rate of 0.015–0.0015/year. Several other approaches produce estimates of current extinction rates that are considerably higher (e.g. Wilson 1992, Stork 1999). Furthermore, impending extinction rates (those expected in the future if current trends continue) are expected to be at least four orders of magnitude faster than the background rates derived from the fossil record (Smith et al. 1993, May et al. 1995, Pimm et al. 1995). Although uncertainty in the analysis may still spawn legitimate debate as to whether current extinction rates are yet as high as those of geological mass extinctions, absolutely no question remains that rates are considerably higher than natural background levels, and if things do not change we will indeed find ourselves presiding over a sixth mass extinction event in the near future.

So what? There are both philosophical and utilitarian reasons to care about loss of species and genetic diversity (see Chapters 9 and 12), as well as the loss in ecologi-

---

[5] Major extinction events are really just the tail of a fairly continuous distribution of extinction magnitudes distributed across geologic history (Groombridge & Jenkins 2002).

cal services upon which other species, including humans, depend. For example, many bird species contribute strongly to ecosystem processes, including decomposition, pollination, and seed dispersal; in turn these services are valued by humans (e.g. commercial plant pollination and control of insect pests that damage plants or spread human disease). If, as predicted, by 2100 as many as one-half of bird species either go extinct or decline to the point where their ecosystem interactions are compromised, humans will likely experience the cascading effects of an altered system (Şekercioğlu et al. 2004).

Humans and sustainable harvest

Obviously, not all species are negatively affected by all human activities. In fact, quite a number of species have reached historically high numbers and are considered pests or overabundant. Certainly, humans have harvested a great many species without negative effects.

The effects of harvest on wildlife populations and the regulation of harvest probably have deeper roots than any other topic in applied ecology. Early recorded history includes Egyptian hunting records tracing back to about 2500 BC (Leopold 1933, Gilbert & Dodds 2001). Graeme Caughley (1985) noted that under Genghis Khan (13th century) the Mongols "conserved wildlife much better than they did people," for example by restricting hunting to the 4 months of winter and allowing some animals to escape hunting drives.

Despite the long history of harvest regulation, many cases exist of overharvest by humans. Most of the US colonies had established closed seasons on some species by the mid-1700s, although the first hunting licenses were not required until 1864 (in New York) and the first bag limits implemented by about 1878 (when Iowa limited prairie-chicken harvest to 25 birds per hunter per day; Connelly et al. 2005). There was little connection between these early laws and enforcement, or between the laws and expected population response. By the late 1800s extinctions due to overharvest loomed for species ranging from passenger pigeons to beaver to bison. Leopold (1933:17) attributed these disasters to an American viewpoint – in rebellion against the European philosophy of wildlife harvest that was perceived to benefit only the wealthy – whereby game laws "were essentially a device for *dividing up* a dwindling treasure which nature, rather than man, had produced."

The reversal of this failing approach to harvest management is usually traced to Theodore Roosevelt. In a passage that typifies his views, Roosevelt wrote in 1903 (Morris 2002:221):

> Every man who appreciates the majesty and beauty of the wilderness and of wild life, should strike hands with the far-sighted men who wish to preserve our material resources, in the effort to keep our forests and our game-beasts, game-birds, and game-fish – indeed, all the living creatures of prairie and woodland and seashore – from wanton destruction. Above all, we should recognize that the effort toward this end is essentially a democratic movement . . . But this end can only be achieved by wise laws and by a resolute enforcement of the laws.

Sportsmen-naturalists such as Roosevelt, dedicated to hunting and fishing, led the movement in the USA to preserve nongame species as well as game species. By the late 1870s these individuals called for preserves to protect against forest decimation and denounced market hunting for food and for fashionable hats adorned with bird feathers. Thus in the late 1800s individuals committed to sport hunting and conservation led the charge to conserve wildlife, "in spite of the utter indifference of a nation seemingly obsessed with economic development" (Reiger 2001:88).

One can point to a number of success stories over the last century where knowledge of population biology has been linked to enforced regulations, leading to sustainably harvested populations (see Chapter 14). For example, wood ducks and sharp-tailed grouse were nearly wiped out by overharvest in the early 1900s, but are now successfully and sustainably harvested in the USA (Bolen & Robinson 2003). Equally striking has been the recovery of white-tailed deer in North America: overhunting and habitat destruction reduced numbers to less than 500,000 by the late 1800s, but by the 1980s – following widespread initiation and enforcement of hunting regulations (as well as habitat change) – deer numbers had increased 100-fold or more.

In Australia, large kangaroos have been harvested by humans for more than 20,000 years (Grigg & Pople 2001). At least 66 million kangaroos were harvested in the 1980s in the state of Queensland alone (Calaby & Grigg 1989), and 1–3 million continue to be harvested annually from about 40% of mainland Australia, even as the most heavily harvested species increase to higher densities than before the arrival of Europeans. In large part the increase is due to tree clearing for agriculture, implementation of watering points for introduced stock, and control of dingo, but clearly credit also goes to a well-managed harvest. The kangaroo is a national symbol for Australia, and so the animals have high conservation status and generate money as tourist attractions. However, their range overlaps with sheep, making them pests in the eyes of the sheep industry. The resolution to this tension between conflicting desires for conservation and extermination may be linked to the commercial value of kangaroos. Leather from kangaroo hide is arguably superior to that of cattle, and the high-quality meat has growing potential on the world market. Thus the conflict between the sheep industry – who would like to vastly decrease or eliminate kangaroos as pests – and conservation may be partially resolved by a truly sustainable harvest. An expansion of the meat market, coupled with the market value of the skin, may give landholders an incentive to reduce sheep numbers and maintain kangaroo numbers as a harvested species, while still providing the draw of live animals for tourists. This vision has been referred to as "sheep replacement therapy for rangelands" (Grigg & Pople 2001).

In the same way that declining or small wildlife populations can be better managed based on population biology principles, harvest management is most sound on a scientific footing. Human removal of individuals via harvest is analogous in many ways to the effects that predators have on their prey populations. Insights into how hunting or predation affects the prey species depends, for example, on how many of which ages or sex are killed, and how much the mortality is added onto other factors that cause death to the prey (see Chapters 8 and 14). Determination of vital rates, monitoring of populations, and incorporation of ecological understanding are all part of harvest management.

The big picture

Harvested species and species in decline are driven by similar ecological processes, and the history of hunting parallels and overlaps conservation of non-harvested species. Thus conservation of harvested species is inseparably related to conservation of non-harvested species. In a broader sense, I am happy to see the dying out of the dichotomy that some have held in the past between "conservation biology" and "wildlife biology," or between "conservation management" and "wildlife management." Ecology, conservation biology, and wildlife biology have merged and intertwined to contribute to the bottom line: that biological knowledge, data, and models help us understand and manage interactions between humans and other species (Thomas & Pletscher 2000). Today's wildlife population ecologist must also master knowledge of experimental design (see Chapter 2) and genetic tools (see Chapters 3 and 9). In short, my attempt in writing this book is to combine demography and genetics, theory and practice, and to apply slices from the conceptual basis of population biology to problems of wildlife conservation.

Further reading

Cohen, J.E. (1995) *How Many People Can the Earth Support?* W.W. Norton & Co., New York. A careful compendium of historical and current human population growth and its consequences.

Groombridge, B. and Jenkins, M.D. (2002) *World Atlas of Biodiversity*. UNEP World Conservation Monitoring Centre, University of California Press, Berkeley, CA. A complete reference addressing biodiversity status, with lots of great graphics.

Lawton, J.H. and May, R.M. (eds.) (1995) *Extinction Rates*. Oxford University Press, Oxford. A collection of articles addressing many aspects of past, current, and future extinctions.

Leopold, A. (1933) *Game Management*. Charles Scribner's Sons, New York. Leopold's insights are as timely now as they were 75 years ago; the book also provides an excellent history of harvest management.

2

Designing studies and interpreting population biology data: how do we know what we know?

It is not enough to say that we cannot know or judge because all the information is not in. The process of gathering knowledge does not lead to knowing. A child's world spreads only a little beyond his understanding while that of a great scientist thrusts outward immeasurably. An answer is invariably the parent of a great family of new questions. So we draw worlds and fit them like tracings against the world about us, and crumple them when they do not fit and draw new ones. The tree-frog in the high pool in the mountain cleft, had he been endowed with human reason, on finding a cigarette butt in the water might have said, "Here is an impossibility. There is no tobacco hereabouts nor any paper. Here is evidence of fire and there has been no fire. This thing cannot fly nor crawl nor blow in the wind. In fact, this thing cannot be and I will deny it, for if I admit that this thing is here the whole world of frogs is in danger, and from there it is only one step to anti-frogicentricism." And so that frog will for the rest of his life try to forget that something that is, is.

John Steinbeck (1960), *Log From the Sea of Cortez*

Introduction

Why learn about study design, data interpretation, and a touch of scientific philosophy and ethics in a book on ecology and conservation of wildlife populations? Simply put, because without reliable knowledge both science and the application of science fall flat. The best scientists, managers, and decision-makers are those who can separate lousy, unreliable, or irrelevant information from important and trustworthy information.

Reliable knowledge is nothing less than the outcome of the quest to judge **truth**. Heavy stuff, for sure, but it has profoundly practical implications. In essence, truth is the correspondence between an idea created in our mind – that is, an **idea**$_{mind}$ – with a referent fact obtained from sensory experience (call this a **fact**). Dr Charles Romesburg – who rattled the field of wildlife science with a classic paper on "Gaining reliable knowledge" in 1981 – has noted that without comparison to a factual referent, an idea of the mind is only an opinion (Box 2.1). Saying it another way, the process of

Box 2.1 Notes on truth, knowledge, and opinion

These notes were modified from unpublished notes by Dr Charles Romesburg, Utah State University.

- There are two kinds of idea:

    1 referent "facts" from sensory experience, and
    2 ideas "from the mind" developed from free creation apart from sensory experi-
      ence ($idea_{mind}$).

- We say $idea_{mind}$ is **true** when $idea_{mind}$ = fact (what is, is). We say $idea_{mind}$ is **false** when $idea_{mind} \neq$ fact.

Without appeal to a referent humans tend to disagree. That is why there is consensus that a given $idea_{mind}$ that has been tested and found to agree with facts is a rightful candidate to the concept "knowledge", as opposed to a given $idea_{mind}$ that has never been exposed to the factual arbitrator.

Opinions are ideas that have been declared as neither knowledge or falsehood. They are in limbo and will remain so until risked in comparison to the factual referent (either direct or indirect comparison). Note that the tolerance for testing truth is an opinion, making all knowledge depend on opinion.

Reason has always been around. Knowledge made little growth up to the 16th century because reason was the sole basis for knowledge. Then came the scientific revolution that blended reason with facts. Galileo, Kepler, and others changed how science was done by risking the predictions from their reasoning with facts.

moving from opinion to **knowledge** is all about coming up with creative ideas ($idea_{mind}$) and comparing them to reliable facts, then revising the $ideas_{mind}$ if they fail to hold up to the facts[1]. Although Steinbeck's frog (in the quote above) recognizes that his idea of the mind (that cigarettes either originated from local materials, or dispersed) fails to match his facts (there is a cigarette in the pond), he denies the facts instead of recognizing that his $idea_{mind}$ should be revised!

So, the scientific process is really just a process of comparing creative and meaningful $ideas_{mind}$ to reliable facts, and then following an objective and orderly process of modifying the $idea_{mind}$ when the two do not match. That's what this chapter is all about.

---

[1] Of course, both $idea_{mind}$ and facts are rich fodder for epistemological and philosophical discussion, including subtleties on how the $ideas_{mind}$ affect our ability to recognize facts. I use the terms to make the simple but profoundly important point that reliable knowledge in wildlife population biology requires deep thought as to how nature works (the $ideas_{mind}$) as well as information gained from study design and data collection (the facts).

Obtaining reliable facts through sampling

How do we obtain the facts from the field against which we compare the ideas of our mind? We do it through appropriate sampling and study design. The facts might be estimated effects of treatments, or they might be estimates of **parameters**. Parameters are quantities of the population for a given area and time; for example, the true survival rate of adults. We almost never know what the true parameters are; rather, we estimate parameters from data. By convention, **estimates** have hats ($^\wedge$) over them; for example, an estimate of abundance (N) is denoted $\hat{N}$. In this section I'll discuss some main points of sampling and study design.

Replication and randomization

When you cook a pot of spaghetti and wonder if it is done, how do you decide? Perhaps you slavishly watch the clock, read the directions, and remove the noodles at just the right time. More likely, though, you sample. But how? Typically, you grab one noodle and taste it. If your pot is big enough and the water has boiled vigorously, one noodle from anywhere in the pot should be enough. But if you have a small pot, or it hasn't boiled very well, you would probably sample a few noodles, perhaps at least one from the bottom and one from the top.

So, in a homogeneous noodle population you might draw inferences from as few noodles as one, but as your noodles become heterogeneous – variable in their doneness – you would **replicate** your sampling. There is almost no meaningful question in our field that involves a population as homogeneous as a pot of noodles, so replication is a cornerstone of reliable sampling. Formally, replicates are the multiple members sampled from a population of interest, or the number of units to which a treatment is independently assigned. The **sample size** for a treatment is the number of replicates. Replication keeps us from making a decision based on a single, potentially unusual, sample. Replication also facilitates an estimate of variation, providing a basis for a statistically sound decision as to whether the population in question – given its variable nature – is really different from another population that we care about (Johnson 2002).

There are a couple of rules about appropriate replication. First, the replicates should be sampled at an appropriate scale to capture the relevant variation over time and space. If you are interested in how owl clutch sizes differ between a single logged and a single thinned forest, then 10 reproducing owls in each forest would constitute the number of replicates. However, if instead you were interested in the general question of whether owl clutch size differs in unlogged and thinned forests, then 10 reproducing owls in one forest of each treatment would constitute only one sample (no replication) with 10 **subsamples**. To treat the 10 subsamples as samples would lead to a mismatch between the number of independent samples (one) and the desired inference, leading to **pseudoreplication** (Hurlbert 1984). To avoid pseudoreplication for the question of owls in unlogged and thinned forests, several replicates of different unlogged and thinned forests would be sampled across the landscape of interest, with one to many owls subsampled within each forest. It makes sense that we would try to avoid pseudoreplication, as it is easy to think of many other extraneous factors other than thinning – including predators, prey, aspect, vegetation, and so on – that would cause the owls in one forest stand to reproduce differently from the owls in another stand even if thinning had nothing to do with it.

A second rule about replicates is that they should be chosen **randomly**; that is, every member of the population should have a chance of being sampled. In a manipulative experiment, the treatment received by each unit should be assigned randomly. Randomization reduces the chance that some ancillary factor will bias our measurements because it makes it less likely that systematic differences other than the treatment could have caused the observed effects[2]. If we picked unlogged forests at high elevation and logged forests at low elevation, and elevational differences led to differences in clutch size, then our inference about the effect of logging would be wrong, even with 100 replicates, because it would really represent an effect of elevation. The bottom line is this: although it may be more convenient to sample nonrandomly – perhaps along a road or a trail – inference becomes limited and much less valuable than with random sampling.

Controls

Controls include a number of ways that scientists ensure that facts are not confounded by some unexpected influence other than the idea$_{mind}$ we are evaluating. For example, when someone refers to **controlled conditions** they mean that they are making sure that the desired test conditions are occurring. In field projects, **control sites** or **control treatments** are those whose only consistent difference from the treated sites is the application of the treatment (chosen randomly from available sites). Similarly, control procedures account for effects that might be caused by methods used to apply experimental treatments; for example, if the question is how leg bands affect bird survival, then some birds might be handled but not marked as a control procedure to identify whether the handling may confound the effect of the leg band on survival (Hurlbert 1984).

Finally, when a degree of subjectivity is involved in measuring a factual referent, **blind controls** can minimize the chance that the observed treatment effect does not carry insidious (and often unconscious) influences from wishful thinking by the observer. In a blind control, the person collecting or analyzing the data does not know the treatment group or identity of the sample they are evaluating. For blind controls, impeccable record-keeping and protocols for conducting the test are especially critical.

---

[2]Although random sampling is usually best, sometimes the nature of the question makes it impossible. In such cases, other approaches, such as systematic, stratified random, cluster, adaptive, sequential, or other sampling methods, may be appropriate (Thompson 2002). The key is to think through how the sampling method may affect estimates of both bias and variance, given the sampled population.

Accuracy, error, and variation

Accuracy and **error** are two sides of the same coin, describing how well the mean estimated from sample observations corresponds to the true mean. Accuracy, or its flip side, error, is made up of two components that can be quite unrelated to each other: bias and precision. Consider an estimate of a population parameter, say abundance, survival, weight, or sprint speed. **Bias** refers to systematic deviation of the estimate from the true parameter of interest. **Precision** refers to the amount of scatter, or repeatability of the estimate when made many times; it is quantified by the variance, with high variance indicating low precision (Box 2.2). Either a large bias or low precision (high scatter) will result in low accuracy[3], as portrayed in the classic bulls-eye diagram shown in Fig. 2.1.

Let's consider in more detail the error that comes from low precision. Lack of precision arises from both process variance and sample variance. **Process variance** is genuine biological variance that arises because conditions vary (e.g. temperature, moisture, diseases; Thompson et al. 1998, Mills and Lindberg 2002). Specifically,

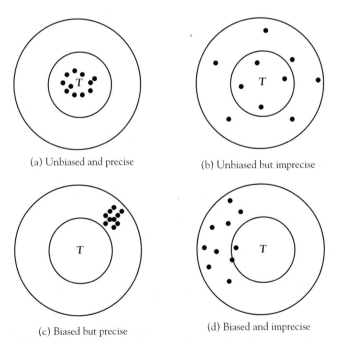

(a) Unbiased and precise

(b) Unbiased but imprecise

(c) Biased but precise

(d) Biased and imprecise

Fig. 2.1 An accurate estimator is one that is unbiased and precise. The center of the circle (the bulls-eye) is truth, denoted by T (for example, a survival rate of 0.85 or an abundance of 220). The dots show sample estimates of the parameter T. Estimates in (a) are accurate, being unbiased and having high precision; all others are inaccurate. Modified from White et al. (1982).

---

[3]For the statistically inclined, overall error is captured by the mean squared error (MSE) for an estimate X: $\text{MSE}(X) = \text{variance}(X) + \text{bias}(X)^2$ (Williams et al. 2002).

Box 2.2 A primer on variance, standard deviation, and standard error

Variance gives an indication of the **spread** of what you measured in your population. Correcting for finite sampling (as we always must when we sample just part of a population) with n observations, the equation for sample variance is the average squared deviations of all of the x_i measurements from the mean ($\bar{x}$):

$$s^2 = \frac{\sum_{i=i}^{n}(x_i - \bar{x})^2}{n-1}$$

Standard deviation (SD) is the square root of the variance. This is useful because it describes the spread of your variable in terms of what you measured (say, animals per hectare), instead of the non-intuitive "squared animals per hectare" from calculating variance. For data with a bell-shaped (normal) distribution, the mean plus or minus two SDs will contain about 95% of the population.

So far we have only talked spread of measurements from a population. But what if we are interested particularly in how well the mean characterizes the sample? The **standard error** (SE) **of the mean** is the estimate, from a single sample, of the SDs of the distribution of means expected if we collected many samples of size n and calculated a mean for each sample. In practical terms, SE quantifies how confident we are that our estimated mean is close to the true mean.

The SE is estimated from our one sample as $SD/\sqrt{n}$. (In some cases, such as for most mark–recapture estimates of vital rates, the SEs are simply the square root of the variance).

It makes sense that a larger sample size (n) will decrease the SE, because more sampling should increase our confidence that the true population mean will be close to the estimated mean. Notice that the word error in standard error is not a statement of mistakes or bad judgment.

An **X% confidence interval** is obtained by adding and subtracting from the mean a value weighted by the SE (for example 1.96 * SE for a 95% confidence interval assuming a normal distribution); informally this is thought of as providing the range in which we suspect the unknown true mean should be found. More formally, that confidence arises from the fact that if we were to repeat the study many times, X% of the confidence intervals constructed in this way would include the true mean. Both the SE of the mean and the confidence interval indicate how precisely the mean is estimated.

The **coefficient of variation** (CV) expresses variation relative to the mean:

$$CV = SD/mean$$

CV is useful in the many cases where variance (and SD) increase with the mean, so we want to know whether some measure (say, offspring production) is relatively more variable for a group with a high mean (say, coyotes) compared to one with a low mean (say, bears). It is often expressed as a percentage.

Box 2.3 An insidious form of error due to being human

Although researchers work hard to avoid bias and increase precision in measurements – for example by calibrating instruments, and by proper replication and randomization – unconscious errors (or lack of accuracy) can sneak up when the study requires a subjective decision. Social psychologists, medical researchers, and educators have worried about this a lot, and regularly use so-called double-blind setups where the person being interviewed does not know the purpose of the study, and the interviewer does not know the identity of the interviewee.

During a study of coyotes in Northern Utah (Mills & Knowlton 1989), Fred Knowlton and I wondered if radiotelemetry error tests really captured the error inherent in estimating azimuths to animals. Field assistants were working under grueling conditions, squashed in a little fixed-station box on the back of a truck for 8 hours or more through the night. Might the error in their telemetry bearings be different in those conditions than in a known telemetry test where they were aware that they were being tested?

To evaluate this possibility, we told field assistants that we had captured some new coyotes over the weekend, when actually the new transmitters were test transmitters placed at known locations. After several nights of collecting data (with known error because we knew the exact location of the fake coyotes), we conducted a traditional telemetry accuracy check where assistants knew they were being tested.

When observers knew they were being tested their precision was quite a bit better than the blind test. Knowing that their diligence and accuracy were being examined, they worked harder and longer to obtain azimuths. Our recommendation to obtain a better estimate of telemetry accuracy is to conduct telemetry accuracy tests without field helpers knowing they are being tested.

---

spatial process variance arises from changes in species present, habitat quality, and habitat heterogeneity over the landscape, which in turn may be related to environmental conditions such as aspect, slope, precipitation, and successional-stage differences. Temporal process variance is often driven by weather, as well as interactions with competition, predation, disease, and human impacts. Even if you know a parameter (say, survival) exactly, it will fluctuate over time and space due to process variance.

In contrast to process variation, which acts directly on organisms, sampling variation is a product of incomplete information from the act of sampling a larger population. One component of **sample variance** comes from human error, as when observers make subjective decisions (Box 2.3), but more fundamentally it is an inevitable result of estimating something by sampling from a population. Although sample variance is present and real in nearly every measurement of a fact, it is a nuisance that inflates total variance artificially. Thus process variance, the actual biological variation inherent in the thing that we measure, has to be teased out of the total variance measured[4]:

---

[4]We'll be returning to process versus sampling variance as components of total variance in later chapters.

Real process variance = total variance measured − sample variance

Process variance is often as important to quantify as the average or mean. As an example, here's the folksy plea of a waterfowl biologist named Johnny Lynch, frustrated by how duck population goals were being set based on a national average that ignored variation (quoted by Ankney 1996:41):

> Did someone say that the "average" numerical standing of the North American mallard population over some period of years would be a good standard for management to try to maintain? Don't let the Old Forecaster hear such talk. Not long ago, he got involved in certain philosophical deliberations regarding the "average" condition of dynamite. Which commodity, in its quiescent state, was a small cylinder having a volume of a few cubic inches, yet at its peak of explosion occupied hundreds of cubic yards. Seemingly, the "average" condition of dynamite could be determined by adding together measurements taken at various levels between these two extremes, and dividing their sum by the number of measurements. The forecaster took one look at the results of all this arithmetic, turned slightly purple, and then decided ruefully that dynamite in its "Average condition" must be one helluva thing to crate, ship, and otherwise handle. He has assiduously avoided "averages" ever since that unfortunate experience.

Without a doubt, variance and extreme events are critical for understanding wildlife dynamics. In fact, throughout the book we will see that the resolution to many applied issues – ranging from expected time to extinction of endangered species, to the number of animals in an area over time, to the consequences of a particular harvest rate – are driven by the extremes, as in Lynch's dynamite example. Variance may be of interest in its own right, so we may compare (for example) a coefficient of variation (Box 2.2) in clutch size among species, or in relative humidity among logging treatments. In other cases the variance of estimated treatment means can warn us to be humble about concluding differences. For example, if density of snowshoe hares in logging treatment A averages 1.8/ha and density in another logging treatment B averages 1.3/ha, can we say that density is higher in A? It depends on the variation around the means. So, embrace uncertainty by describing variation.

Linking observed facts to ideas$_{mind}$ leads to understanding

So far we have thought a bit about how to obtain field observations, or facts, that we can count on. This is critically important, to be sure, but with only a fact in hand we come face to face with the question, "So what?" For a fact to be converted into meaningful knowledge or understanding it must be linked to an idea of the mind that helps to explain a phenomenon. I will touch on some ways to do that via frameworks to evaluate hypotheses using frequentist statistics and P values, information-theoretic criteria, and Bayesian analysis.

The hypothetico-deductive approach

Instead of belaboring philosophy of science or the **scientific method**, I will talk a bit about the hypothetico-deductive method as a workhorse for gaining reliable knowledge. The essential steps of the hypothetico-deductive method are:

- identify the question or phenomenon of interest,
- develop hypotheses that explain the phenomenon[5],
- deduce diagnostic predictions that follow from each hypothesis, and
- gather observations (facts) to test the predictions.

Do not underestimate the importance and difficulty of the first steps of defining an important question and deriving meaningful and testable hypotheses *a priori* (before the study or data analysis begins). One important basis for hypothesis generation is descriptive work, including natural history studies and the accumulation of insights from field observations. I think there is ample reason to mourn the shattered stature of natural history work, both because it helps frame our questions and because our souls are filled when we can be a sponge in the field, absorbing all that we can see, hear, smell, and touch. However, descriptive studies by themselves are not sufficient to quickly advance knowledge in any branch of science. Therefore, while we should be earnest students of natural history, filling our field notebooks with observations and thoughts, realize that this is just a first step in a process that leads to rapid accumulation of reliable knowledge.

Similarly, **induction** – the process of forming general conclusions based on associations in a collection of observations – serves an excellent role in hypothesis generation, but is not an efficient scientific process by itself. Correlations between two variables are a classic form of induction, and the truism that correlation does not equal causation applies to all walks of life. For example, I have a newspaper article that makes a serious attempt to use a rather sketchy correlation to imply causation between values of stocks in the USA and women's skirt lengths! Correlations and associations are good for generating hypotheses to be tested, but weak for concluding mechanisms.

Once a hypothesis is developed, it must be strongly connected to predictions. First, the prediction(s) must logically follow from the hypothesis. In other words, if the prediction is falsified we need to have confidence that the hypothesis is also false; a prediction that may or may not follow from the hypothesis will not reject the hypothesis as false. Second, confounding factors may cause the predictions to be supported even if the hypothesis is false. Avoiding the insidious problem of concluding a hypothesis is supported, when really a confounding factor supported the prediction, is at the heart of most study design and statistical analysis.

---

[5]The distinction between **hypothesis** and **theory** is a topic of much discussion among philosophers of science; typically a theory is a broader conceptual framework from which specific, testable hypotheses are derived. Also, here I'm using hypothesis in the sense of a **research hypothesis** into how nature works, as opposed to **null** or **statistical hypotheses** that investigate specific questions that may or may not be causal (Steidl et al. 2000).

Hypothesis tests can be accomplished either by making observations about the world as it is or by manipulating something and observing what happens. The latter approach is by far the most powerful, because by manipulating the system you make it less likely that the observed results came from something other than your treatment (Romesburg 1981, Johnson 2002). The process of multiple alternative hypotheses being exposed to critical experiments to efficiently weed out non-viable alternatives is known as **strong inference** (Platt 1964).

Obviously, manipulation (and replication and randomization for that matter) can be difficult or impossible for many important questions, especially those on big scales (say, community-level effects of removal of sea otters) or those that happen in just one place (effects of an oil spill, or a dam). Still, such questions can be formally evaluated, perhaps through careful collection of data before and after treatment, and by mini-mizing extraneous confounding factors. Also, whole studies can and should be repeated (**metareplication**; Johnson 2002) and analyzed as a group (**meta-analysis**). If an idea$_{mind}$ is supported in a study repeated in different years, at different sites, with different methodologies, or by different investigators, you are much more likely to believe that it is real.

P values, power, and biologically important differences

In the most common form of hypothesis testing, the final step is to determine a P value, otherwise known as a test for **statistical significance**. By convention, if $P < 0.05$ (or sometimes $P < 0.10$) the null hypothesis is rejected and a **significant** effect is declared[6]. The validity and misuse of null hypothesis significance testing has been vig-orously debated in the last decade (Yoccoz 1991, Anderson et al. 2000, Robinson & Wainer 2002). Here are some main points. First, no heavenly dictum supports a magic threshold of $P < 0.05$. Far too often biological sense has been thrown out of the window with a $P = 0.05$ dichotomy leading to $P = 0.04$ being trumpeted as significant – with all sorts of implications for ecology and management – while the same test with $P = 0.06$ is panned as meaningless and insignificant. We should "stop treating statistical testing as an all-or-nothing procedure and instead use appropriate wording to describe degrees of uncertainty" (Robinson and Wainer 2002:269). For example, differences might lean in a certain direction, or indicate a hint about the true direction, or even indicate simply that differences could not be determined; thus the study needs to be repeated before reliable inference can be made.

Second, the P value does not give the probability that the null hypothesis is true, and smaller P values alone do not necessarily mean a more false null hypothesis. Rather, P values tell you the probability of observing data as extreme (or more extreme) as the observed data given that the null hypothesis is true, with repeated sampling of

---

[6]To be precise, Type I error or α is set at 0.05 as a cutoff, and the P value for the test is compared to the preset α; if P is less than $\alpha = 0.05$, then the test is deemed statistically significant (Type I error is discussed later in this section).

Table 2.1 Possible outcomes of decisions based on frequentist statistics and P values. (a) A medical example where the null hypothesis is that a patient does not have a fatal disease. (b) An example from common endangered species monitoring where the null hypothesis is that a population is stationary (no downward trend in numbers). λ represents the population growth rate. Statistical power is represented by the lower right-hand cell in both panels. Values in parentheses give the probabilities associated with each decision.

(a) Medical example

	Patient actually . . .	
Your decision	**does not have the disease**	**does have the disease**
Fail to detect disease	Correct: state no disease ($1 - \alpha$)	Incorrect: Type II error (β)
Detect disease	Incorrect: Type I error (α)	Correct: state that disease is present ($1 - \beta$)

(b) Endangered species monitoring example

	Population actually . . .	
Your decision	**is stationary, $\lambda = 1.0$**	**is declining, $\lambda < 1.0$**
Population is stationary, $\lambda = 1.0$	Correct: state no decline ($1 - \alpha$)	Incorrect: Type II error (β)
Population is decreasing, $\lambda < 1.0$	Incorrect: Type I error (α)	Correct: decline detected ($1 - \beta$)

the data[7]. A small P value does not necessarily indicate that the effect or treatment was large because small P values can also arise with a small effect size if sample sizes are large or variability small.

When testing a hypothesis using P values, the inference either correctly matches the true state of nature, or is wrong. If wrong, the inference can either conclude a difference between treatments when really there is none, or conclude no difference when really there is one (Table 2.1). The first error, falsely concluding a difference, has traditionally received the most attention and for historical reasons is called a Type I error (symbolized by α); a P value is deemed statistically significant if it is less than α. The second error, concluding no difference when really there is one (or, in null hypothesis jargon, concluding that the null hypothesis of no change is supported when really the null is false) is Type II or β.

We are ingrained, often through statistics classes, to focus primarily on minimizing Type I error to decrease the probability of saying there is a difference or effect when

---

[7]The fact that the statistics are conceptually based on long-run frequencies under repeated sampling explains the moniker of **frequentist** statistics.

really there is not one. That's why the arbitrary threshold of $\alpha = 0.05$ (5%) is so ingrained in our field. But is Type I error always worse than Type II? Consider a medical analogy (Table 2.1a). The null hypothesis is that a person does not have a fatal but treatable disease. If the person really does not have the disease, then we could either correctly detect no disease, or incorrectly detect the disease, thereby falsely rejecting the null hypothesis and committing a Type I error (called a false positive in medical research). On the other hand, if the person actually has the disease, we could either correctly detect the disease or commit a Type II error by failing to detect it (a false negative). Which of these mistakes is worse? A Type I error would upset the patient, and potentially cause them to question the credibility of the test. But follow-up tests would surely occur, indicating no disease. If, however, a Type II error occurred, the disease would not be detected and the doomed patient would head back out in the world, soon to die, a victim of Type II error.

By analogy, an assertion that an endangered species is decreasing when it is really stationary (Type I error) may initiate unnecessary restrictions, leading to some loss of credibility (Table 2.1b). However, a declaration that the population is stationary when really it is declining (a Type II error) means that the population is heading toward extinction while nothing is done! Similarly, when evaluating the effects of exploitation, the consequences of Type II errors (failing to detect a real negative effect) may be of more concern than Type I error (Nichols et al. 1995).

Statistical power is the probability of not making a Type II error, or the probability of correctly rejecting a false null hypothesis. In practical terms, statistical power is the probability of detecting a pre-specified difference or effect that is really there. Power is positively related to α, sample size, and size of effect (e.g. the difference among treatments or the steepness of trend lines), and negatively related to variation. Thus, if sample sizes, effect sizes, or the specified Type I error rate are very small, or if variation is large, then power may be too low to detect a difference that is real. In fact, because studies of trends in endangered species almost always involve subtle changes, small sample sizes and/or large variance, there will be low power to detect real declines (Taylor & Gerrodette 1993, Gibbs et al. 1998)[8]. Beware of the stacked deck of cards presented to biologists assessing declines or other effects in a null hypothesis framework: it may be very difficult to document a real decline when sample sizes are small or variance is large.

Power analysis should be used in the planning stages of an experiment or management action to determine necessary sampling designs and sample sizes, and whether the study is even possible given logistical and financial constraints that limit sample size or lead to power-busting high variation[9]. Such *a priori* analysis greatly improves

---

[8]Sometimes people will increase α to, say, 0.1 instead of 0.05, as a way to increase power.

[9]Power tests typically should not be used to interpret results after a P value has been obtained (so-called *a posteriori*, retrospective, or *post hoc* power analysis). Although many statistical packages offer these *post hoc* power tests, the power estimates are redundant with the P value of the finished study, and are unreliable (Steidl et al. 1997, Gerard et al. 1998). Once you've done your study, just present effect sizes and confidence intervals and let the reader interpret how sample size, variance, α, and effect size affect the interpretation of biological significance.

sampling design because it helps determine: (i) the sample size needed to detect an effect that you feel is biologically meaningful; (ii) the detectable effect for a given sample size, power, variance, and α level; and (iii) the power to detect a certain effect if we were to initiate a study with an expected effect, sample size, variance, and α. Ideally one should consider a range of values, which is relatively easy given the power-analysis modules in data-analysis software (Thomas & Krebs 1997).

The misinterpretations of P values and problems with power in the null hypothesis framework have led some to suggest that the P value framework be abandoned in favor of only presenting effect sizes and confidence intervals (Johnson 1999, Anderson et al. 2000). Certainly presenting a naked P value without the effect size and precision is almost always a bad idea. Presenting the effect (e.g. histograms of means) and precision (e.g. SE) helps clarify the distinction between biological and statistical significance because it lets the reader judge the implications of the observed effect and how well the parameter of interest was estimated (Yoccoz 1991).

Recently, much attention has been paid to alternative frameworks for hypothesis testing that avoid the issues of P values and power entirely. These include information-theoretic methods and Bayesian methods, briefly described below.

Model selection based on information-theoretic methods

Instead of using P values, null hypotheses, significance testing, and α or β errors, a model-selection framework selects among models conceived by the researcher to identify biologically realistic sources of variation (Burnham & Anderson 2002, Johnson & Omland 2004). By determining which models in the candidate set best approximate the data, hypotheses about biological processes (alternative models) can be tested, and parameters can be estimated.

The first step is to build models from biological intuition, knowledge of the system, and previous studies. With these *a priori* models in hand, data are collected and used to test the fit of each model to the data. The comparison among models requires an objective criterion, bringing us to **Akaike's information criterion** (AIC), one of the most famous analytical buzzwords roaring onto the scene of 21st century population ecology. AIC in particular and information-theoretic methods in general operate on the simple principle that ideas$_{mind}$ should be rewarded when they fit the data with the least number of parameters, or pieces. Under the principle of **parsimony** the best model (or models) are those with the highest likelihood[10], given the data, but also those that are the simplest, with the fewest parameters. If a model has too few parameters, or the wrong ones, it will not fit the data well and will lead to biased estimates; adding parameters will almost always improve fit to the data – thereby decreasing bias – but the additional pieces make variance balloon and make spurious factors more likely to

---

[10]In case the word **likelihood** is foreign to you, the key to distinguishing between likelihood and the more familiar probability is "with probability the hypothesis is known and the data are unknown, whereas with likelihood the data are known and the hypotheses unknown" (Hilborn & Mangel 1997:133).

be deemed important. So AIC is calculated for each model; the lower the AIC value the better the fit to data without extra parameters. Formally,

$$\text{AIC} = -2\ln\left[L(\hat{\theta})\,|\,data\right] + 2K \tag{2.1}$$

Where $\ln\left[L(\hat{\theta})\,|\,\text{data}\right]$ is the value of the maximized log-likelihood parameter θ given the data and the model, and K is the number of parameters estimated in that model. All models are ranked, beginning with the model with the lowest AIC considered to be the best one for that set of empirical data[11]. The **Akaike weight** of each model provides a relative weight (w_i) of evidence for each model, interpretable as the probability that model i is the best for the observed data, given the candidate set of models[12].

If a single model is clearly the best approximating of the set – say seven or more AIC units smaller than the next best model – then there is little uncertainty in model selection. In many cases, however, there is not clear support for any single model, and indeed a commonly used convention is to consider any models within two AIC units of the best approximating model as having substantial support. In such cases, one can account for **model-selection uncertainty** (within the candidate models considered) by summing Akaike weights for all the models that contain particular parameters or predictor variables. Similarly, if you are interested in estimating a parameter such as survival, and no single model has overwhelming support (say, with an Akaike weight, w_i, of the best model <0.9), then you could calculate a weighted average of ($\hat{\theta}$) across all R models:

$$\hat{\theta} = \sum_{i=1}^{R} w_i \hat{\theta}_i \tag{2.2}$$

(for variance and confidence intervals see Burnham & Anderson 2002).

There are several important caveats. First, AIC is not the only game in town; alternative model-selection criteria exist (Taper 2004), including the Bayesian approaches

---

[11]Actually, in virtually all applications in our field you should use the small-sample unbiased version, AIC_c. If the data are overdispersed – with inflated variance – a variant called QAIC_c is used. Overall, realize that although I just refer to AIC in the text, you will need to go to more advanced sources to learn the subtleties of which version of AIC to use.

[12]The dirty details: first you calculate a ΔAIC for each model i as the difference between that model's AIC and that of the best model (the one with the lowest AIC):

$$\Delta_i = \text{AIC}_i - \text{AIC}_{\min}$$

Because the likelihood of each model i given the data is $\exp(-1/2\Delta_i)$, the Akaike weight normalizes the likelihoods across all R models in the set, so they sum to 1:

$$\text{Akaike weight} = w_i = \frac{\exp\left(-\dfrac{1}{2}\Delta_i\right)}{\displaystyle\sum_{r=1}^{R}\exp\left(-\dfrac{1}{2}\Delta_r\right)}$$

described next. Second, the concept of sufficient sample sizes and statistical power continues to be relevant here because appropriate model selection depends on the data available.

Third, remember that under this model-selection framework all inferences are dependent on (or, in statistics lingo, are conditional on) both the data and the set of models considered[13]. You should avoid **overfitting** a shotgun blast of arbitrary and potentially spurious models that lead to wrong estimates and indicate support for variables that fit the particular data-set but that are not biologically relevant. Perhaps more serious, however, is **underfitting**, or leaving out important models. In an interesting discussion of the consequences of underfitting in AIC analyses, Beissinger and Snyder (2002) argued that biological inferences – and subsequent management recommendations for snail kite recovery – were fundamentally compromised when a study left out key models when testing for effects of water level on nesting success (see the response by Dreitz et al. 2002).

To minimize the risk of a poor model set, the most parameterized **global model** (the one that includes all potentially relevant effects and causal mechanisms considered) can be tested for **goodness of fit**. Model-fit statistics (e.g. regression residuals, R^2, or formal chi-square tests) assess whether the most complex model adequately describes the variation in the data. In general, the dependence of inferences on *a priori* models developed by the researcher reinforces the points made above about the importance of the critical development of biological hypotheses early in the study (preferably prior to data collection, and certainly before data analysis!).

There are certainly critics of information-theoretic approaches based on AIC (Guthery et al. 2005). However, model-selection approaches are here to stay as facilitators of the estimation of population parameters, and as a complement – or in some cases an alternative – to null hypothesis testing using *P* values.

Bayesian statistics: updating knowledge with probability distributions

Bayesian statistics are another alternative to traditional null hypothesis testing using *P* values, and in fact have much in common with model selection using an information-theoretic framework (Hilborn & Mangel 1997). In essence, Bayesian methods formally incorporate information that has already been acquired (or presumed) to establish a **prior probability** that an idea$_{mind}$ is true; new data update those prior probabilities, leading to a **posterior** probability that a model is true, given the data. This becomes the revised current opinion, to be again modified as more data are collected.

Bayesian statistics trace back to a short memoir published posthumously in 1763 by a preacher and hobby-mathematician named Thomas Bayes (Bayes 1763). Bayes' Theorem quantifies how new evidence changes the probability that the existing belief is correct (see Ellison 1996, Hilborn & Mangel 1997, Wade 2000). The pieces are as follows.

---

[13]A quotable quote from Burnham and Anderson (2002:64): " 'Truth' is elusive; model selection tells us what inferences the data support, not what full reality might be."

- $P(H|D)$ The posterior probability of the hypothesis being true (or of obtaining the specified parameter, such as survival or abundance), given the data at hand.
- $P(H)$ The prior probability before the experiment is conducted or data collected. This is your initial estimate of the weight of evidence in favor of the hypothesis.
- $P(D|H)$ The likelihood function for the data, given that the hypothesis is correct. This is the same as the likelihood $\ln[L(\hat{\theta})|data]$ for AIC in eqn. 2.1.
- $P(D)$ The averaged probability of the data across all hypotheses.

And Bayes' Theorem is

$$P(H|D) = P(D|H) * \frac{P(H)}{P(D)} \tag{2.3}$$

The denominator is basically a scaling constant that can be factored out to give a simple statement of Bayes' Theorem:

$$P(H|D) \propto P(D|H) * P(H) \tag{2.4}$$

This says that the posterior probability for a hypothesis or parameter is proportional to its prior probability multiplied by the degree to which the hypothesis explains the data. Or, what we think now depends on what we thought before, modified by the insight we just got from the new data.

Bayes' Theorem can be extended to assess the relative probabilities of multiple working hypotheses. For example, for two hypotheses the posterior probabilities in favor of each hypothesis would be as follows. For hypothesis 1:

$$P(H_1|D) = \frac{P(H_1)P(D|H_1)}{P(H_2)P(D|H_2) + P(H_1)P(D|H_1)} \tag{2.5a}$$

For hypothesis 2:

$$P(H_2|D) = \frac{P(H_2)P(D|H_2)}{P(H_2)P(D|H_2) + P(H_1)P(D|H_1)} \tag{2.5b}$$

Although there are lots of sophisticated ecological examples of Bayesian analysis (see Ellison 2004), for the purpose of distilling the basic approach here is a simple example (Phillips 1973). Suppose an unscrupulous gambler carries around a biased coin (it favors heads slightly, with a 60% chance of coming up heads and 40% tails[14]), but after getting some change he is suddenly not sure whether a coin in his pocket is fair or biased. How does he decide? First he embraces his uncertainty by setting the probability of either hypothesis (biased or fair coin) as equal:

$$H_1: \text{fair coin}; P(H_1) = 0.5$$

$$H_2: \text{biased coin}; P(H_2) = 0.5$$

---

[14]In case the terms **heads** and **tails** are unfamiliar to you, those are just words we use in the USA to refer to the two sides of a coin.

Now he needs some data. He tosses the coin twice, realizing that the biased coin is more likely to be heads (H). He gets two heads. The likelihoods of two heads with the fair or biased coins are:

$$P(D|H_1) = \text{Probability of H and H given a fair coin} = 0.5 * 0.5 = 0.25$$

$$P(D|H_2) = \text{Probability of H and H given a biased coin} = 0.6 * 0.6 = 0.36$$

Plugging the prior probabilities and the likelihoods from the data into Bayes' Theorem (eqn. 2.5), he gets the following posterior probabilities:

$$P(H_1) = 0.41$$

$$P(H_2) = 0.59$$

Our gambler is now 59% sure that he holds his beloved biased coin. But he wants to be more sure. The posterior probabilities become the prior probabilities, and he continues to flip: he gets TH, HH, HH, and TH. Now, after 10 flips, eight of which came out heads, the gambler is 73% sure that he holds the biased coin. Depending on how sure he wants to be, he can keep going, or just pocket the coin and look for his next gambling victim.

Although the complexities of real-world Bayesian analyses has hampered their application, ever-growing computing power and ability to perform simulations are increasing their application[15]. Proponents of Bayesian approaches in ecology note that probability distributions and their uncertainty are simple to understand, directly show biological relevance, and easily allow comparison of different models. Whereas there are certainly detractors and controversies (e.g. Dennis 1996, Taper & Lele 2004), it seems clear that, like model selection using AIC, Bayesian approaches are established as a viable alternative to traditional *P* value-based hypothesis testing for some applications in wildlife population biology.

Ethics and the wildlife population biologist

It may seem a bit odd to have an ethics section in a book chapter on how we know what we know in wildlife population biology. But think about it: ethical transgressions in the pursuit of knowledge make everything else irrelevant. No amount of statistical rigor or experimental elegance can undo damage made by actions that violate ethical standards.

I will not dwell on the ethical obligations of biologists to address applied issues, or on the implications of the **land ethic** (eg Pister 1999, Leopold 2004). Instead, I will simply reiterate that at its core, ethics has to do with the standard of behavior within

---

[15]Currently, for Bayesian statistics ecologists use software such as WinBUGS (www.mrc-bsu.cam. ac.uk/bugs/winbugs/contents.shtml).

our profession. Following those ethical norms guides the integrity of our inference from idea through collected data, all the way to conclusion and application of a study to real-world wildlife population issues.

Most of the scientific societies in our discipline, including the Ecological Society of America, The Wildlife Society, and the Society for Conservation Biology, have codes of ethics. Every student of applied population biology should know about them. Jack Ward Thomas (1986) has paraphrased, in simple yet eloquent terms, the code of ethics for The Wildlife Society:

> 1 Tell folks that your prime responsibility is to the public interest, the wildlife resource and the environment.
> 2 Don't perform professional services for anybody whose sole or primary intent is to damage the wildlife resource.
> 3 Work hard.
> 4 Don't agree to perform tasks for which you aren't qualified.
> 5 Don't reveal confidential information about your employer's business.
> 6 Don't brag about your abilities.
> 7 Don't take or offer bribes.
> 8 Uphold the dignity and integrity of your profession.
> 9 Respect the competence, judgment and authority of other professionals.
>
> Implied but not specifically mentioned is the requirement simply to tell the truth . . . Tell the truth, all the truth, all the time. It's the right thing, the healthy thing, the professional thing to do.
>
> **(Thomas 1986)**

The rules are simple, but can be hard to follow in a complicated world. Thomas' last comments about truth, in particular, can be difficult when an applied biologist feels outgunned, outspent, or out-politicked. It boils down to the question: "Irrespective of the righteousness of the cause, is distortion of the truth ever permissible?" (Erman & Pister 1989). The short answer, which goes to the heart of the integrity of our profession, is no.

A few years ago I learned first-hand the brutal consequences that can occur when these simple ethical guidelines are violated in a wildlife study. In the USA, Canada lynx became a species of special concern to land managers in the late 1990s, and was listed in March 2000 as a federally threatened species in the contiguous USA. At the time of listing it was not known precisely where lynx occurred. To provide a basis for subsequent monitoring, Kevin McKelvey of the US Forest Service (USFS) Rocky Mountain Research Station and I led a project called the National Lynx Survey to provide a consistent, standardized, reliable process for determining the current range of lynx in the USA. We designed the National Lynx Survey around non-invasive genetic sampling using hair rub pads (to be described more in Chapter 3).

Before initiating the study we carefully developed, validated, and exposed to peer-review a DNA-based species-identification protocol, using both blind and widespread geographic range tests (Mills et al. 2000a, Mills 2002). Although the collection of samples across 16 states in the northern USA was administered through the USFS infrastructure, people from several agencies participated; approximately 800–1000 field helpers deployed and collected the hair pads, then sent them to my laboratory for analysis. Detailed protocols for collecting the hair samples – including discussion of the controls that we had used to develop the species-identification test – were included in all collection kits sent to field workers.

However, a handful of field workers in Oregon and southern Washington (where no actual lynx were detected) took it upon themselves to label some lynx samples as if they had been collected from the field (complete with slope, elevation, location, and vegetation types filled out on data forms) when really they were collected from captive or wall-mounted animals.

These mislabeled samples (which we correctly identified to species) would have been folded into our analysis of samples collected as part of the National Lynx Survey if not for a telephone call months later from one of those who mislabeled samples. When McKelvey and I found out about the mislabeling, our response was to deal with it internally, re-iterating to the hundreds of people collecting data on this project that internal controls were all in place and that the most important thing they could do was to ensure the integrity of samples coming from the field to the laboratory. But before we could do that, a political and media frenzy erupted. In December 2001, the *Washington Times* broke the story of mislabeled samples as a symptom of rampant fraud among applied biologists. Some in the US Congress followed up on the frenzy, saying that all actions on species protection should come into question. These are extreme interpretations to be sure, but in response others at the opposite end of the political spectrum – committed above all else to defending what they perceived as the higher goal of endangered species protection against attacks from Congress – espoused an equally extreme view that mislabeling samples had been an appropriate and even noble thing to do[16]. Three government investigations and one Congressional committee were launched to investigate this matter, and it was covered by dozens of journalists.

In my testimony to the US House of Representatives on March 6, 2002, I noted that we can never know the motivations for those who mislabeled samples. However, I stressed that, although there was no scientifically valid explanation for the mislabeling of samples, the actions of these few should not condemn the credibility of applied biology:

> My experience throughout my career in working with hundreds of biologists and field personnel – including employees of USFWS, USFS, NPS, state wildlife departments, private groups, and several universities – is that they have exceptionally

---

[16]I actually had one activist tell me that I should publicly announce that the field workers had good reason to mislabel samples as a test to expose incompetence because my laboratory was unreliable! He told me that if I did not help him make the argument that the field workers had done the right thing that I would be playing into the hands of those in Congress who wanted to bring down the Endangered Species Act. It horrified me that this person was seriously suggesting that the scientific process by which applied biologists contribute to important policy decisions should be twisted into a political tool.

high ethical standards in their pursuit of knowledge. Although inappropriate actions may occur on an individual and rare basis, my opinion is that these instances do not invalidate the larger body of biology, in the same way that inappropriate actions by a few physicians does not mean that we should shut down the practice of medicine.

In the end, the National Lynx Survey continued intact. However, there is no doubt that for the short term the credibility of our profession was damaged by "Lynxgate." Wildlife and conservation biology are relatively young professions, with credibility hard won on the backs of thousands of professional lifetimes (Thomas & Pletscher 2002). Increasingly, applied population biology work shows up on the front pages of newspapers, is heard in courtrooms, and is considered in the drafting of laws. Trust is everything to our continued relevance as applied biologists. Here are seven lessons from the mislabeled samples in the National Lynx Survey that should be considered action items for all wildlife biologists (quoted from Thomas & Pletscher 2002:1285):

1 Refresh our acquaintance with the ethical standards of our profession.
2 Assure adherence to those standards by bringing attention to actions that are inappropriate.
3 Condemn violations.
4 Consider every action by the standard of whether we would be proud to see it printed in the newspaper – because that is likely.
5 Understand that wildlife biologists now play a significant role in national affairs, and individual and collective actions will be considered in that light.
6 Know that in a mature profession with significant public trusts and responsibilities, there is simply no room or excuse for operating outside the rules of the game.
7 Recognize the responsibility – of teachers, agencies, and the profession – to state, formulate, teach, and continuously reinforce ethical standards and the need for transparent processes.

So, the very fact that all this happened convinced me that it is worth a page or two to remind all of us of our simple and perhaps obvious ethical obligations. Whether we are a scientist designing a study or a technician carrying out a study, our honesty is the bedrock on which lies all else in designing studies and interpreting population biology data.

Summary

By considering **truth** to be when an idea of the mind (idea$_{mind}$) matches a factual referent, it becomes apparent that applied population biology can only move forward by coupling creative ideas, reliable factual referents measured in wild populations, and a formal process to connect the idea$_{mind}$ and the facts. That is how reliable knowledge is gained.

Some of the prerequisites to obtaining trustworthy facts from the field include repli-cation, randomization, and the use of controls. Accurate measurements have low bias and high precision. In the spirit of our **embrace uncertainty** mantra, variance may be more important than the mean. Process variance comes from spatial and temporal variance in nature, while sampling variance is an inevitable nuisance that arises anytime we sample from a population.

The strongest formal approach to connect facts to ideas of the mind is through the hypothetico-deductive approach. Natural history and observed correlations (induc-tion) form an important basis for hypothesis generation, but specific *a priori* predic-tions should then be tested against data. This is the stage for care and forethought in developing the scientific question to be asked, and for specifying both the sampling strategy and how data will be interpreted.

The best way to distinguish among hypotheses is currently a matter of debate. The traditional approach of testing for statistical significance using P values has been crit-icized as being too reliant on an arbitrary threshold of $P = 0.05$. Also, there has often been too little consideration of statistical power – the probability of detecting an effect that is there – a real problem in studies where missing an effect could damage wildlife populations.

Some have argued that alternative frameworks for hypothesis testing should be implemented more widely. Model-selection frameworks using either AIC or Bayesian criteria are rapidly gaining footholds in wildlife population biology. In both cases, data inform the likelihood of particular models (or idea$_{mind}$) being best-suited to explain a particular set of data.

Finally, any rigor in study design or implementation is bereft without a strong foun-dation in ethics. Scientific societies in applied wildlife population ecology have ethics guidelines, and we should all be familiar with them. The integrity of our profession depends on it.

Further reading

Garton, E.O., Ratti, J.T., and Giudice, J.H. (2005) Research and experimental design. In: *Techniques for Wildlife Investigations and Management* (ed. C.E. Braun), pp. 43–71. The Wildlife Society, Bethesda, MD. A well-applied description of key aspects of research philosophy and experimen-tal design.

Hilborn, R. and Mangel, M. (1997) *The Ecological Detective*. Princeton University Press, Princeton, NJ. A lucid overview, with examples, of advanced concepts in maximum likelihood, model selec-tion, and Bayesian analysis.

Krebs, C.J. (1999) *Ecological Methodology*, 2nd edn. Benjamin Cummings, Menlo Park, CA. A much-admired classic that describes the fundamentals of the decisions and analyses made by practicing field ecologists.

Morrison, M.L., Block, W.M., Strickland, M.D., and Kendall, W.L. (2001) *Wildlife Study Design*. Springer-Verlag, New York. A fine compilation of study design with application to wildlife popu-lation and habitat ecology.

3

Genetic concepts and tools to support wildlife population biology

For those not studying biology at the time in the early 1950's, it is hard to imagine the impact the discovery of the structure of DNA had on our perception of how the world works. Reaching beyond the transformation of genetics, it injected into all of biology a new faith in reductionism. The most complex of processes, the discovery implied, might be simpler than we had thought. It whispered ambition and boldness to young biologists and counseled them: Try now: strike fast and deep at the secrets of life.

E.O. Wilson (1995), *Naturalist*

Introduction

Until fairly recently the term "genetics" would have hardly been mentioned in a book on applied wildlife population biology. Now, however, genetic issues related to wildlife populations seem to paint the newspapers almost every day: Is the red wolf taxonomically distinct enough to warrant special management? Are northern hairy-nosed wombats genetically impoverished? Are red-cockaded woodpeckers suffering lower survival or reproductive rates due to inbreeding? Does a sample from meat in a freezer match that of an illegally harvested deer?

Because genetic concepts and tools will increasingly be at the heart of many wildlife management issues in the future, a variety of genetic applications will be considered throughout this book. This chapter will explain what genetic variation is and how it can be described using several common genetic markers, and give a few of the insights into wildlife populations that can be gained by genetic analysis. In short, this chapter will build the foundation for applying genetic approaches to other topics throughout the book.

What is genetic variation?

The **phenotypic expression** of almost all individual traits (ranging from body weight to camouflage pattern to sprint speed to metabolic efficiency) is a function of the genetic makeup, or **genotype**, coupled with the environment to which the individual is exposed. The **genes** that make up the genotype are stretches of DNA along chromosomes in the nucleus of all cells; the location of a gene on a chromosome is called a **locus** (plural **loci**). A gene can vary in the specific sequences of DNA nucleotides that comprise it, and the different forms of a gene are called **alleles**. Sometimes a phenotype is a **single-gene trait**, determined from combinations of alleles at just one locus – for example, whether or not you can roll your tongue, or the dark coat color in the red fox (Våge et al. 1997) – but most often traits are determined by complex combinations of genes, often interacting with the environment.

In diploid organisms, including nearly all vertebrates, one allele at each locus is inherited from each parent. If the two alleles at a locus are the same the individual is **homozygous** for that gene; if they are different then the individual is **heterozygous**. At the population level, the description of heterozygosity relies on the concept of Hardy–Weinberg equilibrium (Box 3.1). While heterozygosity describes variation in how genes are packaged at each locus, several other terms describe the number of alleles at each locus. A gene is considered **polymorphic** if more than one allele is detected at a locus across all individuals sampled[1]; otherwise the gene is **monomorphic**. **Allelic diversity** or **allelic richness** describes the average number of alleles per locus[2]. As a practical aside for wildlife management, allelic diversity is more likely to be lost following a severe population contraction (bottleneck) than is heterozygosity, because heterozygosity is not much affected (at least initially) by the changes in frequencies of rare alleles that are lost during a bottleneck (Nei et al. 1975, Leberg 1992).

So far we have only mentioned measures of nuclear genetic variation based on loci with distinct, identifiable alleles in the nucleus of the cell. However, another form of genetic variation occurs in **mitochondrial DNA** (mtDNA). Mitochondria are organelles often referred to as the cell's powerhouse because they produce energy. The genes in mitochondria are different than those in the nucleus, with mtDNA coding for cell machinery functions and not for phenotypes that we can observe (with a few exceptions). Because mtDNA is haploid (having one form of the gene, not two as in nuclear genes), heterozygosity cannot be measured. However, mtDNA has some important features that make it very useful for applied population biology. First, in contrast to the one copy of each nuclear gene within each cell, mtDNA is present as multiple identical copies – thousands within most mammalian body cells – allowing analysis of mtDNA from very small or poor-quality samples (for example, single hairs). Second, in vertebrates mtDNA is maternally inherited and does not recombine, meaning that sons and daughters inherit their DNA from their mother only. As a result of the maternal inheritance of haploid mtDNA molecules, one breeding pair of parents contains only one transmittable mtDNA genome, in contrast to the four possible nuclear genomes. These features make mtDNA a sensitive marker for detecting

---

[1]Technically, a gene is polymorphic if the most common allele has a frequency of less than 95% (or sometimes 99%).

[2]Estimates of allelic diversity must be adjusted for sample size because fewer samples will tend to have fewer different alleles (Leberg 2002).

Box 3.1 The Hardy–Weinberg Principle and describing heterozygosity

The Hardy–Weinberg Principle forms the cornerstone of conservation genetics used in wildlife application. If two alleles (A_1 and A_2) at a locus have frequency p and q respectively, then after one generation of random mating the frequencies of the three possible genotypes (A_1A_1, A_1A_2, and A_2A_2) are p^2, $2pq$, and q^2 respectively. Since these are the only possible genotypes with just two alleles, then $p^2 + 2pq + q^2 = 1$.

Extending this same idea to multiple alleles, the Hardy–Weinberg frequency of the homozygote genotype for any allele i with frequency p_i is p_i^2. Because an individual that is not homozygous must be heterozygous at a locus, the expected Hardy–Weinberg frequency of heterozygotes, given k alleles at a locus, is:

$$1 - \sum_{i=1}^{k} p_i^2$$

The Hardy-Weinberg allele and genotype frequencies will remain constant over time if they are at equilibrium; that is, if they are unaffected by evolutionary forces such as natural selection, genetic drift, mutation, and gene flow. Populations are of course affected by these processes, but Hardy–Weinberg equilibrium genotype frequencies provide a tremendously useful benchmark; if the population is out of Hardy–Weinberg equilibrium we can ask why, thereby taking the first step toward elucidating mechanisms acting on a population's genetic composition.

Heterozygosity at the population level is typically described as **expected** or **observed**. Expected heterozygosity is that expected under Hardy-Weinberg equilibrium. By contrast, observed heterozygosity is the actual proportion of individuals observed to be heterozygous, averaged across loci. Deviations between observed and expected heterozygosity can be useful for inferring processes that are acting upon wildlife populations, such as genetic drift, selection, and gene flow.

Let's work through a simple example for just one locus based on data from the endangered Hawaiian Laysan finch (Frankham et al. 2002:74); real studies would use multiple loci.

At this one locus there are three alleles, with the following frequencies: $p_1 = 0.364$, $p_2 = 0.352$, and $p_3 = 0.284$. (Notice that the three allele frequencies sum to 1.0.) Using the Hardy–Weinberg Principle, the expected heterozygosity is

$$1 - \sum_{i=1}^{k} p_i^2 = 1 - (0.364^2 + 0.352^2 + 0.284^2) = 0.663$$

In this case, 29/44 sampled finches are heterozygotes, so observed heterozygosity at this locus (0.659) was very close to expected heterozygosity (0.663).

reductions in population size, tracing maternal lineages and sex-specific dispersal, and inferring mating systems.

Finally, **quantitative variation** describes phenotypic traits determined by multiple loci interacting with the environment, including fitness-related characters such as morphology, survival, disease resistance, and reproductive rate. Identifying and tracking quantitative trait loci (QTLs) is more difficult than monitoring single-locus variability, because large numbers of individuals of known relatedness must be followed in intensive field studies or experimental crosses. Despite the fundamental importance of quantitative variation in conservation applications, the effort and expense required to measure it have limited the use of QTLs in applied population management for wild animals (Sherwin & Moritz 2000).

Genetic markers used in wildlife population biology

How do we actually measure genetic variation (heterozygosity and allelic diversity or polymorphism) and use genetic markers to resolve questions in applied wildlife population biology? I will start with a brief sketch of a few of the most commonly used markers, to make understandable the applications discussed throughout this book.

Modern molecular biology has made available a staggering array of tools for applied biologists. Prior to the 1960s, genetic variation was measured almost entirely by observing breeding patterns and observing phenotypic variants, or injecting purified protein into rabbits and observing antigen–antibody reactivities. Both of these procedures had obvious limitations for tracking genetic variation in wild populations of elusive vertebrates. The development of protein electrophoresis in the 1960s provided – for the first time – a direct and easy way to measure genetic variation. Unfortunately, protein electrophoresis usually requires killing the study subject to obtain protein-rich tissue from an organ (e.g. brain, liver, or heart). The next revolutionary wave splashed in the 1980s, with the coupling of the direct analysis of DNA with the polymerase chain reaction (PCR; see Box 3.2). These markers allow genetic sampling of wildlife populations that is **nondestructive** (where a biopsy or other tissue is obtained but the animal is not killed) and even **noninvasive** (where the genetic sample is collected without having to catch or otherwise disturb the animal).

Protein electrophoresis

Protein electrophoresis begins with ground-up tissues from each individual being applied to an electrophoretic gel through which an electrical charge is passed. The protein products of different alleles, called **allozymes**, separate along the gel according to their net electrical charge based on positively and negatively charged amino acids. Once allozymes are visualized on the gel by the application of locus-specific stains, a sample for an individual that is homozygous at a given locus shows up as one band while heterozygotes display as two bands. Although banding patterns can prove more complicated, the key point is that they can be interpreted to infer the underlying genotype at that locus.

Allozyme analysis is relatively inexpensive and requires less training than DNA-based methods, and a rich source of comparative data exists for hundreds of vertebrate species studied since the 1960s. However, allozyme markers have several disadvantages for most wildlife applications. First, the fact that the animal must usually be killed to extract the necessary organ tissue sample is an obvious limitation when the species is rare or when the question being asked necessitates the animal being alive (e.g. a study of survival, or abundance over time). In addition, tissue must be analyzed when fresh or after quick-freezing, confronting researchers with the prospect of trans- porting liquid nitrogen or freezers into the field. Third, allozymes have low resolution: only a small portion of the genome can be examined, only about a quarter of the loci examined for vertebrate species so far are polymorphic, and heterozygosity tends to be less than 0.1 (Hartl & Clark 1997). The low resolution means that for questions related to recent, subtle changes in genetic variation – for example, due to human- caused population fragmentation – allozymes typically have low power to detect differences.

Box 3.2 How PCR turns tiny samples of DNA into larger samples

PCR is a process of amplifying samples of DNA that are low in quantity or low in quality. The specific steps for conducting PCR can be simplified and summarized as follows (figure modified from Frankham et al. 2002:58 after Avise 1994).

Step **a** is to obtain a sample and extract (i.e. isolate) DNA from it. Put the isolated DNA in a plastic tube, along with **primers** (DNA fragments of about 20 bp) that attach to each strand of the DNA at specific places outside the locus to be amplified. Add some synthetic nucleotides, remembering to marvel at the fact that you are working with the building blocks of life: the four nucleotides that make up all DNA on Earth! Finally, add *Taq* polymerase, an enzyme that attaches the nucleotides to the synthesized DNA strand. Place the tubes into a thermal cycler machine.

Next (step **b**), for each PCR cycle the machine first heats the DNA up to about 94°C to separate (denature) the two strands of DNA, and then cools to 55–65°C to bind (anneal) the primers to their target sequences on each strand. Next the temperature increases slightly to about 72°C to extend the primer into double-stranded DNA, with the *Taq* polymerase attaching the nucleotides to their respective complementary bases on the other strand. In one PCR cycle you have just doubled the number of DNA strands for your target region!

In step **c** the PCR cycles about 30 more times, nearly doubling the DNA each cycle, ending with millions of copies of each original DNA strand.

Without *Taq* polymerase, PCR would not happen. Importantly, *Taq* polymerase was syn- thesized from *Thermus aquaticus*, a bacterium that lives in hot springs in Yellowstone National Park in the USA. The fact that prior to the 1960s *Taq* was just unknown slime in a hot pool reminds us of a benefit of biodiversity (Varley 1993:14):

> Here in the world's most popular geothermal region, an obscure, primitive, hot spring bacterium is discovered that contains an even more obscure enzyme that in turn establishes a procedure that promises to change the world for the better . . . The fact is that [Taq] was available for discovery there in Mush- room Pool because the feature and its basin were not available for more destructive, short-term uses . . . Our celebration of Taq is thus tinged with a vague sense of waste: what else, around the world, have we lost already, and how much more can we afford to lose?

Box 3.2 Continued

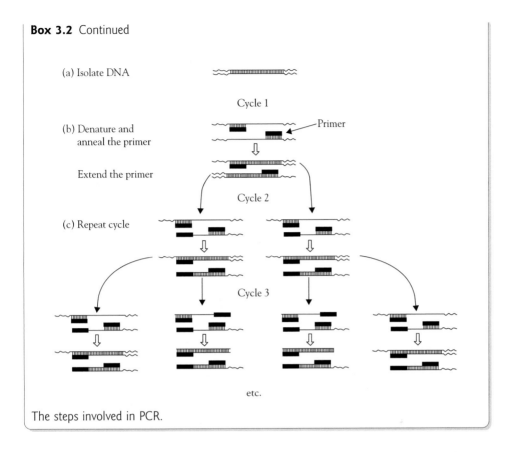

(a) Isolate DNA

Cycle 1

(b) Denature and
anneal the primer

Primer

Extend the primer

Cycle 2

(c) Repeat cycle

Cycle 3

etc.

The steps involved in PCR.

DNA-based markers

One benefit of DNA-based methods as genetic markers is that they typically have high resolution for distinguishing individuals and populations[3]. Another huge benefit is that PCR (Box 3.2) facilitates detection of a genetic signal even from samples that are poor in quality or tiny in quantity. Thus PCR allows analysis of ancient DNA from museums or archived samples, including 20,000-year-old saber-toothed cats (Janczewski et al. 1992) and a 120 million-year-old weevil (Cano et al. 1993)! In the field, DNA can be extracted from hair, feathers, feces, urine, blood, ear punches, toe clips, eggshells, and deer antlers (Morin & Woodruff 1996, Taberlet et al. 1999). Freezing of field-collected samples for DNA analysis is not immediately necessary if the sample has been stored in a proper container and dried or preserved (common options include silica gel and alcohol; see Murphy et al. 2002).

---

[3]Recall that DNA codes for amino acids, which in turn form chains that make up proteins. Not all DNA changes result in amino acid changes, and not all amino acid changes alter the protein structure. So by looking directly at the DNA you have the potential to pick up many genetic changes that are missed by protein electrophoresis, which detects only changes in the protein products.

Fragment analysis

A wide array of techniques fall into this category, whereby the size of DNA fragments (mtDNA or nuclear DNA) is compared among individuals: I will describe just a couple. When DNA fragments are run through a gel in a process called gel electrophoresis, smaller pieces of DNA migrate faster, ending up further down the gel than the longer pieces. When the fragments of DNA are illuminated, they show up sorted by size, with big pieces near the top of the gel and smaller pieces near the bottom.

Restriction fragment length polymorphism (RFLP) markers are produced when restriction enzymes recognize specific 4–8-bp (base pair) sequences of DNA, and break the DNA (or **restrict** it) at that spot. Different-sized fragments are produced depending on whether and how mutations have changed the DNA sequences recognized by the restriction enzyme. For example, if an individual has only one of the restriction sites, then only one cut occurs and one linear DNA fragment will be created from a circular mtDNA molecule; two fragments would be created from two mtDNA restriction sites, and so on. If the illuminated fragment bands are characteristic for a species or individual, they can be used for diagnostic identification at the species or individual level. In some applications, specific sections of DNA are first amplified via PCR and then restricted to produce banding patterns (Fig. 3.1).

The variation in RFLP band number comes only from mutations at the restriction site. Other approaches take advantage of the fact that the DNA in between restriction sites may also differ among individuals due to stretches of DNA being repeated for varying numbers of times. For example, **DNA fingerprints** are based on stretches of DNA 10–100 bp long repeated up to several hundred times; because different animals typically have different numbers of repeats (in addition to different restriction sites), the fragments on the gel appear as 10–30 individual-specific bands. These fingerprints, analogous to a bar code on a grocery product, are formally called **multilocus minisatellite** markers.

The multiple bands in multilocus minisatellite DNA fingerprints cannot be assigned to specific loci, preventing direct measures of heterozygosity and allelic diversity. Also, the relatively large size of minisatellite markers means that PCR amplification is usually not possible, especially if the quality of the sample is poor (Bruford et al. 1996). Nevertheless, DNA fragment analysis markers including minisatellites are excellent choices for many questions of individual and/or species relatedness (Burke et al. 1996).

Microsatellite DNA

Technically, **microsatellite markers** belong in the fragment analysis class as cousins of minisatellites. However, their properties are different enough, and their use in wildlife population biology widespread enough, to give them their own section heading. Each microsatellite locus contains short (1–10 bp, usually 2–5 bp) sequences of nuclear DNA repeated between five and 100 times (for example, the two nucleotide bases cyto-

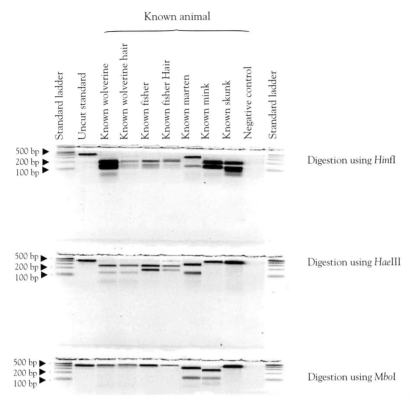

Fig. 3.1 An example of RFLP fragment analysis of mtDNA to distinguish different forest mustelids of the northern USA using single hairs from non-invasive snags (from Riddle et al. 2003, with kind permission of Springer Science and Business Media). After amplifying the cytochrome *b* region of mtDNA with PCR, the DNA was digested with three different restriction enzymes, creating species-specific fragments that collectively distinguish among different species. The first and last lanes are a molecular ladder that helps to determine size of the bands, and the uncut standard contains a PCR product from a wolverine not subjected to the restriction digests; the negative control is pure water to check for contamination. An example for practice: the first restriction digest (*Hin*fl) distinguishes between marten (with two fragments, of 329 and 113 bp in size) and wolverine (with three fragments, of 212, 132, and 98 bp), but wolverine has exactly the same bands as fisher. So the next digest (*Hae*lll) distinguishes between wolverine (259, 140, and 43 bp) and fisher (259 and 183 bp). Thus multiple restriction enzymes are like multiple morphological characteristics that we might use to tell different bird species apart.

sine and adenine, or C and A, repeated 17 times)[4]. Microsatellite loci are amplified using PCR, with the size of the amplified alleles determined by the size and number of repeats (so an allele with CA repeated 17 times will be 4 bp smaller than an allele with CA repeated 19 times). As with other fragment analysis, smaller alleles run further

[4]The nature of this marker explains its other names: simple sequence repeats (SSRs) and variable number of tandem repeats (VNTRs).

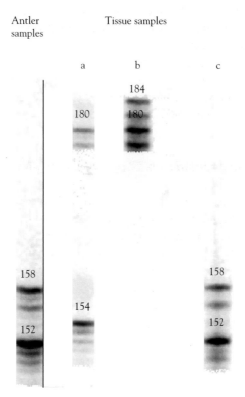

Fig. 3.2 An example of a forensic application using microsatellite DNA (modified from Blanchong et al. 2002, copyright The Wildlife Society): which deer did the antler come from? Shown here is one microsatellite locus analyzed from an antler sample and from tissues (a–c) from three different white-tailed deer. The size of each allele (in base pairs) in each sample is written above the allele (larger alleles have a higher number of the repeat). In this case each individual sample is a heterozygote. Notice that the antler sample matches tissue sample c. Although just one locus is shown for demonstration, actual applications use multiple loci to minimize the likelihood that two individuals share the same genotype.

down the gel. A homozygote individual displays only one band, whereas a heterozygote displays two bands (Fig. 3.2).

Microsatellites are well suited to traditional population genetic models because each locus is codominant, with alleles displaying Mendelian inheritance. In this sense, microsatellites produce similar sorts of information (including heterozygosity and allelic diversity) to allozyme electrophoresis. Unlike allozymes, however, microsatellites have very high levels of variation and are PCR-based, facilitating their use with nondestructive sampling (Luikart & England 1999).

DNA sequencing

Almost certainly, the DNA-based approach of the future is direct DNA sequencing. Sequencing is the most highly informative, highest-resolution technique, and with the growing popularity of automated DNA sequencers is becoming cheaper and easier to implement. Specific regions of the nuclear or mitochondrial genome can be amplified with PCR, and then the exact sequence of DNA nucleotide bases – adenine (A), guanine (G), thymine (T), and cytosine (C) – is revealed when base pairs are illuminated, read by a laser, and interpreted by computer software. Currently, sequencing is the most expensive of the DNA techniques, and the statistical framework for analyzing the data is the least developed. These drawbacks are fading rapidly, however, as the equipment, protocols, and analytical framework improve.

Insights into wildlife population biology using genetic tools

Molecular biology and noninvasive sampling have truly awesome potential for population analysis (Table 3.1). Most of the chapters in this book will contain some application of genetic tools to wildlife population biology questions, for example in quantifying connectivity and isolation among populations, estimating abundance, and solving forensics cases in harvest violations. Instead of elaborating all of these myriad applications in this chapter, I will introduce just a few uses of genetic tools that relate to some of the most basic tasks in population biology: identifying important taxonomic units and distinguishing among individuals.

Taxonomy and hybridization

How individuals are grouped into taxonomic levels determines the fundamental units of conservation and management. Genetic characteristics supplement morphology and other information (e.g. life history, geographic range) to determine taxonomic affiliation. In so doing, genetic information may reveal that groups historically lumped into one species are actually multiple species, with potentially different conservation needs. Conversely, multiple species or subspecies actually may not be distinct; recognizing the similarity may release resources that could be spent on taxa with more critical needs. Box 3.3 describes case studies of each of these scenarios. Genetic markers can also supplement other information to help resolve important taxonomic affiliations below the species or subspecies level, including **evolutionarily significant units** and **management units** (Box 3.4).

Genetic information can also help in detecting and interpreting the consequences of **hybridization**. Hybridization – defined broadly as the interbreeding of individuals from genetically distinct populations – is an enormously complex topic with difficult biological issues and perplexing management implications (for excellent overviews see Rhymer & Simberloff 1996, Allendorf et al. 2001). Prior to 1990, interpretation of the US Endangered Species Act 1973 reflected the widespread view that hybrids were impure, so that protection under the US Endangered Species Act should be discouraged for hybrids between species or subspecies.

In an excellent example of biological information directly influencing policy, a paper by O'Brien and Mayr (1991) helped overturn the hybrid policy. Although there is currently no formal policy on hybrids, federal agencies recognize the following (US Department of the Interior, Department of Commerce 1996).

Table 3.1 The table shows a few examples of wildlife population ecology insights that can be gained from the use of genetic markers and analysis. In most cases nondestructive or even non-invasive sampling can be used with both nuclear and mtDNA markers. Applications a–e are described primarily in this chapter, whereas applications f–h are primarily described in other chapters of this book.

Application	Examples
(a) Species identification	Determination of carnivore species based on hair-rub pads or scats; determination of what prey species are eaten based on remains in owl pellets.
(b) Taxonomic relationships	Tuatara across New Zealand should be managed as multiple taxonomic groups instead of just one; seaside sparrows should be managed as just a couple of taxonomically important units instead of nine.
(c) Determination of hybrids	Detection of lynx–bobcat hybrids, or barred owl–spotted owl hybrids.
(d) Determination of individual identity and sex	Determination from puncture wounds of the sex and identity of coyotes that kill sheep; estimation of the abundance of humpback whales in the North Atlantic or wombats in parts of Australia.
(e) Determination of parentage	Identification of which wolves in a pack mated to produce pups; determination of which male rhinoceros breed and which do not.
(f) Rates of movement of individuals among populations	Estimation of the number of skinks moving between rocky outcrops; determination of whether female white-toothed shrews are more likely to disperse than males.
(g) Levels of genetic variation and size of historical populations	Evaluation of how much heterozygosity has been lost in Florida panthers; determination of how numerous northern elephant seals were before being decimated by hunting.
(h) Forensic applications	Determination of what species were killed to produce so-called whale meat in restaurants; using blood samples taken from a poacher's gun to find out what species have been poached.

- Occasional hybrids are to be expected between species, and "natural occurrences of hybrid individuals or hybrid zones between recognizable species do not disintegrate the genetic integrity of the species" (O'Brien & Mayr 1991:1187–8). For example, occasional hybrids with bobcat should not influence the threatened status of Canada lynx in the contiguous USA. As long as the hybrid offspring more closely resemble the listed species – based on morphological, behavioral, ecological, and molecular data – US Endangered Species Act protection extends to those offspring.
- Hybrid lineages between species usually die out, but they will sometimes establish themselves as a breeding population with their own adaptations and evolutionary history worthy of conservation (assuming the lineage was developed outside of confinement and is self-sustaining and naturally occurring). Thus red wolves should receive protec-

Box 3.3 Genetic information clarifying taxonomy and improving wildlife management

Case study 1: a group managed as a single species is actually multiple species with distinct conservation requirements

The New Zealand tuatara (*Sphenodon*) is the only surviving genus of one order of reptiles and is probably the most distinctive surviving reptile genus in the world, with a morphology nearly unchanged over the last 100 million years. Tuatara have been protected fully since 1895, with the focus on a single species (*Sphenodon punctatus*) throughout New Zealand. Subsequent genetic and morphological analyses, however, have determined at least two different regional taxonomic groups that warrant separate management (Daugherty et al. 1990). Neglecting these distinctions could lead to the extinction of evolutionarily distinct groups, or lead to inappropriate mixing during translocations.

The Stephens Island tuatara, an as-yet-unnamed subspecies of *Sphenodon punctatus*.

Case study 2: multiple taxonomic units are recognized to actually be just one (from Avise & Nelson 1989)

Historically, nine subspecies of the seaside sparrow (*Ammodramus maritimus*) were recognized based on plumage and subtle morphological characteristics. One of these subspecies, the dusky seaside sparrow (*A. m. nigrescens*) was listed as endangered in 1966, as it dwindled in number due to habitat change. In 1980, only six males remained (demographic stochasticity in action; see Chapter 5), and the subspecies was considered extinct by June 1987. However, subsequent mtDNA analysis indicated that the dusky seaside sparrow was not a unique subspecies after all: there was no basis for phylogenetic distinction of *A. m. nigrescens* from other Atlantic coastal populations of *A. maritimus*. Because all Atlantic populations

(Continued)

Box 3.3 Continued

shared one mtDNA genotype and all Gulf Coast populations shared another (see figure), the major conservation focus should be on two subspecies – Atlantic coastal populations and Gulf Coast populations – instead of nine. Recognizing that a taxonomic revision is strongest when supported by a combination of approaches, Avise and Nelson (1989) also found morphological and ecological support for their thesis. In short, there is no question that the habitat loss in this case has been disastrous and that it wiped out the local population formerly known as the dusky seaside sparrow. However, in retrospect the dusky sparrow probably did not warrant the conservation attention that we might provide for more unique lineages, and conservation would be better served by focusing on just two forms: the Atlantic and Gulf Coast forms.

Geographic distributions in the eastern USA of the nine originally recognized subspecies of the seaside sparrow. Open and closed circles represent birds carrying distinctive Gulf Coast and Atlantic Coast mtDNA genotypes respectively (from Avise & Nelson 1989. Copyright (1989) AAAS).

tion under the US Endangered Species Act, even if they originated as wolf–coyote hybrids (see Box 3.5).

- At the subspecies level, hybridization naturally occurs and may have adaptive benefits. In cases where genetic variation is low and new genetic variation is brought in to combat inbreeding depression, offspring should receive protection under the US Endangered Species Act and the population status should not be compromised. The breeding of Texas panthers with highly endangered and inbred Florida panthers (see Chapter 9) is an excellent example.

Box 3.4 Evolutionarily significant units and management units

Evolutionarily significant units (ESUs) and management units (MUs) describe distinctive populations for management purposes at a finer level of resolution than that of species and subspecies concepts. An ESU has been defined as "a population (or group of populations) that (1) is substantially reproductively isolated from other conspecific population units, and (2) represents an important component in the evolutionary legacy of the species" (Waples 1995:9). In other words, ESUs are considered to be the ecological and evolutionary building blocks of the species, whose conservation will allow the continued evolution of the species. ESUs are relevant to legal and policy frameworks, including the Endangered Species Act (1973) in the USA, the Species at Risk Act (2003) in Canada, and the Environment Protection and Biodiversity Conservation Act (1999) in Australia.

The scientific criteria for defining ESUs are evolving, with much discussion on the appropriate role of genetic markers. For example, some scientists argue for an operational definition based strictly on molecular phylogenies to discern historical isolation and evolutionary potential: ESUs are reciprocally monophyletic for mtDNA alleles (i.e. all members are descended from a single common ancestor unique for each ESU) and differ significantly in allele frequency at nuclear loci (Moritz 1994, 1995). In contrast, others call for incorporation of geographical, life-history, habitat, behavioral, and morphological differences to identify both isolation and the degree that the population represents an important part of the species' evolutionary legacy (Waples 1995, Crandall et al. 2000).

Sometimes these different approaches will not point toward the same ESU designation. For example, one small group of populations of Cryan's buckmoths is geographically separated from other populations in North America. A strictly molecular ESU criterion would suggest that this is not an ESU, as there are no significant differences in allele or haplotype frequency between these isolated populations and others. However, host-plant performance experiments indicated that Cryan's buckmoth larvae consume and grow on a unique plant host compared to other populations, implying an ESU based on genetic differences and potentially isolation (Fraser & Bernatchez 2001). In other cases, molecular criteria will provide insights – particularly about the historical legacy of populations – when there are neither resources nor time to conduct experiments of adaptive differences among population groups. In summary, differing ESU approaches can complement each other, focusing on the common goal of addressing the protection of evolutionary potential.

The MU is analogous to the stock concept in fisheries, and refers to population groupings based on restricted demographic interchange. Just what restricted means depends on management objectives, so MUs may have little resemblance to ESUs. Although MUs may have diverged allele frequencies in nuclear DNA, they are not expected to show reciprocal monophyly for mtDNA alleles (Moritz 1994). Thus MUs are unlikely to have different evolutionary potentials: the target level of distinctiveness will be driven largely by policy or management needs and explicit consideration of risks (Taylor & Dizon 1999). For example, the relatively well-connected populations that make up a MU may collectively sustain a higher harvest rate than if harvest focused on a more isolated target (Brook & Whitehead 2005). In contrast to ESUs, where translocations are generally avoided, translocations among MUs will generally not be detrimental, and may even be advantageous for maintaining genetic variation.

Box 3.5 The red wolf (*Canis rufus*) as a case study in detecting and interpreting hybridization

The red wolf was once distributed throughout the southeastern USA. In the early 1900s its numbers plummeted as a result of predator control, habitat destruction, and hybridization with coyotes. The species was listed as endangered in 1967. Free-ranging red wolves were rare and becoming hybridized out of existence by coyotes (which were expanding eastward), so the last red wolves were removed from the wild to use as breeding stock for eventual reintroduction. In 1973 a captive breeding program was begun with 14 of the most "pure" red wolves out of 400 animals captured in southwestern Louisiana and southeastern Texas. Although the wild population was considered extinct in 1980, in 1987 reintroductions began into a 680,000 ha peninsula in eastern North Carolina. The reintroduction appears to be a success: by 2002, all red wolves in the population were wild-born, and the population consisted of at least 100 animals distributed in 20 packs.

Within the scientific community, red wolf conservation has been controversial, largely because it touches on so many of the vexing issues related to hybridization. Hybridization with coyotes (which expanded into the eastern USA only in the 1990s) is the biggest threat to red wolf persistence, so current management requires that hybrids be identified using molecular methods (Adams et al. 2003) and eliminated or sterilized. Although some argue that red wolves may have originated as hybrids between coyotes and gray wolves, such an event would have pre-dated modern human activities. As Dowling et al. (1992:602–3) note: "Genetically distinct taxa of hybrid origin must not be denied protection [under the US Endangered Species Act 1973] due to mixed ancestry. If the red wolf proves to represent an historically stable entity generated by long past (maybe even ancient) hybridization between gray wolf and coyote, then it is a taxon of hybrid origin that clearly should be protected."

Source: M.K. Phillips et al. (2003).

A red wolf pup being held by a US Fish and Wildlife Service biologist. Photograph courtesy of Chris Lucash.

- By contrast, hybrid progeny (among species or distinct subspecies) that arise from human actions that do not target recovery should be discouraged, and removed as appropriate. This is especially true when the intercross progeny jeopardize the persistence of a listed species. Rhymer and Simberloff (1996) provide an avalanche of instances of this problem. As just one example, mallard ducks have been introduced around the world and hybridize readily with narrowly distributed endemic species; introduced mallards have been implicated in declines of New Zealand grey ducks, endangered endemic Hawaiian ducks, endemic Florida mottled duck, and the native Australian black duck. To hint at the complexity of the hybrid issue, Allendorf et al. (2001) point out that when such introgression becomes nearly complete, as it has in the case of the New Zealand grey duck, conservation (instead of elimination) should be considered because there may be no other option to avoid complete loss of the hybridized species.

mtDNA can determine the direction of hybridization based on its maternal inheritance. For example, because coyote mtDNA is found in gray wolves but not vice versa, hybridization between coyotes and wolves occurs solely by way of a male wolf mating with a female coyote (Lehman et al. 1991). Similarly, lynx were the mothers of Canada lynx–bobcat hybrids (Schwartz et al. 2004), sage grouse the mothers of sage- and sharp-tailed grouse hybrids (Aldridge et al. 2001), and barred owls the mothers of barred owl–spotted owl hybrids (Haig et al. 2004).

In summary, the appropriate way to deal with hybrids once again invokes the nondemocratic adage trumpeted throughout this book: not all hybrids are created equal. The hybrids that are most important to eliminate can also be those that are hardest to detect: for example, hybrids derived from domestic animals or from human-induced habitat changes (e.g. barred owls moving west with logging and other habitat changes and mating with threatened spotted owls; Haig et al. 2004). For small populations, hybridization from human-induced changes may be an underappreciated threat, as sterile hybrids can lead to demographic dead ends for population growth of the species of concern, and fertile hybrids can lead to hybrid swarms that also threaten persistence of the pure species (Chapter 11); hybrids also may be especially vulnerable to parasites (Sage et al. 1986).

Determining species identity

Species determination using the morphology of hairs or scats is notoriously unreliable (Piggot & Taylor 2003). Now, however, species can be reliably identified through DNA analysis of evidence that they leave behind. mtDNA is the usual marker of choice for species identification based on small or degraded samples, primarily due to the multiple mtDNA copies in each cell (typically 100–1000 or more) compared to the one copy of nuclear DNA. The researcher identifies regions of mtDNA that are variable among species but conserved (constant) within species. Using RFLPs or sequencing of the PCR product, diagnostic signals identify a sample with a species.

For example, the distribution of Canada lynx and other forest carnivores across the entire northern USA has been evaluated noninvasively by sampling hairs left behind on rub pads (Box 3.6). Similarly, in cases where feces cannot be determined based on

Box 3.6 The National Lynx Survey as a case study for species identification using noninvasive genetic sampling

In Chapter 2 I described the rationale behind the National Lynx Survey. The basic idea behind the survey, including the use of noninvasive sampling to assess distribution of lynx across 16 states, proved to be a reliable and informative approach.

The sampling device for sampling elusive and low-density lynx was a 10-cm × 10-cm carpet pad with nails sticking out, smothered in a beaver castoreum and catnip oil scent lure. Lynx (and other species) rub against it and leave hairs behind (McDaniel et al. 2000). At each sampling site, 125 rub pads were placed in a systematic grid: 25 transects 3.2 km apart, with each transect consisting of five rub pads 100 m apart. Pads were checked after 2 weeks.

Species identification of the collected hairs relied on PCR amplification of short (about 400 bp) segments of mtDNA, coupled with the use of restriction enzymes to produce species-specific fragments of DNA (Mills et al. 2000a). These fragments are consistent across the range of a species and are not shared by other species (see figure). Importantly for identification of species of political concern, exhaustive tests to validate the species-identification protocol were conducted prior to initiating the survey (Mills 2002).

After 3 years of sampling, more than 21,000 pads had been placed in the field and from these approximately 7000 samples were processed (McKelvey et al., unpublished data). About 67% of the hair samples – including single hairs or fragments of hair – could be identified to species. Although the sampling method was designed to target lynx, and 96 rubs from lynx were recorded (mtDNA only identifies species, not number of individuals), similar approaches facilitated the identification of other forest carnivores that happened to rub on the pads (Riddle et al. 2003; see Fig. 3.1). For example, 2040 rubs occurred from black bears, 414 from bobcats, 109 from cougar, 25 from domestic cats, and 383 from coyotes.

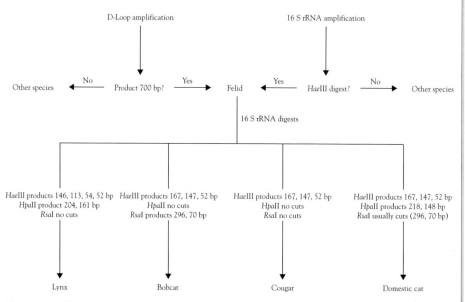

The approach used to distinguish felids in the National Lynx Survey of the USA. The diagnostic test for felid species identification was a 360-bp section of the 16 S rRNA amplified from mtDNA and subsequently digested with a battery of three restriction enzymes to produce species-specific fragment patterns. From Mills et al. (2000a). Reproduced with kind permission of Springer-Verlag.

morphology, "molecular scatology" (Kohn & Wayne 1997) can come to the rescue. In Venezuala, scat sizes overlap for sympatric species of carnivores (puma, jaguar, ocelot, and crab-eating fox), so scats could not be used to determine species-specific food habits until the development of a diagnostic mtDNA test to distinguish the scats by species (Farrell et al. 2000; see Paxinos et al. 1997 for US carnivores). Excreted material can even yield the identity of the species that were eaten, such as small mammals identified from material in owl pellets (Taberlet & Fumagalli 1996).

Of course, the plethora of new molecular techniques has not gone unnoticed by the wildlife law-enforcement community, with **forensic** applications identifying species or population of origin, sex, and individual identity. Although we'll return to examples throughout the book, one of the earliest cases of DNA forensic work involved investigating whether whale meat was from species that could be legally harvested. Baker & Palumbi (1994) perused Japanese retail markets and restaurants and purchased whale meat ranging from unfrozen sliced meat to dried and salted strips marinated in sesame oil and soy sauce. Because international laws prohibited them from transporting tissue samples to their laboratories in New Zealand and Hawaii, they set up a mobile PCR laboratory in their hotel room, amplifying mtDNA so it could later be identified to species via sequencing[5]. By comparing the species identity and geographic origin obtained from DNA analysis with catch records, they were able to conclude that several species of protected whales were being hunted, processed, and imported illegally and sold openly on the Japanese retail market. Subsequent repeats of this approach during 2000 in both Japan and South Korea (Lento et al. 2001) led to the purchase of over 1000 samples. Of these, 61 turned out to be from internationally protected whale species whereas more than 140 were not true whale meat at all, but rather were porpoise or dolphin, or – in four cases – sheep or horse!

Determining sex and individual identity

Once the species is identified, sex and individual identity may be identified with other genetic markers. In mammals, males carry DNA markers associated with the Y sex chromosome, whereas females do not (female sex chromosomes are XX and males are XY), so sex of the animal can be determined from a scrap of skin, a bundle of hairs, or feces (Woods et al. 1999, Shaw et al. 2003, Pilgrim et al. 2005). Sex-determination techniques are often used in forensic work, especially for ungulates, where game laws tend to be strongly sex-specific. The use of PCR with Y-chromosome markers has successfully determined the sex of killed ungulates based on bloodstains (from knives and rifle bolts), hair, and meat (Gilson et al. 1998). In many other animal species – including birds, snakes, and some turtles and lizards – females have the heterogametic sex

---

[5]It is important to underscore both the innovation and integrity of this project in complying with international law under the Convention on International Trade in Endangered Species (CITES). No DNA from the original purchased meat was transported out of the country, because the researchers used a clever molecular approach (Palumbi & Cipriano 1998): the DNA synthesized during PCR attaches to magnetic beads that get pulled out of the tube with a magnet, separating the synthetic product entirely from native whale DNA, which is left behind.

chromosomes (e.g. females are ZW and males are ZZ; Griffiths et al. 1998, Modi and Crews 2005), so DNA markers can again be used to identify sex. However, sexing other reptile, amphibian, and fish species using sex chromosomes is complicated by the influence on sex determination of other factors such as temperature, pH, or social conditions (Chapter 4).

Individual identity is usually based on microsatellite DNA, both because its high variability provides high power for individual identification, and because Mendelian codominant expression provides insight into other characteristics such as population structure and connectivity. With sex and individual identity determined, a plethora of incredible forensic applications become possible. As one example, Blejwas et al. (2006) obtained salivary DNA from puncture wounds on sheep carcasses and determined not only the species (primarily coyotes) but also the sex and individual identity of the attacking coyote (primarily breeding males).

In addition to forensic applications, molecular individual and sex identification can provide population-level insights into sex ratio, abundance, geographic origin of individuals, and potentially even survival. In one of the first applications of these approaches, humpback whales in the North Atlantic Ocean were sampled noninvasively (from sloughed whale skin) and nondestructively (from biopsy darts): an abundance of 4894 (95% confidence interval, 3375–7123) male and 2804 (1776–4463) female whales were estimated, with local and migratory movements of up to 10,000 km and genetic mixing in winter breeding areas. Grizzly bear abundance was determined from hairs left behind on barbed wire (Woods et al. 1999, Mowat and Strobeck 2000), coyote abundance from feces (Kohn et al. 1999), and highly endangered northern hairy-nosed wombat abundance from single hairs collected at burrow entrances (Sloane et al. 2000). The potential for future applications is tremendous, although as with any new technique pitfalls may manifest in both the molecular and the mark–recapture analyses (Box 3.7).

In addition to estimates of abundance, genetic markers can facilitate estimates of N_e, the genetically effective population size, thereby helping to determine the rate of loss of variation (inbreeding) in small populations (see Chapter 9). Similarly, the genetic composition of a contemporary population can carry a signature that indicates whether the population has undergone severe contractions, or **bottlenecks**, in size (e.g. Schwartz et al. 1998, Spencer et al. 2000).

Finally, **parentage** can be assessed with genetic markers by comparing alleles of an individual to alleles of putative parents. Genetic analysis of parentage can help determine reproductive success, including the proportion of parents that breed, and mean and variance in number of offspring per parent. Genetic measures are most useful when direct behavioral observations are either impossible or potentially misleading. For example, little was known about the mating system of the highly endangered black rhinoceros before Garnier et al. (2001) used microsatellite analysis of feces to document strong polygyny and skewing of reproductive success (of 19 offspring, more than half were sired by one male, whereas seven of 11 adult males had no offspring over 10 years). These findings are important both for captive breeding and for deciding appropriate translocations among the small, scattered remnants of the wild population.

Box 3.7 Abundance estimation using non-invasive genetic sampling

Noninvasive genetic sampling has revolutionized the possibilities for estimating abundance of wildlife species formerly considered too elusive or expensive to sample using traditional approaches. In other words: "Relief from sampling despair has an unexpected source in feces" (Kohn & Wayne 1997:226). Without the animal knowing it, multiple samples can be collected and individually genotyped, just as individual animals are often marked in traditional mark–recapture studies (Chapter 4).

However, this revolutionary approach to marking animals noninvasively has limitations (Waits & Leberg 1999). Low quantities or quality of template DNA may indicate nonexistent individuals through several mechanisms (Broquet & Petit 2004). Obviously, contamination of the sample can give false signals. Even without contamination, however, slippage of DNA polymerase during PCR can create **false alleles** if the size of an allele is scored incorrectly. Also, **allelic dropout** can occur when one or more alleles at a heterozygous locus fail to amplify during PCR, so a heterozygote is scored as a homozygote (genotype *ab* is scored as either *aa* or *bb*). These genotyping errors will tend to cause a positive bias in abundance estimates, because the same animal captured multiple times may appear to be different animals (see Creel et al. 2003).

On the other hand, an opposite set of challenges could create a negative bias in the abundance estimate. We have called this the **shadow effect** (Mills et al. 2000b), because multiple different animals could be indistinguishable genetic shadows of each other. In mark–recapture studies, the shadow effect means that a sample may be recorded as a recapture when, in fact, different animals were captured.

These concerns extend beyond abundance estimation and are applicable to most uses of noninvasive sampling. Are they insurmountable problems? In most cases, probably not. Laboratory techniques are constantly improving, and in many cases a pilot study using high-quality tissue samples will allow identification and correction of the problems (e.g. Lukacs & Burnham 2005, Kalinowski et al. 2006).

Summary

Describing, measuring, and interpreting genetic variation has become a central component of modern wildlife population biology. A major reason for the explosion of genetic techniques and applications for wildlife has been the development of the PCR, which allows genetic information to be obtained from unfathomably old or tiny tissue samples obtained noninvasively. A few of the applications of DNA markers to wildlife population biology include determination of the hybrid status or taxonomic affiliation, and identifying the species, sex, and individual identity of animals that have been difficult to sample in the past. With this background information we are well poised to consider other genetic applications throughout this book, including estimating abundance and reproduction, measuring isolation and connectivity of populations, predicting inbreeding depression, detecting diseases, and solving forensic cases.

Twenty-five years ago it would have been impossible to even begin to imagine the insights into wildlife population biology that genetic markers would provide. Surely the same will be true for the next 25 years. Therefore, it makes sense that wildlife biologists understand basic genetic concepts and tools, and that genetic samples be collected and archived as a matter of course whenever studies involve handling animals. Because all the problems have not been worked out in the analysis of poor-quality DNA inherent in noninvasive samples, the most prudent approach is to archive higher-quality samples from nondestructive sampling (for example, blood or ear-tissue punches) when animals are handled, instead of just relying on noninvasive sampling. These archived samples will provide baselines, for example, in assaying future changes in abundance or connectivity (say, in parks), and for building databases crucial for forensics cases.

Many of these techniques and applications are in their infancy. The burgeoning of these applications means inevitable mismatches will occur between technique, analysis, and application. That is why the strongest applications of genetic tools are those that are accompanied by demographic, ecological, and field data.

Further reading

Avise, J.C. (1994) *Genetic Markers, Natural History, and Evolution*. Chapman and Hall, New York. Although markers have changed over the last decade, this book remains a classic explanation of fundamental genetic concepts and techniques.

Hedrick, P.W. (2005) *Genetics of Populations*, 3rd edn. Jones and Bartlett, Sudbury, MA. A rigorous source covering both the breadth and depth of population genetics, with many applications.

Oyler-McCance, S.J. and Leberg, P.L. (2005) Conservation genetics in wildlife management. In: *Techniques for Wildlife Investigations and Management*, 6th edn (ed. C.E. Braun), pp. 632–57. The Wildlife Society, Bethesda, MD. An up-to-date and complete reference on the application of genetics in wildlife population ecology.

4

Estimating population vital rates

If I were an animal, I'd choose to be a skunk: live fearlessly, eat anything, gestate my young in just two months, and fall into a state of dreaming torpor when the cold bit hard. Wherever I went, I'd leave my sloppy tracks. I wouldn't walk so much as putter, destinationless, in a serene belligerence – past hunters, past death overhead, past death all around.

Louise Erdrich (1993), in *The Georgia Review*

Introduction

How many leopards are in a National Park, and how fast are they dying and reproducing? Are deer in Colorado few and declining, or many and increasing? How is the proportion of male and female turtles changing due to global warming?

Population characteristics estimated from field data form the skeleton on which the body of applied population biology rests. These **vital rates**, within and among populations, are united by the famous BIDE equation:

$$N_{t+1} = N_t + B + I - D - E \qquad (4.1)$$

Abundance (N) at time $t + 1$ equals the abundance the previous time step, t, plus the number of animals that arrive due to birth (B) or immigration (I), and minus those dying (D) or emigrating (E). Because males and females often have different dispersal and mortality patterns, sex ratio also affects and is affected by these vital rates.

In this chapter I will discuss within-population vital rates, including abundance and density, survival, reproduction, and sex ratio. Later in the book (Chapter 10) I will describe how to estimate immigration and emigration, key vital rates for the dynamics of multiple populations.

Estimating abundance and density

Abundance is probably the piece of information most sought after in wildlife population biology: it is key for determining harvest regulations, for deciding on protection

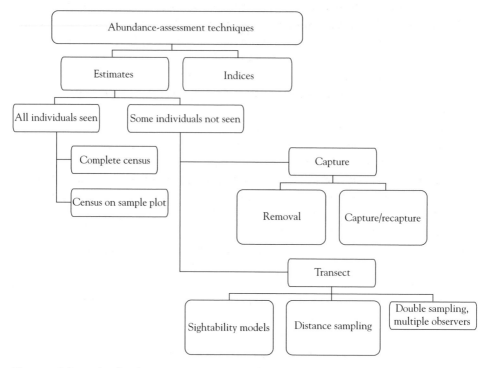

Fig. 4.1 Schematic of various measures mentioned in text to assess abundance of wildlife populations. The major groupings are based on indices or estimators, and whether all animals can be seen ($\hat{p} = 1.0$) or not ($\hat{p} < 1.0$). Modified from Lancia et al. (2005). Reproduced by permission of The Wildlife Society.

for species, and for evaluating the effects of predation or human actions, to name just a few. On the face of it, animal abundance might seem to be trivial: why not just count them? And yet, abundance estimation is one of the most mathematically sophisticated and conceptually challenging components of wildlife population biology. I will cover the basics of some of the most widely used approaches (Fig. 4.1). Throughout, I will use **abundance** and **population size** interchangeably to refer to the number of animals, and **density** to refer to abundance per unit area.

Abundance estimate versus census versus index

Here is one of the most important take-home messages of the chapter: all estimates of animal abundance, no matter how wild the math or intimidating the equations, can be reduced to a simple equation with profound implications:

$$\text{Estimate of abundance} = \hat{N} = \frac{\text{Count of animals}}{\text{Estimated probability of detection}} = \frac{\text{Count}}{\hat{p}} \quad (4.2)$$

When all animals are detected ($\hat{p} = 1$), the count equals the estimate. But as the probability of detection declines, our estimate will increase.

This equation for abundance estimation has been called canonical (Williams et al. 2002), in the sense that it is a simple yet axiomatic and universally binding principle. The estimated probability of detection ($\hat{p}$) is a function of both detectability (e.g. capture, sighting) within the sampled area and the proportion of the area sampled relative to the total population. Setting aside for the moment the second of these components – the fact that time and money often limit us to sampling only subsections of a population's range (Chapter 2) – we will focus on the implications of detecting fewer than 100% of the animals within a sampled area.

In nature animals are elusive and sometimes downright uncooperative when we try to detect them. They avoid traps, turn their reflective eyes from spotlights, hide under trees when biologists fly overhead, and pass rub pads without leaving hairs behind. In short, detection probabilities will nearly always be less than 1.0 in wildlife studies, so eqn. 4.2 shows that if we use the count alone without accounting for detection probability the resulting measure of abundance will be too small. For example, if we count 40 rabbits on a spotlight transect and ignore the fact that detection probability is, say, 0.5, then we would report 40 animals when really $\hat{N}$ should be $(40/0.5) = 80$. Detection probability can change over time, space, and even among individuals, and these complexities are why so many articles and books describe mathematical ways to estimate this probability. As we will see, accounting for detectability is also relevant for estimating other vital rates such as survival and movement.

In wildlife applications, the word **census** is reserved for the special and unusual case where detection probability equals 1.0; that is, all animals are counted. This might occur when studying a visible species on a small island, or with surveys in a narrow open transect where all individuals are seen. Botanists quite appropriately census numbers by counting all the plants in a plot, but a true census is rare for wildlife (and sometimes even for plants, where seeds or young plants can be missed). In fact, even the census of humans conducted by many governments is actually not a census at all but rather a count index with unknown detection probability (Box 4.1).

An **index** is a field count of animals or their sign that (hopefully) contains information about the relative number or density, but is not in itself an abundance or density estimate. Examples include mammal captures uncorrected for detection probability, as well as pellet counts, bird-call counts, track counts, numbers of burrow entrances or lodges, harvest numbers at check stations, questionnaires of wildlife sightings, and many others. Because they are typically cheaper to implement than formal estimators of abundance or density, indices are usually favored when money is tight, the species is difficult to observe directly, and/or when the questions are of such broad scale that more intensive estimators are impractical.

As an indirect assay of abundance, the utility of indices must be judged against how well they track changes in absolute or relative abundance across time, space, habitat types, or management treatments. An index can reliably indicate trends over time or relative difference across space only if its relationship to true abundance remains linear and constant, or at least does not change systematically (Bart et al. 2004).

If the relationship between the index and abundance does vary, you will not know whether you are seeing real changes in abundance or changes in the index/abundance relationship (Nichols & Pollock 1983, Tallmon & Mills 2004). For example, fur-

Box 4.1 Example of a non-census: the US "census"

The US Constitution mandates a count of the population in each state every 10 years to apportion the 435 seats in the US House of Representatives and to distribute federal funds to the states. Although it is called a census, the logistics are just too daunting for it to have a detection probability of 1.0; even Thomas Jefferson, who initiated the first census, noted that some persons had been missed. Ignoring that detection probability is less than 1.0 and relying on the count alone means that it is really a US index, with no known relationship to true census population size.

Statisticians proposed for the year 2000 census a transition from index to population estimate. Prior to the traditional census (where an intensive count is followed by random sampling of non-responding housing units) a totally independent set of 750,000 housing units would be picked. After determining the number of housing units present in both counts, the abundance estimate would be:

$$(Count\ 1 * Count\ 2)/number\ of\ matching\ housing\ units$$

The census bureau calls this the one number census or dual system estimation, but we will recognize it as the Lincoln–Petersen estimator described later in this chapter.

The fascinating part of this story is that this straightforward proposal to move beyond an uncorrected count index has stirred enormous political debate, because of the possibility that different groups of citizens may have differing detection probability, which would adjust the estimates of numbers and thereby shift political power. Even the US Supreme Court ruled that a 1976 federal census law "directly prohibits the use of sampling in the determination of population for the purposes of apportionment." Apparently we have a way to go to educate some that a complete count – a census – is impossible with hundreds of millions of people, and that a solid sampling strategy is the best way to move from an index to a reliable population estimate.

Source: Wright (1998).

trapping data are probably a poor index of abundance because the number of trapped animals will in large part reflect trapper effort, which in turn will be driven by economics and social norms. Obviously, some control on the constancy of the relationship between the index and abundance can be exerted by the researcher; in a bird-call index survey, for example, sampling could be restricted to the same time of day or year, and steps taken to minimize observer bias. Also, some indices better lend themselves to testing for constancy of the relationship between index and abundance; for example, bird calls or counts of captured animals can be tested statistically for constant detection probability (MacKenzie & Kendall 2002).

Clearly, then, some indices will perform better than others at portraying changes over time or relative differences across habitats or treatments. However, a growing movement argues that instead of hoping that an abundance index will reliably indicate relative differences in abundance over time and space, it is better to either directly estimate abundance (see below) or to switch to a different state variable such as proportion of area occupied (MacKenzie et al. 2005). Certainly, indices will almost always fail as descriptors of absolute abundance, because the relationship between index and abundance will rarely be both constant and known. In short, if the goal is to estimate how many individuals are actually in a population, an index will not do it; you need to use a statistically based estimator (Fig. 4.1), perhaps based on transect sampling or capture–mark–recapture (CMR).

Transect methods for estimating abundance

It is easy to envision counting animals on either side of a line that you sample by walking, driving, riding a horse, or flying. If all animals in a series of known width were detected, the abundance in the study area would simply be the number counted divided by the proportion of area sampled in all the transects (Thompson 2002). But if some animals are likely to be missed in a transect, the probability of detection must be estimated (eqn. 4.2). One way to do this is **double sampling** (or ratio estimates): incomplete counts are made over an extensive area (e.g. counts on transects in a helicopter or airplane) while a simultaneous complete census on the ground at a subset of the transects provides an estimate of detection probability (which equals mean aerial count/mean ground count) for the aerial count. Similarly, **multiple observers** can count animals, with the estimated detection probability based on overlap in observations (Lancia et al. 2005). Two of the most widely used approaches to estimate abundance on transects include distance sampling and sightability models.

Distance sampling

Distance-sampling techniques are based on the idea that detection probability decreases with distance from the observer, so detections at various distances can be used to estimate detection probability as well as abundance and density (Buckland et al. 2001). The most common applications of distance sampling include sighting animals at various distances from a line during a transect count, and sighting (or listening to calls) from the center point of a circle. The essential data needed to conduct distance sampling are the measured perpendicular distances from the center line of the transect to each animal seen. To minimize flushing animals as the observer draws close, you can estimate the straight-line distance from where you first see the animal, then use trigonometry to calculate the perpendicular distance (Fig. 4.2).

The distance data are used to estimate the probability of detection ($\hat{p}$). It is easiest to first see how this works graphically (Fig. 4.3) and then summarize the calculus. If all animals could be seen equally at all distances, we could draw a **perfect sightability rectangle** with a $\hat{p}$ value of 1.0 across all distances. Because the sightability is assumed to decline with distance, the curve showing animals sighted at different distances would only occupy a portion of the perfect sightability rectangle. Therefore, the overall $\hat{p}$ in Fig. 4.3 is the proportion of the perfect sightability rectangle under the curve.

Mathematically, we first estimate the probability of observing an animal, given that it is found at distance x from the line [$g(x)$]. Then the computer software (such as the

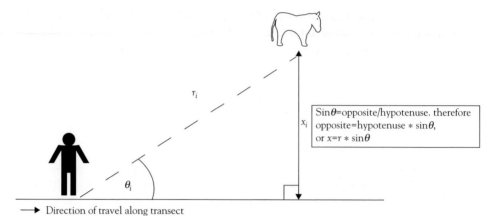

Sinθ=opposite/hypotenuse. therefore
x_i opposite=hypotenuse $*$ sinθ,
or $x=r * \sin\theta$

→ Direction of travel along transect

Fig. 4.2 How perpendicular distance for line-transect sampling can be calculated if the observer is not able to directly measure it. As the observer moves along the transect, they record the distance r_i from the line to where the animal is in the field. They also record the angle θ_i formed by the line of sight and the transect line. From these measures, the perpendicular distance (x_i) is $x_i = r_i(\sin \theta_i)$. Trigonometry really does have some interesting applications in applied biology!

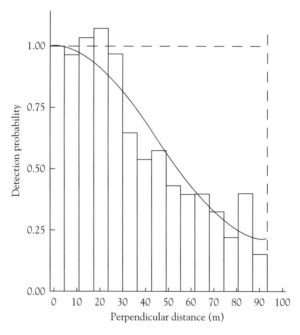

Fig. 4.3 Detection probabilities calculated from perpendicular distance-sighting data using line-transect sampling of wood ducks in forested habitat. The curve shows the fitted Fourier series estimator from the program DISTANCE. The dashed line shows what the overall detection probability would be if it were perfect at all distances (the perfect sightability rectangle); an intuitive estimator of overall detection probability for these data is the proportion of the perfect sightability rectangle under the curve. The increased detections at 15 and 20 m are assumed to be sampling anomalies. Modified from Kelley (1996). Reproduced by permission of The Wildlife Society.

program DISTANCE) integrates the detection function across all distances. In symbols, the probability of detection $\hat{p}_w$ is roughly the average of the detection probabilities across all distance categories from 0 to the maximum distance w:

$$\hat{p}_w = \frac{\left[\int_0^w g(x)dx\right]}{w} \tag{4.3}$$

With an estimate of detection probability, abundance ($\hat{N}$) may be estimated using the canonical formula (eqn. 4.2), and with the known transect length (L) and width ($2w$; because you are sampling both sides of the transect line) density, D, can be estimated[1]:

$$D = \hat{N}/(2 * L * w) \tag{4.4}$$

Here are some key assumptions for distance sampling.

- All animals directly on the line are seen (this assumption can be relaxed).
- Animals are counted only once, and do not move before being sighted.
- Perpendicular distances are measured exactly. It is permissible to lump estimates into categories (for example, 10-m intervals) to deal with uncertainty in distance estimation[2].
- Sightings are independent, such that one animal does not cause others to be more or less likely to be sighted (more complicated models can account for this, as in herding animals).

As a rough guideline for required sample sizes, Buckland et al. (2001) recommend measuring distance to at least 40, and preferably 60–80 animals.

A common variant of distance sampling uses a single point instead of a line. For example, in point counts of birds the observer records over a specified time period (e.g. 5 minutes) all birds that are seen or heard, along with their estimated distance. Rather than perpendicular distances, radial distances are recorded, but the approach again assumes a monotonic decline in detection with distance. The detection curve is then exactly like line-transect sampling, with similar assumptions and analysis. Considerable error can arise in estimating distance when you can hear but not see the birds (Nichols et al. 2000a).

Sightability or observation probability models

Given that accounting for detectability is critically important yet hard to do for every time and place, sightability models can be developed and the resulting detection probabilities used to estimate abundance in future surveys. Often coupled with aerial

---

[1]Note: I have derived the density estimate in a way that connects distance sampling to the canonical approach to estimating abundance. Buckland et al. (2001) provide an excellent treatment of variance estimators, and show how density can be directly estimated without going through these steps.
[2]In practice, the most important part of the sightability curve is near the center line of the transect. As such, Buckland et al. (2001) recommend having smaller or more numerous distance categories nearer the line than away from the line.

transect surveys, a known population (usually radio-marked) is observed while additional variables likely to influence sighting probability (for example, snow conditions, group size, animal activity, vegetation type, etc.) are recorded simultaneously. The sightability model is developed from the relevant variables coupled with the proportion of known animals detected. In subsequent surveys, observers record data on the sightability variables and use the formula to convert the raw counts into an abundance estimate with variance (Unsworth et al. 1994).

As an example, Samuel et al. (1987) used radio-marked elk to develop a sightability model for aerial surveys in Idaho. Observers flew over the sampling area and documented whether they actually observed radio-marked individuals. Sightability for a group of a given size was modeled with logistic regression, where the radio-marked elk were either observed or not, with covariates that affect the sightability. They found that sightability increased with group size and decreased with vegetation cover. Specifically:

$$\text{Sighting probability} = \hat{p} = \frac{1}{1 + e^{-[1.22 + 1.55\ln(\text{group size}) - 0.05(\%\,\text{vegetation cover})]}} \tag{4.5}$$

Thus for a future survey following similar protocols and in similar conditions to the Idaho study, a survey detecting a group of five elk in 70% vegetation cover would have a sighting probability of 0.55 (arrived at by putting 5 and 70 into the equation above). Each group count divided by its sighting probability (eqn. 4.2) gives an estimate of abundance for that group, and the sum of all abundances gives the overall abundance estimate for the sampled area.

The appeal of this approach is that after the sightability model has been developed and tested, future efforts require only counts and data on the model variables, without the need to directly estimate detection probability. Although one must be aware that the sightability model may only work well under the particular conditions for which it was developed, a sightability model thoughtfully applied from one place to another is better than a guess at numbers not framed in statistical sampling (Box 4.2).

CMR methods for estimating abundance

A different class of abundance estimators relies on capturing and marking, and recapturing again (with **capturing** and **marking** defined broadly and not necessarily literally; see Box 4.3)[3]. A **capture history** for each animal is generated, with a 1 denoting capture on a sample occasion and a 0 denoting no capture.

Two broad classes of CMR model may be distinguished: **closed-population models**, where the population is assumed to experience neither losses (by death or emigration) nor additions (by birth or immigration) during the period sampled, and **open-population models**, where losses and additions occur and can, in fact, be measured. The more complicated open models build on concepts and definitions from closed models.

---

[3] I will not cover **removal** abundance estimators whereby captured animals are permanently removed from the population (these are often used for fisheries or game animals where the harvest can be used to help estimate abundance).

Box 4.2 Application of a sample-based population survey to resolve a debate over numbers of mule deer

The Colorado Division of Wildlife (CDOW) used harvest, sex, and age data to estimate a relatively stable population of approximately 7000–9000 mule deer through the 1990s for a population of deer residing in northwest Colorado. Some sportsmen in Colorado believed that mule deer in the state were in serious peril, and used casual methodology (e.g. personal observations, outfitter guesses) to estimate the number of deer in this particular population at closer to 1750. The discrepancy led some sportsmen to accuse the CDOW of misleading the public by inflating estimates of deer population size. In a mediation process, it was agreed that an intensive aerial survey system developed in Colorado would be used to estimate numbers of deer in this population and, additionally, a sightability model developed for mule deer in Idaho would be applied to counts of deer using the CDOW survey system (Freddy et al. 2004). The CDOW system used intensive censuses of deer on randomly selected sample units or quadrats and assumed 100% detectability of deer on each sampled quadrat. In the area having the contested numbers of deer, the CDOW method led to an estimated population size of 6782 ± 2497 (90% confidence interval), whereas adjusting the counts using the Idaho sightability model led to an estimate of 11,052 ± 3503. Thus the original CDOW estimate of approximately 7000 deer was supported by two approaches to estimating deer population size, but no statistical support could be found for the sportsmen's estimate of 1750. This case study demonstrates that intuitive guesses at wildlife abundance without a formal sampling framework can be very wrong. It also shows the potential of under-estimating numbers of deer even when intense counts are conducted on relatively small parcels of land having complex cover and terrain features, underscoring the need to estimate detectability through sightability models or other approaches.

Closed CMR models of abundance

The simplest and most well-known closed-population model is the Lincoln–Petersen (LP) estimator of abundance based on two sampling periods. The LP method has a deep history, with applications stretching back to estimates of the human population of France in 1786 and of waterfowl of North America in 1930 (Williams et al. 2002:290). I will explain the LP in depth, both because it is commonly used to estimate abundance and because it forms the basis for understanding most other CMR estimators.

LP estimator with two samples

Suppose you want to estimate the abundance (N) of mice on a grid. You open traps in the afternoon, and the next morning you check traps and mark some mice uniquely (n_1 marked mice). You release them, and then a short time later (say, the next day) repeat the process. In the second sample you capture a total of n_2 mice, of which m_2 of these are marked. The best way to understand the LP abundance estimator without having to memorize it is to remember eqn. 4.2 and think of the first capture and marking session as the count and the second as the means for estimating the probability of detection:

Box 4.3 Methods of marking animals

Individual marks for wildlife studies usually conjures images of bird leg bands, turtle shell notches, and mammal ear tags. Although these traditional methods continue to be useful and widely applied, a number of other approaches are now available to mark animals (Silvy et al. 2005). Radio transmitters can be implanted in animals as small as shrews, worn as backpacks in birds, and equipped with global positioning satellite location monitors for larger animals. Telemetry has the benefit of being able to help a researcher differentiate movement from mortalities. Passive integrated transponder (PIT) tags are rice-grain-sized glass and metal cylinders that are injected under the skin; they do not transmit a signal but individual tags are recorded by an electronic reader passed over the animal. For amphibians, elastomer (rubbery paint) of different colors can be injected under the skin. If the mark does need not be permanent or individuals do not need to be distinguished (e.g. in short studies using two-sample Lincoln–Petersen methods), options could include paint balls, dyes, or hair clipping.

In some cases animals can be distinguished with noninvasive methods whereby the animal does not have to be captured. Animals with stripes, spots, or other patterns can be individually identified (e.g. body patterns on species ranging from salamanders to tigers), as can animals with distinctive scars from wounds (e.g. manatees scarred by boat propellers; Langtimm et al. 1998). As discussed in Chapter 3, noninvasive individual identification can also be obtained via genotype marks from bits of hair, feather, feces, or other material.

In all cases the choice of tag should result from careful deliberation, weighing possible negative effects on the animal against the efficiency of the method and the information to be gained from the particular study design. The ideal mark is humane, does not affect the response being studied (e.g. abundance, survival, reproduction, or behavior), is not prone to misidentification, and lasts reliably for the length of the study. Some can be used for multiple purposes; for example, toe clips provide an instant DNA sample as well as a mark that is permanent for small mammals and for some (but not all) amphibians.

$$\hat{N} = \frac{n_1}{\hat{p}} = \frac{n_1}{\left(\dfrac{m_2}{n_2}\right)} \tag{4.6}$$

This rearranges to give the intuitive abundance estimate:

$$\hat{N} = \frac{n_1 n_2}{m_2} \tag{4.7}$$

Although eqn. 4.7 shows the intuitive form of the LP estimator, it turns out that it is negatively biased, so the operational forms of the LP formula for abundance (the one to use in actual application) and its variance are:

$$\hat{N} = \left[\frac{(n_1 + 1)(n_2 + 1)}{(m_2 + 1)} \right] - 1. \tag{4.8}$$

$$\mathrm{var}(\hat{N}) = \frac{(n_1 + 1)(n_2 + 1)(n_1 - m_2)(n_2 - m_2)}{(m_2 + 1)^2 (m_2 + 2)} \tag{4.9}$$

The square root of the variance gives the SE of the estimate, and the 95% confidence interval of the abundance estimate is:

$$\hat{N} \pm 1.96(\text{SE}) \hspace{3cm} (4.10)$$

An example of abundance estimation using the LP method is shown in Box 4.4.

Three key assumptions underlie the LP and other closed-population estimators of abundance.

1 The population is closed, so that $\hat{N}$ applies to both capture occasions. If deaths or emigration occur between the two periods, the LP estimate refers only to the popula-tion at the time of the first sampling period[4]. Immigration or births can also violate closure, in which case the LP estimate refers to population size at the time of the second sample (see Kendall 1999). Closure in CMR studies is usually biologically reasonable if the trapping sessions are close together in time (e.g. consecutive nights of trapping).

2 Marks are not lost or overlooked by observers. If tags are lost between the two samples, the LP estimate will be positively biased (because fewer of the n_1 animals will be avail-able to become m_2 animals, so $\hat{p}$ will be biased low and therefore $\hat{N}$ will be biased high).

3 All animals are equally likely to be captured in each sample. For any CMR study, capture probability might vary in three primary ways: over time, among individuals, or in response to the animal having been trapped (Box 4.5). For LP, a change in capture prob-ability over time is not a problem (the first sample merely introduces marked animals to the population to facilitate an estimate of $\hat{p}$ at the next session), but individual het-erogeneity or trap response can bias the LP estimator. As an example of the effects of individual heterogeneity, remember from Chapter 3 that noninvasive genotyping of hair or scat to mark individuals can in some cases lead to a shadow effect whereby certain individuals share the same genotype (the same mark). These shadows are essentially genotypes more likely to be detected, biasing $\hat{p}$ high which causes a negative bias in $\hat{N}$ (Mills et al. 2000b). Finally, a trap-shy response leads to a positive bias in $\hat{N}$ while a trap-happy response leads to a negative bias.

A few other practical points about the LP estimator are worth mentioning. First, because it is a two-sample estimator, marks are only applied once and checked once which means that animals do not have to be individually identified. This makes LP unusual among CMR estimators in that simple batch marks – perhaps a dab of paint on the back – are sufficient to identify the marked animals. The one-time mark also means that the second capture session can be based on any method that obtains an estimate of (m_2/n_2), including the use of hunters or anglers to obtain the sample.

A related feature arising from the two-sample construction of the LP estimator is that the mark–recapture can often be improved by using different methods for the two

---

[4]Deaths during the second trapping event will not affect LP. If a death occurs during the first session (due to trapping or handling), the dead animal(s) should not be included with the n_1 animals but rather added to the abundance after N is estimated; the variance estimate is unaffected (Williams et al. 2002:293).

Box 4.4 An example of abundance estimation using the LP method and noninvasive sampling

Eastern North Pacific humpback whales can be uniquely (and noninvasively) identified from natural markings including pigmentation, scars, and ridging of the flukes. Population closure over 1–3 years can be assumed because the whales have high site fidelity to distinct feeding aggregations off the coast of California, Oregon, and Washington before migrating to wintering grounds off Baja California, mainland Mexico, and Central America. Some humpback whale data and abundance estimates are shown below. Many animals were captured (photographed) several times in a year, so the number of photographs is much greater than the number of uniquely identified whales (n_1 and n_2; Calambokidis & Barlow 2004).

Years for n_1 and n_2	Number of identification photographs in first year	n_1	Number of identification photographs in 2nd year	n_2	m_2	$\hat{N}$	95% Confidence interval around $\hat{N}$
1991 and 1992	668	269	1023	398	188	569	537–601
1992 and 1993	1023	398	512	254	173	584	547–620
1993 and 1994	512	254	402	244	108	572	512–633
1994 and 1995	402	244	661	331	100	804	704–904
1995 and 1996	661	331	564	331	144	759	690–829

n_1, The number of individuals identified in photographs in the first year; n_2, the number of individuals identified in photographs in the second year; m_2, the number of individuals in the second year that had been identified in the first year.

capture sessions. With **mark–resight** methods, potential logistical advantages accrue with using a different method (sighting) for the recapture, and trap response and individual heterogeneity should be reduced if capture probabilities of the two samples are independent. Finally, two samples does not necessarily mean only 2 days of trapping. Multiple days can be collapsed into two samples of unequal length, so for example the first three nights of trapping could be sample one and the second two nights sample two. Collapsing multiple days into two samples can help deal with population closure (Kendall 1999), and has the benefit of increasing sample size per trapping event. However, if you can conduct sampling over more than 2 days, other CMR models such as those discussed next may be more appropriate.

Box 4.5 Three ways that CMR studies can violate the assumption of equal catchability (and how field researchers try to minimize the violations)

1 **Time** may change capture probabilities for all animals in different capture sessions. Weather, moon phase, or time of year could cause capture probability to change. Researchers attempt to minimize this violation of equal catchability by closing down traps when weather patterns are likely to affect capture probability.

2 **Heterogeneity among individuals** indicates that different animals have different capture probabilities. It can be caused by an animal's age, sex, dominance status, home-range location relative to trap locations, and so on. A possible solution to heterogeneity is to estimate abundance separately for classes of animals thought to have different capture probabilities (e.g. males and females). Individual heterogeneity can also be minimized by making sure that each animal is likely to encounter more than one trap during the course of daily movements; although the traps do not need to be in a uniform grid pattern (e.g. Karanth & Nichols 1998), regular trap spacing is usually used with at least four per home range. It can also be minimized by using different techniques – such as marking, telemetry, sighting, and so on – on different occasions (for example, mark the first session, resight the second).

3 **Behavioral response** arises when an animal becomes more or less likely to be captured after the first capture. Trap-happy animals are more likely to be captured again (perhaps due to the novelty or security of the trap, or the allure of free food). Trap-shy animals are less likely to be trapped after first exposure. Trap happiness can be decreased by pre-baiting traps (which is also a good idea because it tends to increase capture probability in general), trap shyness by minimizing the time and severity of handling.

Closed-population estimates requiring three or more samples

Although the LP method is robust to some forms of unequal trappability it is not robust to all. If you are able to employ more than two capture sessions, then you will have a sufficient data stream to first test which forms (if any) of unequal trappability are apparent in the data, and then be able to employ an estimator robust to the identified deviations from equal catchability. You will not be doing this by hand; the approaches were originally codified in the computer program CAPTURE in the late 1970s (Otis et al. 1978, White et al. 1982), and more recently in the program MARK (Cooch 2001). For a particular data-set, the model whose assumptions of capture-probability structure most closely approximate the trapping data is chosen to provide the most accurate (least biased, most precise) estimate of abundance. Briefly, here are the models:

- Model M_O: equal catchability. Every animal has the same probability of capture for each sampling period in the study (no behavioral response, individual heterogeneity, or temporal variation).

- Model M_h: individual heterogeneity. Individuals have different capture probabilities.
- Model M_b: behavioral response. All animals initially have the same capture probability but after first capture may become trap happy or trap shy.
- Model M_t: time-variation model. Probabilities of capture change from trap period to trap period but within a period all animals have an equal chance of capture. Essentially an extension of the LP model.
- Models M_{bh}, M_{th}, M_{tb}, and M_{tbh}: various forms incorporating multiple deviations from equal trappability.

Open CMR models of abundance

In many cases, the length of the study makes it impossible to assume that the population is closed to additions and losses. In the mid 1960s Richard Cormack, followed closely and independently by George Jolly and George Seber, developed a modeling framework to provide estimates of vital rates in open populations. Abundance estimates in open populations are therefore based on **Jolly–Seber** (JS) models, while survival estimates are called Cormack–Jolly–Seber (CJS) models.

The Jolly–Seber model for estimating abundance is analogous to the LP estimate, except that the pieces are generalized to more than two sessions (with subscript i referring to the session) and capture probability focuses on marked animals only, to account for the open population (Pollock et al. 1990). As with LP, a total of n_i animals are caught at time i. Although m_i marked animals are captured at time i, in an open population with a probability of detection of less than 1.0 we know that some of the animals previously marked have died or emigrated. Therefore $\hat{M}_i$, the number of marked animals alive and in the population just before occasion i, is estimated using information on animals captured both before and after i (and therefore known to be alive at i)[5]. Thus, the Jolly–Seber abundance estimate for sample i is:

---

[5]While I am trying to avoid gory details I do not want to leave a big black box. $\hat{M}_i$, the number of marked animals in the population just before time i, is estimated based on the following (Pollock et al. 1990).

- R_i, The number of the n_i animals caught at i that are successfully released with marks (this could be less than n_i if there were losses during capture).
- r_i, The number of the R_i subsequently recaptured after i.
- m_i, The number of marked animals recaptured in the ith sample.
- z_i, The number of marked animals not captured at i but recaptured after i.

If we assume the probability of ever seeing again a marked animal that has just been released is the same as the chance of seeing again a marked animal that was not captured at time i then $\dfrac{r_i}{R_i} = \dfrac{z_i}{(M_i - m_i)}$, which rearranges to: $\hat{M}_i = m_i + (R_i z_i / r_i)$. So in words, we estimate the number of marked animals making it to occasion i based on both the marked animals captured (m_i) and the number of animals not captured but known to be alive because they were captured later ($R_i z_i / r_i$).

$$\hat{N}_i = \frac{n_i \hat{M}_i}{m_i} \tag{4.11}$$

Robust design

In 1982, wildlife biometrician Ken Pollock made the simple but profound observation that closed- and open-population models could be combined to take advantage of each of their strengths. Closed-population models can be robust to unequal capture probabilities among individuals or over time, but are only valid over relatively short periods of time during which no additions or losses to the population occur. Open-population models allow estimates of abundance in the face of gains and losses, but are not as precise, or as easily accommodating to unequal catchability in the form of individual heterogeneity or behavioral responses (Lebreton et al. 1992, Burnham & Anderson 2002). Pollock's suggestion was to use a **robust design** such that a long-term study of an open population is implemented as a sequence of short-term studies of closed populations (Pollock et al. 1990). The robust design is implemented with several primary sampling occasions, between which the population is likely to be open to gains and losses (Fig. 4.4). Each primary session includes several secondary sampling periods (preferably four or more), analyzed using closed models. For example, there may be four nights of mark–recapture trapping once a month for 5 months (Fig. 4.4). Abundance is estimated with closed-population models within each month. Cormack–Jolly–Seber survival estimates (described below) are based on pooled data across the secondary periods (e.g. each animal is recorded if it was captured at least once during a four-night set of secondary periods).

   The robust design provides a powerful engine for estimating vital rates (Box 4.6). In addition to good estimates of survival and abundance, with two age classes you can estimate new individuals entering the population from both immigration and *in situ* reproduction (Nichols & Pollock 1990, Nichols & Coffman 1999), as well as temporary movement into and out of the study area (Kendall et al. 1997, Bailey et al. 2004a, 2004b; Chapter 10 will show the application of this method to estimating dispersal).

A note on density estimation in capture-mark-recapture studies

Often the density of animals per unit area is of more interest than abundance per se. Intuitively, density is simply the abundance estimate divided by the area of the trapping grid. However, animals whose home range barely overlaps the trapping grid, or animals that come onto the grid from outside its perimeter, make the effective size of the trapping grid larger than the actual grid. The size of the **effective trapping grid** depends on the animal, the grid shape, and the study design, so it should be estimated for each study.

   A practical approach assumes that animals off the grid are just as likely to move toward the grid as away, and that the distance moved can be indexed by the maximum distance between captures for animals on the grid. A boundary strip equal to half

Box 4.6 Some insights using the robust design

The robust design provides a powerful framework for estimating abundance and survival within populations, and for connectivity (movement) among populations. I will save examples of quantifying movement for Chapter 10. Here are two case studies to show how detection and survival can be quantified with robust design.

Case study 1: how many salamanders are missed during monitoring counts? (Bailey et al. 2004a, 2004b)

Concerns about global amphibian declines have led to widespread efforts to monitor amphibians over time and space. Plethodon salamanders have received special attention because they may be particularly susceptible to human-caused stressors, and therefore may be good indicator species (Chapter 13). But how good are raw counts of salamanders as a tool for population monitoring? Working in Great Smoky Mountains National Park, Larissa Bailey and colleagues were interested in the likelihood of being able to actually detect salamanders present in an area, and in the stability of detection probability over time and space. For salamanders that spend a lot of time underground, the probability of capture or detection is a product of the probability that the salamander is near the surface and thus available to be captured multiplied by the probability of catching a salamander given that it is near the surface during a set of secondary samples. The first bit is the probability that the salamander has not temporarily emigrated, or is otherwise temporarily unavailable for sampling; this is a problem for lots of species that pop up briefly but then seem to disappear for a while (e.g. marine mammals that only visible when they come to the surface or snow geese that are only detectable when they are breeding). The second bit is simply the capture probability.

  The robust design of sampling involved four primary periods 6–10 days apart and lasting for 3–4 consecutive daily secondary samples, repeated for 3 years. What did they find? The average probability of a salamander being available near the surface (that is, not temporarily emigrated to the soil depths) was only 13%, and the average probability of catching a salamander near the surface was 30%. Together, that means that the probability of detecting a salamander that occurs in a particular plot is only 4%. Furthermore, the capture probability varied across years, across habitat type, and across species. Therefore, actual population size could decline quite a bit with relatively little change in a count-based index. Conversely, a count index could fluctuate a lot because of changes in detection probability, while actual abundance (what we care about) changed very little. Therefore, count indices are not reliable in this case and if abundance is of interest, formal CMR estimators should be used.

Box 4.6 Continued

Case study 2: survival of voracious deer mice on clearcuts and forest fragments (Tallmon et al. 2003)

Deer mice can be voracious seed predators, and are known to respond positively to many human perturbations, so it is of interest how forest fragmentation affects their density and survival. In a study in southwest Oregon, David Tallmon and colleagues used a robust design to trap four primary periods in each of two summers. The first primary session consisted of eight consecutive nights of secondary samples, whereas the other three primary periods were four consecutive nights each. A 16-day interval separated primary sessions (except for one year where 20 days separated two sessions). From this robust design coupled with 340 mouse captures it was possible to calculate density in forest fragments compared with unfragmented controls during the last primary session of each summer (closed models), as well as apparent survival in different fragmentation habitat types over 20-day intervals (see figure). So, deer mice love forest fragmentation!

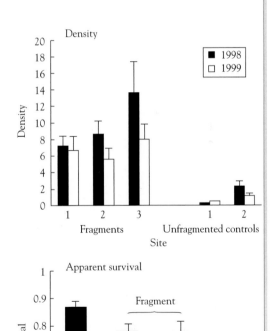

Estimates of deer mouse population density and survival in a fragmented landscape. From Tallmon et al. (2003). Reproduced by permission of the ESA.

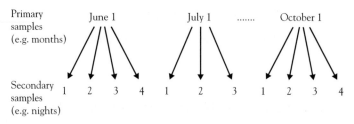

Fig. 4.4 Pollock's robust design. Here I use five monthly primary sampling periods (June–October, with the dots indicating August and September) and three- or four-night secondary periods. Notice that the number of secondary periods can differ among primary periods, which is nice because rain shuts down trapping efforts, trucks break down, field assistants get sick, and so on. Apparent survival (and recruitment and movement) can be estimated between primary periods (e.g. monthly survival using Cormack–Jolly–Seber methods), and abundance and capture probabilities can be estimated using closed models across secondary periods (e.g. abundance for each month).

the mean maximum distance moved is added all the way around the grid to provide an effective grid size (Wilson & Anderson 1985, Karanth & Nichols 1998)[6]. The density estimate is based on estimated abundance divided by estimated effective area, with its variance accounting for uncertainty in both abundance and effective grid size.

Survival estimation

Three main classes of survival estimator can be distinguished by whether or not all animals can be relocated (known fate) or whether survivors are recorded (CMR) or deaths are recorded (band recovery or return)[7]. The computer program MARK (Cooch 2001) is the workhorse for all of these analyses.

Known-fate models

In cases where a method such as radiotelemetry allows for certainty in relocating or detecting an animal that is alive and part of the study, **known-fate models** or **complete follow-up models** can be used. The simplest way to think about survival estimation using known-fate models is to start with the ideal case where the surviving animals (x) relative to the total number (n) are known unambiguously. The estimated

---

[6] Alternative approaches for estimating effective grid size include **nested grid** analysis and direct measurement using radio-collared animals (Lancia et al. 2005).

[7] Many other methods combine and extend the categories (see Further reading at the end of the chapter, and Lebreton et al. 1992, Winterstein et al. 2001).

proportion surviving is like a coin-tossing game (with life and death on each side of the coin):

$$\hat{S} = \frac{x}{n} \tag{4.12}$$

The estimated variance (based on a binomial probability model) is:

$$\hat{var}(\hat{S}) = \frac{\hat{S}(1-\hat{S})}{n} \tag{4.13}$$

This simple bionomial model can be elaborated for time intervals of different lengths, and for cases where animals must be **censored** because their fates become unknown when a transmitter stops working or falls off, or a marked animal leaves the study area, or the study ends. For example, some extensions take advantage of **failure-time** or **hazard** methods used in engineering (estimating the lifetimes of machine components) and medicine (estimating survival time of patients). One of the most widely used failure-time models is the **Kaplan–Meier** method (Pollock et al. 1989, Winterstein et al. 2001). The Kaplan-Meier approach accommodates limited censoring, as well as staggered entry whereby animals are released gradually into the sample over time. The estimate of survival for t units of time from the start of the study is:

$$\hat{S}(t) = \prod_{i=1}^{t} \left[1 - \left(\frac{\text{Number of deaths at time } i = d_i}{\text{Number at risk at time } i = r_i} \right) \right] \tag{4.14}$$

where the number at risk in each time step (r_i) includes everybody tagged, alive, and not censored at the start of an interval (remember that Π means product).

The variance is[8]:

$$\text{var}\left[\hat{S}(t)\right] = \frac{\left[\hat{S}(t)\right]^2 \left[1 - \hat{S}(t)\right]}{r_t} \tag{4.15}$$

An example of the Kaplan–Meier estimator for bobwhite quail with both censoring and staggered entry is given in Box 4.7.

Like all estimates of survival based on marked animals, the Kaplan–Meier method assumes that marked animals are representative of the population (for example, across sex and age classes, habitat types, or other characteristics), and that the mark does not affect survival. Also, it is assumed that whatever reason caused the animal to be censored is not related to their fate; this assumption will be violated if, for example, radios are destroyed only when animals are killed. The final assumption is that survival times are independent for the different animals (one animal dying does not

---

[8]As with abundance (e.g. eqn. 4.10), the 95% confidence interval is $\pm 1.96\sqrt{\text{var}\left[\hat{S}(t)\right]}$.

Box 4.7 An example of how to use the Kaplan–Meier method to estimate survival from the start of a study through time period t

The data used here are from a study of northern bobwhite quail (*Colinus virginianus*) radio-tagged in North Carolina (Pollock et al. 1989). As an example, let's estimate survival through week 15 using the data in the table. The quail had survived from the start of the study through week 14 with a probability of 0.235. During week 15, an additional four quail died out of 23 at risk. Therefore:

$$\hat{S}(15) = 0.235 * \left(1 - \frac{d_{15}}{r_{15}}\right) = 0.235\left(1 - \frac{4}{23}\right) = 0.235 * 0.826 = 0.194$$

with variance of 0.0013 (95% confidence interval, 0.123–0.265). Because six new animals were added to the study during week 15, but one animal was censored and four died, the number at risk going into week 16 was $(23 + 6 - 1 - 4) = 24$.

Week (t)	Dates	No. at risk (r_i)	No. of deaths (d_i)	No. censored	No. of new individuals added	Survival ($\hat{S}[t]$)	95% CI
1	17–23 Nov	20	0	0	1	1.000	1.000–1.000
2	24–30 Nov	21	0	0	1	1.000	1.000–1.000
3	1–7 Dec	22	2	1	0	0.909	0.795–1.024
4	8–14 Dec	19	5	0	0	0.670	0.497–0.843
5	15–21 Dec	14	3	0	0	0.526	0.337–0.716
6	22–28 Dec	11	0	0	0	0.526	0.312–0.740
7	29 Dec–4 Jan	11	0	0	0	0.526	0.312–0.740
8	5–11 Jan	11	2	0	0	0.431	0.239–0.623
9	12–18 Jan	9	1	0	0	0.383	0.186–0.579
10	19–25 Jan	8	0	1	0	0.383	0.174–0.591
11	26 Jan–1 Feb	7	0	0	3	0.383	0.160–0.606
12	2–8 Feb	10	0	0	6	0.383	0.196–0.569

Box 4.7 Continued

13	9–15 Feb	16	4	0	10	0.287	0.168– 0.406
14	16–22 Feb	22	4	0	5	0.235	0.149– 0.321
15	23 Feb–1 Mar	23	4	1	6	0.194	0.123– 0.265
16	2–8 Mar	24	4	0	0	0.162	0.103– 0.221
17	9–15 Mar	20	2	0	0	0.146	0.087– 0.205

CI, confidence interval. Data from Pollock et al. (1989). Reproduced by permission of The Wildlife Society.

affect others). If this assumption is violated – for example when predators tend to kill off entire litters of newborn radiocollared snowshoe hares (O'Donoghue 1994) – the Kaplan–Meier estimate of survival is not biased but the variance will be too low.

Notice that survival does not need to be constant among individuals or over time, making this method a good choice when survival probabilities change due to hunting pressure, weather, and other events. Precision of the estimate is a function of the number of animals at risk, so it is important to maximize the number of animals followed and to minimize censored animals, with a suggested minimum sample of 25, and preferably 40–50, marked animals in the population at all time steps (Pollock et al. 1989, Winterstein et al. 2001). The Kaplan–Meier approach is easily extended to relate survival to covariates, or to statistically test for differences in survival among treatment (e.g. harvest compared with no-harvest) or other classes (sex, age, etc.; see White & Garrot 1990, Winterstein et al. 2001).

CMR using the Cormack–Jolly–Seber method

Suppose you have animals that are marked but whose fates cannot be known with certainty: the probability of detection complicates assessment of whether they are dead or alive but not captured. A naïve estimate of survival using such marked animals would be the number of marked animals captured at time $i + 1$ (m_{i+1}) divided by the number of animals released with marks at time i (R_i). By now I hope that such a naïve estimator for survival, $\dfrac{m_{i+1}}{R_i}$, sends shivers down your spine, because you know that it will almost always be biased low. Why? Because some of the marked animals not caught at $i + 1$ are very much alive, just not captured!

So we need to estimate the number of marked animals that make it from one interval to the next, following an approach similar to that used for the Jolly–Seber abundance estimate (see footnote 5). At time i, the number of marked animals include the R_i animals caught in i and released with marks back into the population, plus the marked animals alive but not caught in $i (\hat{M}_i - m_i)$. Of these, a total of $\hat{M}_{i+1}$ survive through to the next interval. Therefore, apparent survival (ϕ_i) is the simple proportion of marked animals that have survived and remained in the population from the end of one trapping event to the start of the next[9]:

$$\hat{\phi}_i = \frac{\hat{M}_{i+1}}{\hat{M}_i + R_i - m_i} = \frac{\text{Estimated number of marked animals that make it to } i+1}{\text{Number marked animals leaving trapping event } i} \qquad (4.16)$$

Variances for all of these estimators are available, as are power curves showing expected precision of estimates for various population sizes and capture probabilities (Pollock et al. 1990, Krebs 1999).

Band-return approaches

Band-return approaches are conceptually closely related to CMR methods in that a form of detection probability is accounted for and estimated as part of estimating survival. The big difference, of course, is that you are not tracking marked animals that live to each time step, but rather those that die with bands being returned to the researcher. Because the band returns come from people hunting or fishing, a recovery depends on the animal being killed and reported. The recovery rate becomes loosely analogous to detection probability, in that we must tease its effects out of the data in order to estimate what we care about: survival (Williams et al. 2002).

Other approaches

Many approaches extend the methods described above, such as the development of known-fate telemetry models with a relocation probability less than 1.0 (as might occur if movement or topography precludes detection of telemetry signals on certain sampling occasions; Pollock et al. 1995). Also, approaches have been combined and incorporate auxiliary information. For example, survival of an open population of wood thrushes that were both radiocollared and banded for recapture was estimated by linking Kaplan–Meier estimates with Cormack–Jolly–Seber methods (Powell et al. 2000); elk survival and abundance was estimated by combining known-fate survival models for radiocollared elk with age-at-harvest data from hunter check stations (Gove et al. 2002).

In contrast to the approaches discussed so far, one common approach derives survival estimates solely from **life-table** survivorship curves or other age-distribution

---

[9] $\hat{\phi}_i$ is called **apparent survival** because it does not separate death from the probability of permanently leaving the area. The probability of survival alone is harder to estimate, requiring telemetry or sampling at different spatial scales (Citta 2005).

methods based on unmarked animals. A cohort is followed through time, with the drop in numbers indicating survival through each age class (the horizontal life-table approach). For example, a cohort of 100 2-year olds that is reduced to 60 by the time they become 3-year olds implies a survival of 0.6. It can be logistically very challenging to follow a cohort from birth to the death of the last individual, and the process must be repeated to capture the variation in survival that occurs through time (i.e. good and bad years). Alternatively, if the current age structure is assumed to reflect the past (and population growth is either stationary or known), then the changes from age class to age class could again indicate survival through each age (the vertical life-table approach). A serious limitation of life-table or age-distribution methods is that we are not accounting for age-specific detection probabilities and so must assume that all individuals are equally detected across age classes and time. This problem, coupled with the restrictive assumptions, makes this a much weaker method for estimating survival than radiotelemetry or CMR-based approaches (Lebreton et al. 1992, Williams et al. 2002).

Estimation of reproduction

Reproductive output is another essential within-population vital rate estimated from field data. Sometimes reproductive output refers to females alone, and sometimes to either-sex offspring born to either-sex adults; the important thing is to keep the accounting straight. To be clear in meaning about various terms related to reproduction, I'll begin with some pedantic definitions.

- **Maternity** and **paternity** refer to the mother and father of an animal, respectively.
- **Natality** is the average number of live offspring born (or eggs laid) per female that reproduces, often called **litter size** for mammals and **clutch size** for birds, amphibians, reptiles, and fish. (Again, sometimes only female offspring are counted.)
- **Fecundity** is the average number of offspring born per individual of a given age (or stage) in one time step, so it is a product of natality and the proportion of the cohort that breeds (note that if focusing only on females, fecundity is the product of female natality multiplied by the proportion of females that breed).
- **Recruitment** often gets tossed in with reproduction, but I will use this term (in later chapters) to refer to net population production after both births and deaths have been taken into account.
- Finally, the **average reproductive contribution** of individuals of a given age or stage to the population next year is a product of fecundity and either the survival of young to be counted the next year or the survival of parents to have the young[10] (Box 4.8 shows an example of calculating reproductive rates from field data).

Natality of females has traditionally been determined by observation (for a review see Harder & Kirkpatrick 1994). In some cases, mammal litter size is inferred from counts

---

[10]Some call this **fertility**, but often in the ecology and wildlife biology literature fertility and fecundity are interchanged and used differently. That is why I am making what might seem to be gruesomely detailed and even arcane distinctions for these terms. The meaning of the terms are especially important in population-projection models (Chapter 7).

Box 4.8 Example of calculating reproductive rates from field data

For adult common frogs (*Rana temporaria*) in Europe, reproductive data may be assembled from a number of studies (see Biek et al. 2002). Here we are evaluating reproduction by females of females.
  Measured from the field:

- Clutch size = natality: 650 female eggs per female that reproduces (SD, 133)
- Probability of females laying eggs: 1.0, all females breed annually
- Sex ratio: 0.5 (SD, 0)
- Adult survival: 0.43 (SD, 0.035)

Calculated from field data:

- Fecundity: 650 × 1.0 = 650 female eggs laid per adult female in the population

We'll come back to this example in Chapter 7.

---

made *in utero*. Mammals have corpora lutea scars that form in the ovaries after release of the egg; a count of corpora lutea scars – for example, in an ungulate brought in to a hunter check station – reveals the number of ova shed per estrus for that female. Fetal counts *in utero*, as well as uterine scars marking sites of previous placental attachment, can also be counted in necropsied mammals. Ultrasound is emerging as an increasingly field-friendly tool to count embryos *in utero* (Griffin et al. 2003). Birds do not have corpora lutea or uterine scars but do have postovulatory follicles that sometimes persist long enough to be counted.

Remember that measures of reproductive output taken at different times in the birth process must be scaled by survival: an ova must be both fertilized and implanted in the uterus to become a fetus, and a fetus must not be reabsorbed or aborted to be born alive. Thus to serve as an estimate of natality a count of corpora lutea scars or fetus numbers must account for the probability of making it from that stage to birth.

The proportion of the population likely to breed can also be determined in the field. For mammals, vaginal smears assessing the proportion of epithelial cells and leukocytes can indicate female mating receptivity, fecal steroid hormones can indicate pregnancy (Berger et al. 1999), and lactation can indicate likely nursing. Reproductive status of males is more difficult, with possible assessment based on sperm counts, hormonal levels, or testis size. In birds, the same approaches apply to males; for example, the testis of a mature male white-crowned sparrow goes from less than 10 mg to more than 600 mg during the height of the breeding season. For females, a laparotomy incision to expose the left ovary can indicate the proportion of birds nearing the egg-laying stage, and also confirm the sex of live birds where the sexes cannot easily be distinguished morphologically. Also, necropsied female birds in some families (such as the Columbidae, doves and pigeons) will have distinguishable crop glands filled with a curd-like material fed to young. For a harvested bird such as mourning doves, crop glands can indicate the proportion of birds incubating or rearing young.

Genetic markers can clarify maternity and paternity, and establish reproductive patterns that are not intuitive from observation alone (Jones & Ardren 2003). For example, over 4 years in Yellowstone National Park there was usually concordance between parentage estimates from nuclear DNA markers (microsatellite loci) and visual observation or telemetry data for 11 wolf packs (K. Murphy, personal communication). However, the DNA matches showed two surprising deviations from the behavioral observations of maternity and paternity. First, a putative mother identified via observation was not the biological mother of pups that she nursed; second, a 10-month-old male and an 11-month-old female wolf reproduced, although wolves of less than 1 year old had never been known to reproduce in free-ranging gray wolves. Similarly, in African lions, DNA fingerprinting complemented intensive behavioral observations by determining that a single male was the father for the entire litter in 23 out of 24 cases despite the fact that females had been observed accepting multiple copulatory partners (Gilbert et al. 1991).

Sex ratio

Marilyn Monroe, the famous American actress, is said to have quipped, "What do I think about sex? Oh, I think it's here to stay." As long as Marilyn Monroe is right, as long as we applied biologists deal with creatures that have sexual reproduction (as virtually all wildlife populations do), then the ratio of males to females will always be an important population descriptor.

Although the sex ratio of a population is most commonly given as the ratio of males to females (e.g. 60 : 40) or the percentage of males (e.g. 60%), it may also be described as the ratio of females to males or the percentage of females. The sex ratio often changes with time, with the following conventional distinctions: (i) **primary sex ratio** means conception[11]; (ii) **secondary sex ratio** means birth; (iii) **tertiary sex ratio** means some later date, usually at the end of parental care (weaning/fledging); and (iv) **quarternary sex ratio** means the older breeding adult population. Finally, one should be clear on whether the ratio is referring to a head count or to a breeding sex ratio. The two can be very different, depending on the mating system. For example, in a **polygynous** system, where one male mates with more than one female, a head-count sex ratio may be 1 : 1 (males/females) while the breeding sex ratio may be as extreme as 1 : 20 (for example, in the case of elephant seals where males violently maintain exclusive harems). The flip side of a polygynous mating system is **polyandry**, where one female mates with multiple males, again skewing the breeding sex ratio compared with the head count (this is relatively rare, but can be found, for example, in emus). In a

---

[11]Sometimes researchers use the primary sex ratio to refer to the youngest juveniles they can easily observe and sex.

monogamous mating system, a pair bond between one male and one female leads to more similar breeding and head-count sex ratios.

As an aside, when there is a skewed sex ratio (polygyny or polyandry) there will tend to be increased competition within the sexes for mates. This is called **sexual selection** and it explains crazy things like huge antlers on ungulates which seem impossibly maladaptive (try walking through dense trees with an elk rack on your head) except in the light of increased mating opportunity. Sexual selection also helps to drive selection of sexual dimorphism, as with the elephant seals mentioned above: competition for their large harems leads to selection for massive males that are perhaps five times as large as females.

Sex ratios in the wild

What drives sex ratios in wildlife? The mechanisms leading to primary sex ratios are complex (see Hardy 2002). One much-discussed branch of **sex allocation theory** posits that, although selection should favor equal primary sex ratio under most conditions, females in better physical condition (with more access to resources) should produce more offspring of the sex that would most benefit from the improved condition (Trivers & Willard 1973). For example, consider polygynous systems where males compete for mates or territories and where big strong males mate the most: when food resources are plentiful mothers might invest more in males, increasing their reproductive success; when resources are poor mothers would invest more in females because females would have mating opportunities even with poor food availability. A fair amount of data now support these theoretical predictions (Hardy 2002, Suorsa et al. 2003), including a fascinating application where supplemental feeding affects sex ratios in the critically endangered kakapo (Box 4.9).

In many reptiles, amphibians, and fish, primary and secondary sex ratios may depend on temperature (temperature-dependent sex determination). With a 4°C increase during incubation, painted turtles and loggerhead turtles may produce nearly all females (Mrosovsky 1982, Janzen 1994), whereas a mere 1°C shift can push tuatara to produce all males (Cree et al. 1995). Ignorance of temperature-dependent sex determination can undercut conservation efforts: for years, well-meaning people took the eggs from threatened marine turtles out of the sand and put them into Styrofoam boxes so they could hatch away from predators. It turns out that the boxes were a few degrees cooler than the beach sand, leading to severe masculinization of the hatchlings in boxes (Mrosovsky 1982).

Quarternary sex ratios differ from primary ratios in fairly predictable patterns for mammals and birds because survival between the sexes changes with time. The primary drivers of changes in sex ratio over time arise from behavioral, physiological, and genetic differences between the sexes. As an example of these factors in play, let's get personal and consider human sex ratios (Holden 1987, Williams 2003). At conception the primary human sex ratio is skewed toward males, with 53% : 47% males/females. But males are slightly more likely to be lost *in utero* due to spontaneous abortions, miscarriages, and stillbirths, so by birth the secondary sex ratio is less male-

---

Box 4.9 Sex ratios and kakapo conservation

The kakapo is a large flightless parrot once common in New Zealand but nearly driven to extinction by introduced black rats and weasels. By 1997 only about 60 kakapo remained in the world.

The exceptionally small numbers and vulnerability of this flightless, ground-nesting bird to ever-increasing introduced predators led managers in 1982 to capture and move all remaining birds to mammal-free islands. Adult survival increased substantially, but reproduction and chick survival was poor. Kakapo have a polygynous mating system, with males congregating in leks every 2–5 years in synchrony with the episodic mass fruiting of podocarp trees. Males display vocally to attract females and compete intensely among themselves; larger males are more likely to mate.

To increase reproduction, supplemental food (nuts, apples, and sweet potatoes) was provided to the kakapo beginning in 1989. Paradoxically, this reasonable and intuitive attempt to save the species by supplemental feeding appears to have backfired. Females provided with supplemental food produced 67% males, compared with 29% males for birds not given supplemental food. Thus kakapo sex allocation followed the Trivers–Willard hypothesis, with females producing more sons than daughters when resources were high. The male bias under supplemental feeding is problematic not only because females are needed for the population to grow, but also because the population was already heavily skewed toward males (41 : 21 males/females) because introduced predators kill females as they incubate the nest.

In a terrific example of ecological theory guiding management, in 2002 the New Zealand Department of Conservation acted on predictions of sex allocation theory. While food was provided to all females after mating to increase chick survival, only lighter females (which might most need food and yet would tend to produce more female young than heavier females) were given supplementary food before mating. So far, the strategy appears successful: all but one of the 21 adult females laid eggs, leading to the highest kakapo recruitment in at least 20 years; furthermore, of the 24 young fledged that year, more than half (15 of 24) were females.

Sources: Clout et al. (2002), Sutherland (2002), Robertson et al. (2006).

---

dominated, at about 51 : 49. Males have higher death rates, so by age 30 the sex ratio is even (50 : 50) and by age 65 it is 45 : 55. Why the strong skew toward females between childhood and older age? In large part, male behaviors are to blame. Male humans kill each other and themselves more than females, and are twice as likely as females to die from lung cancer, pulmonary disease, accidents, liver damage, heart damage, and suicide. Males are more likely to drink heavily, use seat belts less often, and die in motor vehicle accidents. In addition to behavioral differences, human sex ratio may be affected by physiological factors including the potential role of estrogen in lowering so-called bad cholesterol levels and reducing heart attacks. And finally, recall that sex determination in mammals depends on the sex chromosome, with females being XX and males XY. An X-linked gene that has negative effects will automatically be expressed in a male (as men only have one X-chromosome), while the same deleterious gene could be masked by the second X in a female. Thus, for example, males are far more prone to X-linked genetic problems such as muscular dystrophy and color blindness.

What do we see in wildlife populations? Many mammal populations follow the trends described for humans, with the same general mechanisms likely to come into play. Parasitism may also play a role, as male-biased parasitism is the general rule among mammals (Moore & Wilson 2002). The heavier parasite load for males is most pronounced in species where male–male competition for mates or territories is most severe, leading to reduced immunocompetence to defend against parasites (perhaps because testosterone is an immunosuppressant or perhaps just because intraspecific competition leads to larger male size, which in turn attracts a larger parasite load). By the way, the parasite-burden hypothesis can also be linked to humans, as men are about twice as vulnerable to parasite-induced death in the USA, UK, and Japan, and more than four times as vulnerable in countries with higher parasite-induced death, such as Kazakhstan and Azerbaijan (Owens 2002).

Birds show a pattern opposite to mammals, with females tending to die earlier and sex ratios becoming more male-biased with age. Behavioral risks typically differ from mammals, most obviously when egg-laying exposes females to high predation risks and energetic drains. Females are also the heterogametic sex in birds (ZW; see Chapter 3), exposing them to deleterious alleles expressed on the sex chromosome. And in some species, notably raptors and polyandrous bird species, size dimorphism is reversed, with females being larger and therefore potentially carrying heavier parasite loads.

Over time, if the myriad interacting factors affecting sex ratios stay relatively constant, then we would expect to see a stable and characteristic sex ratio for particular species. Despite considerable debate as to the adaptive significance of stable sex ratios, for applied purposes knowing why sex ratios are as they are allows us to consider what might happen when we intentionally alter the sex ratio under different scenarios of harvest, translocation, or intensive management (Wedekind 2002), or when the sex ratio is inadvertently altered when environments change due to fragmentation (Suorsa et al. 2003) or global climate change (Janzen 1994).

Summary

Vital rates are the skeleton upon which the body of population biology hangs. Both models and management actions dealing with factors such as harvest, endangered species, and control of exotics rest on estimates of abundance, survival, reproduction, and sex ratios (as well as connectivity – to be discussed in a later chapter).

Although indirect indices of abundance (e.g. snow tracks, call counts, or fur returns) may have some utility for assaying relative abundance over time and space, they are constrained by the fact that a change in the index could be caused either by a change in true abundance (which is what we care about) or by a change in the relationship between the index and true abundance (a bothersome confounding effect). The reliability of information about populations will be improved by estimating abundance. The dizzying blizzard of abundance estimators is unified by the simple idea that all involve some count of animals divided by an estimated probability of detecting (or capturing) an animal during the count period. In rare circumstances all animals are counted, a true census. In most other cases, however, probability of detection is less than 1.0 and must be estimated; two common approaches include transect-based and CMR approaches.

Survival can be based on models that either can reasonably assume a probability of detection of 1.0 – that is, that the fate of the animal is known at all times, usually with the help of radiotelemetry – or that the detection probability is less than 1.0 so that survival must be estimated from recorded deaths (band-return models) or recorded live marked animals (Cormack–Jolly–Seber models). These approaches and others can be mixed and matched to estimate survival, but the commonality is to avoid the naïve estimator of (number encountered/total), which ignores the fact that animals not captured or sighted may be very much alive but simply not detected.

Reproduction is often easier to determine than survival, with a suite of methods to detect reproductive output both before and after birth. Determination of maternity and paternity of offspring, and therefore reproductive output of parents, can benefit from DNA markers.

Because there is a mortality component from conception through old age, and this mortality can be sex-specific, sex ratios should specify the time at which they are taken. Primary (at conception) and secondary (at birth) sex ratios may be affected by resources and in some amphibians, reptiles, and fish by temperature; human-caused changes in resources (through habitat fragmentation) or temperature (under global warming) could severely affect populations by skewing sex ratios. Later in life the sex ratio in mammals tends to become more female-biased because male mammals tend to die faster due to behavioral, physiological, and genetic differences; birds tend to follow the opposite trend.

Further reading

Lancia, R.A., Nichols, J.D., Pollock, K.H., and Kendall, W.L. (2005) Estimating the number of animals in wildlife populations. In: *Techniques for Wildlife Investigations and Management*, 6th edn. (ed. C.E. Braun), pp. 106–53. The Wildlife Society, Bethesda, MD. Describes the fundamentals and wide variety of approaches to estimating animal abundance.

Thompson, S.K. (2002) *Sampling*, 2nd edn. John Wiley and Sons, New York. A good overview of sample design, plus information on a variety of abundance estimators. A good overview of sample design, plus information on a variety of abundance estimators.

Williams, B.K., Nichols, J.D., and Conroy, M.J. (2002) *Analysis and Management of Animal Populations*. Academic Press, San Diego, CA. The most detailed, complete single source for the theory and mathematics of vital-rate estimation.

II

Population processes: the basis for management

5

The simplest way to describe and project population growth: exponential or geometric change

> *I think I may fairly make two postulata. First, that food is necessary to the existence of man. Secondly, that the passion between the sexes is necessary and will remain nearly in its present state . . . Population, when unchecked, increases in a geometrical ratio: Subsistence increases only in an arithmetical ratio. A slight acquaintance with numbers will show the immensity of the first power in comparison of the second.*
>
> **Thomas Malthus (1798),** *An Essay on the Principle of Population*

> *The buffalo herds also vanished before the great pilgrimage . . . Scarce by 1867, those "wild cattles of the prairies" were all but gone by the 1870's. . . . The prairies were repopulated with domestic animals brought in by the settlers. In 1833 Charles Larpenteur drove four domestic cows and two bulls into Montana from the Green River country. A conservative count in 1880 showed 48,287 horses, 1,632 mules, 249,888 sheep, and 274,321 cattle in Montana territory.*
>
> **Bud Moore (1996:55),** *The Lochsa Story*

> *The burial mound was outside the city walls, in a field dotted with cow pies and large stones. The stones had been arranged geometrically in patterns that were supposed to mean something to the gods; presumably, the cow pies had fallen at random, although then, as now, the division between what is random in nature and what is purposeful is extremely difficult to determine.*
>
> **Tom Robbins (1984:28),** *Jitterbug Perfume*

Introduction

The quotation by Thomas Malthus describes a profound paradox between the increases in human population size and the ability to harvest enough food to feed us; the paradox catalyzed Charles Darwin's theory of natural selection as a driver of evolutionary change. Whether or not Malthus was right is a matter of hot debate that we will not wade into (at least 20 books address the topic in my local university library). Rather, the quotation helps us in the transition from considering conceptual pieces (e.g. study design, genetic tools, and measurements of vital rates) to understanding the process of population biology.

In the last chapter we saw that at the most basic level change in abundance (N_t) over time is driven by dynamics both within populations (births and deaths) and among populations (immigration and emigration). In this chapter we will examine the simplest way to quantify population growth: geometric or exponential change unaffected by density, individual qualities, or other factors. Later chapters will add more complexity, including changes in population growth due to density (density dependence) and characteristics of the individuals in the population (e.g. age structure). Importantly, I should note at the outset that **population growth** refers to any trajectory in abundance over time, including increases, decreases, or no change.

At the heart of Malthus' concern is that an arithmetic process generates a straight line over time whereas multiplicative (geometric or exponential) growth creates a curved line (Fig. 5.1). **Arithmetic change** involves adding or subtracting a fixed number at each time interval (or step). Malthus posited that food supply ("subsistence") increases by a fixed absolute amount each time step; for example, an additional 900 tonnes of wheat available to humans each year.

In contrast, **geometric** or **exponential growth** means that a constant fraction of the current number is added to the population each time step. Said differently, this means that the current population is multiplied by a constant number each time step. Money in your interest-bearing bank account grows exponentially (or, equivalently, geometrically) because each dollar contributes to future income; more dollars multiplied by a fixed rate means more of an increase. Likewise, population growth is a multiplicative process which can lead to geometric or exponential increase (or decrease). Geometric and exponential are best considered as the same type of multiplicative growth: the only difference is that one occurs on a discrete time scale while the other happens on a continuous scale (more on this later in the chapter).

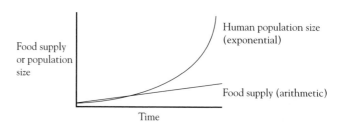

Fig. 5.1 The distinction between arithmetic and geometric (or exponential) growth posited by Thomas Malthus in his famous 1798 essay. His key claim was that food supply was growing arithmetically (the straight line) while the human population was growing exponentially (the curve).

Fundamentals of geometric or exponential growth

Discrete (geometric) growth

Suppose we want to characterize population growth over time for a species that breeds annually and is sampled annually. The growth rate of the population over discrete time steps, otherwise known as the population multiplication rate or **geometric growth rate** (λ; called **lambda**), describes abundance next year as a multiple (or proportion) of the abundance this year. A population with $\lambda = 1.0$ is **stationary**, meaning that it neither increases nor decreases. A population with a λ value of 2.0 will double in 1 year. If $\lambda = 0.75$ the population will be three-quarters the size that it was last year; that is, the population will decline by 25%. Mathematically, the abundance of a population (N) at time $t+1$ is a function of both abundance at time t and the population growth rate:

$$N_{t+1} = N_t \lambda \qquad (5.1)$$

If the population continues to grow at the rate λ for t time steps from an initial abundance at time 0 (N_0), then at time t we would expect N to be:

$$N_t = N_0 * \lambda_1 * \lambda_2 * \ldots \lambda_t$$

$$N_t = N_0 \lambda^t \qquad (5.2)$$

So to estimate population growth, λ, as a function of a change in abundance over one time step, we just rearrange eqn. 5.1:

$$\lambda = N_{t+1}/N_t \qquad (5.3)$$

And to estimate the constant annual growth rate over t time steps, rearrange eqn. 5.2:

$$\lambda = \sqrt[t]{\frac{N_t}{N_0}} = \left(\frac{N_t}{N_0}\right)^{\frac{1}{t}} \qquad (5.4)$$

As an example, consider wolves (specifically, the population in the US northern Rocky Mountain states after reintroduction). Minimum counts for 3 years were $N_{1998} = 275$, $N_{1999} = 322$, and $N_{2000} = 433$ (US Fish and Wildlife Service et al. 2001). Using eqn. 5.3, annual λ estimates were: $\lambda_{1998-9} = 322/275 = 1.17$; $\lambda_{1999-2000} = 433/322 = 1.34$. These can be interpreted as 17 and 34% increases per year. If we only had the abundances in 1998 and 2000 (spanning 2 years) and wanted to calculate the average annual λ value, we would use eqn. 5.4: $\lambda_{1998-2000} = \sqrt[2]{\frac{433}{275}} = 1.25$, which gives an average 25% increase per year.

Because λ is a *rate* of change, two different-sized populations with the same λ will have different *absolute* changes in abundance. For example, if $\lambda=0.80$ a population of 1000 rabbits will change by 200 rabbits in one time step (from 1000 to 800) whereas a population of 100 will decrease by only 20 rabbits.

Continuous (exponential) growth

In some ways it would be easier just to stop at the discrete form of population growth described by geometric change. It is, after all, an intuitive and easy way to explain population growth: "My newt population decreased by 3% per year ($\lambda=0.97$) from 1999 to 2001." But suppose the species of interest does not reproduce seasonally, or we want to compare population growth in an elephant population (which undergoes yearly intervals of population change) with that in a mouse population (monthly or bimonthly intervals of population change). It turns out that discrete growth represented by λ has some awkward mathematical properties, so to really understand population growth requires understanding the calculus-based continuous-time analog of λ, defined by r and interchangeably called the **exponential growth rate** or the **instantaneous per capita growth rate**.[1]

With your help, our little venture showing how r relates to λ will be fairly painless. On Fig. 5.2(a), sketch lines up from 2007 and 2008 on the x axis to the curved line, then from each of these points across to the y axis; the lower intersection with the y axis is N_{2007}, the upper is N_{2008}. The lines show the change in abundance (ΔN) over a 1-year change in time (Δt) from 2007 to 2008. Now, we can shorten the interval, making Δt smaller and smaller, so ΔN gets smaller too (try it on the graph). Make a smaller and smaller interval, converging to a point on the curve. At this point, the change in the population per unit time has gone from the world of discrete time to the world of continuous time described by a derivative, where a tiny (infinitesimal or instantaneous) change in population size (dN) occurs over a tiny interval of time (dt):

$$dN/dt = rN \tag{5.5}$$

where r is the instantaneous growth rate per capita (per individual).

Because dN/dt is a derivative, you can also think of rN as the slope of the tangent of the curve of N plotted against time (Fig. 5.2a). To isolate r, abundance over time can be plotted on a logarithmic scale, which makes the curved line of abundance straight, with the slope of the line equaling r (Fig. 5.2b).[2]

---

[1]The instantaneous per capita growth rate, r, is sometimes called the **intrinsic** growth rate because it represents growth of the population without density dependence.

[2]A reminder: logarithms, regardless of what base they are in (e.g. base 10 or base e) convert multiplicative processes (such as exponential growth) to additive processes. Thus, a multiplicative geometric growth curve becomes a straight line when plotted as the logarithm of abundance.

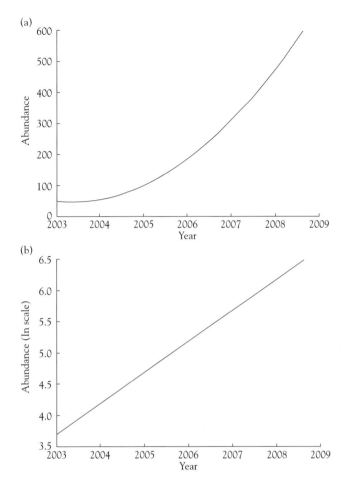

Fig. 5.2 Constant exponential growth of a hypothetical population ($r=0.5$ or the equivalent $\lambda=1.65$), showing how the curve of abundance over time (a) becomes a straight line when plotted against the natural logarithm, ln, of abundance (b). The slope of the line in (b) is equal to r in a constant environment.

How does the instantaneous per capita growth rate, r, relate to λ? The two are interchangeable, after a simple conversion:

$$r=\ln\lambda$$

or

$$\lambda=e^r \tag{5.6}$$

The ln is the **natural logarithm**, with the base e (which is about 2.718).

The property of being able to substitute λ for r, or r for λ, means that eqn. 5.2 can be rewritten as:

$$N_t = N_0 e^{rt} \qquad (5.7)$$

A population with an $r>0$ is increasing, and one with $r<0$ is decreasing.

To get familiar with converting between λ and r, let's convert the previous calculations of wolf λ values to values of r:

$$\lambda_{1998-9} = 1.17 \text{ converts to } r_{1998-9} = \ln 1.17 = 0.16$$

$$\lambda_{1999-2000} = 1.34 \text{ converts to } r_{1999-2000} = \ln 1.34 = 0.29$$

As a convenient shortcut for converting between λ and r, it turns out that when r is close to 0, as it usually is for most wildlife populations, $\lambda \cong 1+r$. This means that when population growth is a relatively small increase or decrease you can approximate λ by adding 1 to whatever r is: an r value of 0.2 gives a λ value of about 1.2, and an r value of −0.2 gives a λ value of about 0.8. This rule of thumb breaks down as the absolute value of r gets further from 0 (e.g. try it for $r=0.6$ or −0.6); in that case you need to use your calculator and eqn. 5.6: $\lambda = e^r$.

Overview of λ and r

Exponential or geometric population growth, described by λ or r, find application in an enormous number of management applications ranging from harvest or endangered species programs to monitoring regimes. These approaches are based on two important assumptions. First, the growth rate over time is assumed to be independent of density. Second, growth is based on population sizes that can be adequately represented by the total number of individuals, ignoring differences between sexes, ages, and so on.

Which is better to use, λ or r? The answer depends on what you are doing. If you are just describing constant growth over time for a population with regular breeding seasons, λ is intuitively easier because it is based on familiar percentage changes. On the other hand, r has many useful properties, as listed here.

- The parameter r is centered on 0, whereas λ is centered on 1. Therefore population change is more intuitive with r, because a positive value indicates an increase and a negative value a decrease. Also, being centered on 0 leads to a symmetry that is lacking for λ: for example, a population going from 100 individuals to 165 and back to 100 over 2 years first has an r value of 0.5 and then an r value of −0.5; the same abundance change described by λ would be 1.65 (=165/100) followed by 0.61 (=100/165).
- If you want to compare species whose biology is based on different time steps, r transforms to other time scales easily whereas λ does not. For example, suppose we want to compare the growth rates of a tortoise and a hare. The tortoise has a growth rate of $\lambda=1.3/2$ years, whereas the hare has $\lambda=1.03/$month. Which is increasing faster? To make the comparison on the same time scale, we'll use a monthly growth rate. First convert both λ values to r ($r=\ln \lambda$): 0.26/2 years for the tortoise and 0.029/month for

---

Box 5.1 Analogy between λ versus r and annual yield versus annual rate in financial circles

For better or worse, few of us in wildlife or conservation biology got into this field because we want to make a lot of money. Still, you can use your knowledge about λ and r to help you make more sense of advertisements for savings and money market accounts, just in case you do make some money and wonder how to save it. Notice that the annual rate (sometimes called the interest rate) is always lower than the annual yield (sometimes called the annual percentage yield). That's because the annual rate means that your money will be compounded instantaneously over the whole year, just like r compounds population size. The annual yield is compounded just once a year, like λ.

So, suppose the annual rate is 5.75%; a population of money growing at $r=0.0575$ (they state r as a percentage to make it more palatable to people). Using $\lambda=e^r$, that is the same as a λ value of 1.0592, or an annual yield (proportional change) of our money of 5.92%. Just as wildlife population growth can be described interchangeably with r and λ, so too can your riches. Investing $100 today at these rates would give you, 1 year later, $100*e^{0.0575}=\$105.92$, which is the same as $\$100*1.0592=\$105.92$. Just as the annual rate and annual yield produce the same amount of interest earned, population growth using r and λ give the same rate of population change after you have transformed them appropriately.

Source: (after Case 2000).

---

the hare. Converting the tortoise r to a monthly rate (0.26/24) we get 0.011/month, much slower than the hare. Note that if we had simply and incorrectly divided the $\lambda_{tortoise}$ by 24 months (without first converting it to r) we would have calculated $\lambda=$ 1.3/24=0.05, an absurd answer implying a screaming decline of 95% per year!
- Similarly, you can add r values of a population over successive time intervals to get total growth over the whole period, but you cannot add λ values.
- Finally, if the growth rates over time are variable, the expected growth rate can be estimated by a simple average, or arithmetic mean, of the r values in consecutive time periods, but not of the λ values. We'll develop this idea much more in the upcoming section on the implications of variation in population growth (see below).

In short, Caughley (1977:52) states it nicely: "Although the replacement of the simple statistic λ by the more complex e^r may seem barbaric, it leads to simplified algebra and a better appreciation of the nature of a rate of increase." To make this comparison less abstract, Box 5.1 shows that the advertisements for money market or savings interest rates also use the concepts embodied by λ and r.

Before wrapping up this overview of geometric and exponential growth, we should address a common mistake, where exponential or geometric growth gets confused with "really big growth." Without a doubt, constant geometric or exponential change can propel abundances rapidly upward. For example, suppose you offered to give your roommate $1 million at the end of the month if he or she will give you just 3 cents today, then double the amount each day for the next 30 days. The population of pennies begins with $N_0 = 3$ and $\lambda = 2$. After $t = 30$ days, at the end of the month your roommate would have paid you (eqn. 5.2):

$$\text{Pennies}_{(\text{day } 30)} = 3 * 2^{30} = 3{,}221{,}225{,}472 \text{ pennies}$$

After paying back your roommate the $1 million you promised, you would still make more than $31 million!

Notwithstanding such striking examples, exponential or geometric growth is characterized not by how big the population gets or how rapidly it increases, but rather by population change being unaffected by density. For example, a population with $\lambda = 1.001$ could grow exponentially for 100 years and only increase from 10 individuals to 11. Simply put, exponential growth may or may not be really large, and a huge population may or may not be growing with exponential growth.

Doubling time

As a way of reviewing and synthesizing the use of simple exponential or geometric growth models, consider how the equations discussed so far can help predict the time it would take a population increasing at a certain exponential rate to double in size. The formula can be memorized, but instead let's derive it from first principles based on eqn. 5.2 ($N_t = N_0\lambda^t$) or eqn. 5.7 ($N_t = N_0 e^{rt}$). Given an observed λ or r value, how many time steps (t) would it take for N_t to be twice N_0 if that exponential growth continued? Going through the procedure for λ and using eqn. 5.2:

$$N_{t(\text{double})}/N_0 = 2 = \lambda^t \qquad (5.8)$$

To solve for the number of time steps (t) we get t out of the exponent by taking the ln of both sides:

$$\ln 2 = t(\ln \lambda) \qquad (5.9)$$

Then solve for t:

$$t = \ln 2/\ln \lambda \text{ (recall that } r = \ln \lambda)$$

$$t_{\text{double}} = 0.69/r \qquad (5.10)$$

Thus the time that it takes to double in size is shorter when population growth is larger. To see how doubling time is affected by growth rate, calculate for yourself the time it would take to double if $r = 0.10$ (6.9 years), and compare that with the doubling time for a population in which $r = 0.05$ (13.9 years). Once you see how it works for doubling time, you should be able to plug in the right numbers to calculate tripling time, the time to decrease by half, and so forth.

Box 5.2 Doubling time of Olympic mountain goats

Approximately 12 mountain goats were introduced to the Olympic Peninsula of Washington state during the late 1920s (Houston et al. 1994). In 1935, the central mountainous portion of the peninsula (including most of the prime goat habitat) was designated as Olympic National Park, with a mission to conserve the natural biota. The estimated exponential growth rate per year (r) of goats from the late 1920s to the early 1980s was 0.08. In the early 1980s, with more than 1100 goats distributed across the mountains of the park, biologists documented damage to native flora from the goats and the park authorities initiated exploratory efforts for decreasing or removing this introduced species. Suppose that in 1980 we were asked how long it would take for the goat population to double in size if no efforts were taken to reduce numbers and the past growth rates continued.

We're assuming no control in numbers and that population growth will continue as it had in the past. Using eqn. 5.10 from the text:

$$t_{double} = 0.69/0.08 = 8.6 \, years$$

Therefore, our starting point under the assumptions above would be that the population would double in size from 1100 to 2200 goats in about 9 years. In this case, that doubling time did not happen because the park began removing goats in the early 1980s (see Chapter 14).

Box 5.2 gives an example of calculating doubling time in a real population. It also reminds us that by projecting into the future we must assume that conditions in the past hold into the future.

Causes and consequences of variation in population growth

Factors that cause population growth to fluctuate

Population change in the real world will never be constant. Consider the time series in Fig. 5.3. Why aren't the changes in abundance constant over time? Obviously, a big reason will be steady impositions on population growth arising from other interacting species (predation, competition, parasitism, habitat loss due to humans, etc.), or from within the species (e.g. competition and other forms of density dependence). These are **deterministic factors** that change population growth in predictable ways (or at least somewhat predictable; see the Robbins quote at the start of the chapter).

But plots of wildlife abundance over time bounce around a lot more than can be explained by deterministic factors increasing or decreasing population growth rate. Some of the bounce is an artifact of **sample variance**, arising from the process of sampling abundance (see the SE bars in Fig. 5.3, which quantify the precision

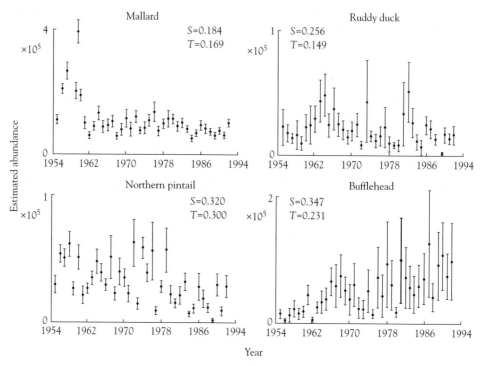

Fig. 5.3 Plots of estimated abundance over time (called **time series**) for four waterfowl species (from Link & Nichols 1994; reproduced by permission of Oikos). Abundance estimates ($\times 10^5$) during the period 1955–92 were obtained by aerial survey corrected for visibility bias, for a stratum of the May Breeding Waterfowl Survey in southern Manitoba. The bars around each abundance estimate represent 1 SE. Total variation in the time series is given by S for each species. This includes true process variance (environmental stochasticity over time) as well as sampling variance of the individual abundance estimates (see Chapter 2). The smaller value, T, for each species represents true temporal (process) variance, after sample variance is subtracted out. Note that total variation is up to 50% larger than true process variance, reminding us that sampling variance makes the trend in the populations look more variable through time than they really are.

of the annual abundance estimates). Notice, for example, that if I had just plotted the mean abundance estimates (without error bars) for the bufflehead in Fig. 5.3, it would appear that the population bounced around a great deal. But the overlap of the standard error bars arising from sample variance means that true population size actually may have been nearly constant across the time series, especially after 1980. Fortunately, there are ways to mathematically remove the noise contributed to time series by sample variance (Meir & Fagan 2000, White 2000, Holmes 2001).

Thus, deterministic changes could modify population growth over time for any population, and sampling error could make a time series appear to have fluctuations simply because we are making estimates of abundances. But imposed on these are true random fluctuations, otherwise known as **stochastic process variance**, affecting population growth[3]. The two main forms of stochasticity in population growth are demographic and environmental stochasticity.

Demographic stochasticity includes the inevitable deviation in birth and death rates arising from the mean rates being probabilities across a population. For example, if the mean annual survival rate is 0.8, each animal can only live through a given year, or die. It cannot 0.8 live. For small populations, demographic stochasticity causes variation in population growth even when mean birth and death rates remain absolutely constant.

One of the easiest ways to understand demographic stochasticity is by example. If you toss a coin, you would reasonably expect the probability of heads to be 50%. But if you toss the coin only three times, you cannot possibly get 50% heads: you will get 0, 33, 67, or 100% heads, by chance, even when the mean expectation is 50%. Even if you tossed it 10 times you would not be terribly surprised to get 3, 4, 6, or 7 heads, all of which are deviations from the expected 50:50. However, if you tossed the coin 100 times you would expect the percentage of heads to be much closer to 50%, and with 1000 tosses you would expect it to closely converge on 50%, the expected probability of heads (see Box 5.3).

Consider how this analogy would apply to sex ratios, which are often expected to be close to 50:50 at birth. In a small population, say with only 20 births, by chance only five or six females might be born in a given year. Such chance events – demographic stochasticity – can hamper the ability of a population to increase. Similarly, when numbers are small there could be large deviations from the expected survival rate or birth rate. So demographic stochasticity can affect sex ratio, reproduction, and survival, causing each to be more or less than the mean expectation.

The result of demographic stochasticity is that a population trajectory for a small population will vary even in a constant environment, with no change at all in mean birth or death rates. In particular, even if average population growth should be positive, demographic stochasticity in a small population could cause a decline towards extinction. An often-cited example is the extinction of the dusky seaside sparrow, for which all of the last five survivors happened to be males. Just as you would be more confident that your coin toss would be close to 50:50 when repeated many times, demographic stochasticity is minimized when abundance exceeds about 100 individuals (see Box 5.3).

In contrast to demographic stochasticity, which produces random variation around the mean rates, **environmental stochasticity** produces random changes in the mean vital rates for the population. Environmental stochasticity arises from extrinsic factors, often driven directly or indirectly by weather (Fig. 5.4). For example, if a salamander requires a certain level of moisture to breed, then a particularly hot dry spring may lower the mean reproduction of adults and survival of juveniles across the population. For many wildlife populations, food supplies may affect average survival of young in

---

[3]To be precise, a lot of the stochastic or random, fluctuations we see in populations may in fact actually be driven by deterministic factors, but can more easily be treated as stochasticity. Given how limited our knowledge is of many things impacting most wild populations, we are usually left treating a lot of variability as randomness even though it might be predictable with more complete information. See the quote about stones and cow pies at the start of the chapter.

Box 5.3 An example of demographic stochasticity based on reintroduced woodland caribou

Although woodland caribou formerly ranged over much of the northern USA, by the early 1980s habitat degradation, poaching, and vehicle collisions had reduced their US distribution to only about 25 animals in the Selkirk Mountains of northern Idaho and northeastern Washington. One recovery strategy for this federally listed endangered species was to translocate caribou back into their historic range (see Chapter 10 for more on caribou translocations). The estimated survival for 60 caribou translocated over 5 years was 0.74 (Compton et al. 1995).

Although of course survival varies over space and time from environmental stochasticity, let's see how much variation we might expect from demographic stochasticity alone. If only two caribou (one female and one male) were released and the environment was constant (environmental stochasticity was 0), the expected number in the next year would be 2 * 0.74 = 1.48. Of course, as in the coin-toss example there would be three possible outcomes: both caribou survive (survival = 1.0), only one does (survival = 0.5), or neither does (survival = 0). The binomial probabilities of each of these outcomes can be calculated[1], and the table shows that for two individuals we would expect neither caribou to survive about 7% of the time, one of two to survive (for a survival rate of 0.5) about 38% of the time, and for both to survive about 55% of the time. A survival rate of 0.74 is not even possible! Even with the actual reintroduction sizes of 12 and 24, lots of survival rates other than 0.74 are possible. As the population gets close to 100 individuals, the variability due to demographic stochasticity decreases and the expected survival rate converges on the annual survival rate of 0.74 for all individuals.

Percent chance that different population sizes would have particular observed survival rates. (Survival rates centered on the true mean of 0.74 are highlighted.)

		Chance of observed survival rate (%)									
Population size	**Survival rate . . .**	**0–0.1**	**0.1–0.2**	**0.2–0.3**	**0.3–0.4**	**0.4–0.5**	**0.5–0.6**	**0.6–0.7**	**0.7–0.8**	**0.8–0.9**	**0.9–1.0**
2		7	0	0	0	38	0	0	0	0	55
6		0	1	0	4	14	0	30	0	35	16
12		0	0	0	0	6	12	20	26	22	14
24		0	0	0	0	1	6	20	52	18	3
48		0	0	0	0	0	1	24	59	16	0
96		0	0	0	0	0	0	20	70	10	0

[1]The binomial probability of y successes in n trials with a probability of success of z per trial is $\left[\left(\frac{n!}{y!(n-y)!}\right)z^y(1-z)^{n-y}\right]$ where ! refers to the factorial (e.g. 4! = 4 * 3 * 2 * 1 = 24).

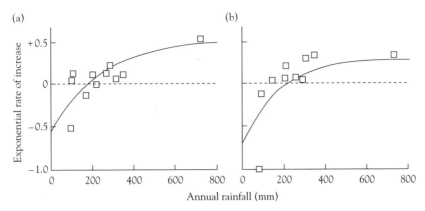

Fig. 5.4 Stochasticity in rainfall patterns explains variation in population growth rates in kangaroos. (a) Red kangaroos; (b) western grey kangaroos. Modified from McCallum (2000). Reproduced by permission of Blackwell Publishing Ltd.

a given year, as do random changes in predators or parasites. In all of these cases, mean vital rates for animals in the population – and therefore population growth rate – vary over time and space in unpredictable ways. Unlike demographic stochasticity, the occurrence of environmental stochasticity is not necessarily dependent on population size: for example, a year where λ fluctuates from 1.2 to 1.1 will reduce population size by about 10% for 50 or 500 animals.

A population exposed to environmental stochasticity (as all are) will inevitably have some very bad years and some very good years (Shaffer 1987). In many cases these stochastic fluctuations may seem extreme to us (perhaps because homes are damaged or human lives are lost), but they are really just somewhere away from the mean. In other cases true outliers can be referred to as **catastrophes** or **bonanzas** because they are outside the range (or at least far out in the tails of the distribution) of normal environmental stochasticity. Volcanic eruptions, avalanches (in some cases), and the plague come to mind as factors leading to catastrophes. Here are some specific wildlife-related examples of catastrophes: in 1973 a hurricane decimated the last Laysan teal population; volcanic eruptions in Japan almost destroyed the last colony of the short-tailed albatross; and a severe winter coupled with overbrowsing wiped out all but 50 of the 6000 reindeer on St. Matthew Island (Simberloff 1988).

Implications of variation in population growth

What happens when numbers of an exponentially increasing population bounce around through time due to stochastic variation? One outcome is fairly obvious: as variance in population growth increases, future population size outcomes become more uncertain and more variable. You can see this if you let a computer play "what if" scenarios of population growth many times (Fig. 5.5). Greater variation in future population sizes leads to an increase in extinction probability (Boyce 1977), even if average population growth is positive. In Fig. 5.5, you can see that a few replicate populations increase to very large sizes while most end small, or even extinct; from this

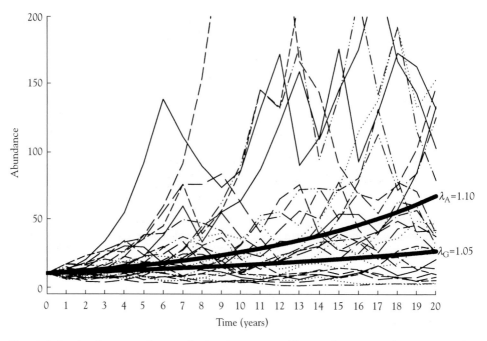

Fig. 5.5 Stochastic geometric growth showing 25 possible population growth trajectories for hypothetical snail kite populations (e.g. Beissinger 1995) beginning with 10 individuals. For each replicate, λ at each of 20 time steps varied randomly between 0.5 and 1.7 ($\sigma_\lambda^2 = 0.12$). λ_A is therefore 1.1 and λ_G approximately 1.05 (represented by thick lines). Because λ_G represents median population growth, about half of the final abundances fall above the λ_G line and half below.

we can take home the generalization that the greater the environmental stochasticity relative to mean population growth, the higher the chance of extinction.

The second outcome of stochastic variation in population growth is related but less intuitive. In a stochastic environment, the likelihood (or probability distribution) of any particular population size at time t in the future becomes more and more skewed, with most populations being relatively small (their numbers can't go below zero) but with a tiny fraction having huge abundance. The few really big populations inflate the average population size, making it much larger than the most likely future population size. Another way to say this is that without variation, the most likely outcome is the same as the average, but with stochasticity the few big winners make the average become larger than the most likely population size. With stochasticity, the most likely trajectory of population size over time is captured by the geometric mean growth rate (λ_G) – which is mathematically identical to the arithmetic mean, r – not the arithmetic mean growth rate (λ_A; Morris & Doak 2002).

Before exploring this idea with wildlife populations, let's think about uncertainty in outcomes using the stochastic world of playing the stock market (Fig. 5.6). Suppose you were offered a stock that each week either gains 80% of its value or loses 60% of its value: in half of all weeks it gains and in half it loses. You could estimate the average change in value as the **arithmetic mean**, which is a 10% gain (calculated from [80+−60]/2=10). Wouldn't that be an awesome buy for a starving wildlife population

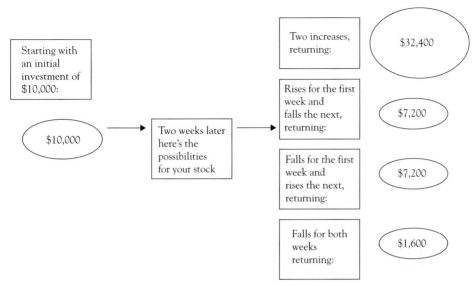

So, the average return per week is still 10% for an arithmetic mean return of $ 12,100 after 2 weeks),
but there is a loss in three of the four possible scenarios, indicating the most likely outcome is that
you will lose money.

Extending the outcome of the investment to 1 year:

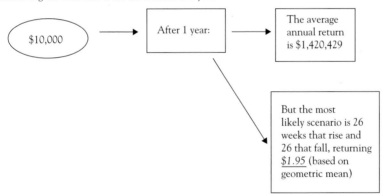

Fig. 5.6 The impacts of stochasticity. This is a deal too good to be true (Chang 2003, Paulos
2003). Suppose I offered you a stock that is updated in value each week: half the time it increases
by 80% of its current value, and half the time it loses 60% of its current value. Although the arithmetic mean indicates a 10% gain, if you bought it you would most likely lose a lot of money. In a
stochastic world with multiplicative growth (of dollars or populations), go with the geometric
mean, not the average from the arithmetic mean.

ecologist? After a year of average growth of 10% per week, a $10,000 investment would
be worth more than $1.4 million dollars! So why not quit your job and just buy this
stock? Because the arithmetic mean, or average, is driven by the few lucky winners who
make a lot of money (due to a rare string of many weeks with 80% gains), while most
investors will lose in a big way. If you want to know the most likely outcome you would
calculate the **geometric mean** (λ_G or $\bar{r}$; see Box 5.4): in this case you would learn that

Box 5.4 Calculating the geometric mean

The geometric mean differs from the arithmetic mean because instead of adding a total of "t" numbers and dividing by t, you instead multiply the t numbers and take the tth root of the product. To put these words into an equation for the geometric mean population growth rate (λ_G) over time:

$$\lambda_G = \sqrt[t]{(\lambda_1 * \lambda_2 * \lambda_3 * ... \lambda_t)}$$

or equivalently

$$\lambda_G = (\lambda_1 * \lambda_2 * \lambda_3 * \ ...\ \lambda_t)^{\frac{1}{t}}$$

The geometric mean will be less than the arithmetic mean when there is stochasticity. Let's run through an example. Suppose an endangered population grows at a constant $\lambda = 1.05$; we would expect a 5% increase per year, so that in 16 years a population of 100 would have an expected size (from eqn. 5.2) of

$$N_{16} = 100 * 1.05^{16} = 218$$

Now suppose instead that the population growth alternated each year between $\lambda = 1.55$ and $\lambda = 0.55$. The arithmetic mean of the growth rate is still 1.05 [from (1.55+0.55)/2]. But the growth of the average population is governed by the geometric mean, which is

$$\sqrt{1.55 * 0.55} = 0.923$$

After 16 years the expected population size would be

$$N_{16} = 100 * 1.55^8 * 0.55^8 = 28$$

This is the same as projecting all 16 years with the geometric mean: $100 * 0.923^{16} = 28$.

A population of 28 is a lot less than the 218 expected from the arithmetic mean! The variation in population growth leads to a likely decline for the population, even though the deterministic growth rate implies that the population should increase substantially.

Similarly, look back at Fig. 5.6, the stock market example, with a variable growth rate of 80% increase ($\lambda = 1.8$) and 60% decrease ($\lambda = 0.4$). The arithmetic mean is (1.8+0.4)/2=1.1, which is what led to the predicted average of

$$\$_{52\ weeks} = \$10,000 * 1.1^{52\ weeks} = \$1,420,429$$

By contrast, the geometric mean ($\sqrt{1.8 * 0.4} = 0.849$) gave the predicted most likely earning of

Box 5.4   Continued

$$\$_{52\,weeks}=\$10{,}000*0.849^{52}=\$1.95$$

An equivalent way to calculate the geometric mean population growth rate from a time series takes advantage of the mathematical properties of $r=\ln\lambda$:

- calculate r for each interval by $\ln(N_{t+1}/N_t)$;
- take the arithmetic mean of all of the r values to obtain $\bar{r}$;
- convert the $\bar{r}$ back to λ (by way of $\lambda=e^r$) and you've got your λ_G.

you and the majority of other buyers will lose almost all of your money, ending the year with $1.95 of your original $10,000 (Fig. 5.6). This is barely enough for a big mug of coffee with which to console yourself. The average population growth leading to the average earnings ($1.4 million) is much less informative than the geometric mean leading to the most likely (median) earnings over time.

Bringing this back to wildlife population biology, how would you feel about the fate of a population that increased by 80% half the time and declined by 60% half the time? Just as mean stock market earnings are dominated by a very few big winners, the average population size in a stochastic environment with multiplicative population growth is driven by the likelihood that a few populations would by small chance become very large while most populations would not. Typically we are interested in the geometric mean – indicating the most likely trajectory and its associated population size – not the growth rate of the average population size.

Thus, the real biological variation around λ will make populations grow more slowly than they would under a constant arithmetic mean growth rate. This leads to the counterintuitive finding that a population whose average growth rate is positive could be most likely to decline over time. The magnitude of the decrement in the long-term growth rate of a population depends on how variable the growth rates are (Fig. 5.7).

Specifically, increasing the variance in growth rate (σ_λ^2) causes the geometric mean (λ_G) to be less than the arithmetic mean (λ_A; Case 2000, Lande et al. 2003)[4]:

$$\lambda_G\cong\exp(\ln\lambda_A-[\sigma_\lambda^2/(2*\lambda_A^2)]) \tag{5.11}$$

In short, λ_G (or its cousin $\bar{r}$, which we'll discuss next section) tells us what the typical population with stochastic growth is likely to do in the future[5]. Because λ_G predicts the

---

[4]If you like thinking in terms of $\bar{r}$, an analogous approximation uses variance in the r values (σ_r^2) instead of σ_λ^2 (Lande et al. 2003:11): $\bar{r}\cong\ln\lambda_A-\sigma_r^2/2$. The applied result is simple: if variance in r is more than twice as big as $\ln\lambda_A$, then $\bar{r}<0$, and the long-term growth rate becomes a decline towards extinction.

[5]Often in the literature on population viability analysis you will see λ_G or $\bar{r}$ denoted with the symbol μ and called the long-run growth rate.

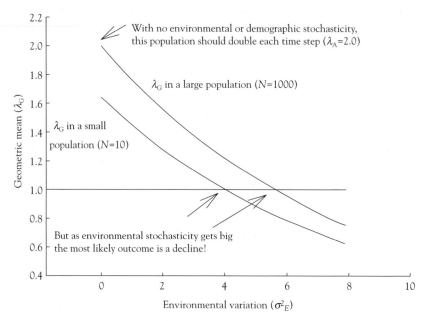

Fig. 5.7 As environmental stochasticity increases there is a reduction in geometric mean growth rate (λ_G). The upper curve shows the reduction in a large population ($N=1000$). The lower curve shows that in a small population ($N=10$) the added fluctuations due to demographic stochasticity further decrease expected population growth. The arithmetic mean growth rate (λ_A) is 2.0. Notice that the most likely population trajectory is a decline ($\lambda_G<1.0$; horizontal line) as environmental variation increases. Modified from Case 2000.

median population size along the curve (i.e. at any point approximately half the replicates have population sizes above and half below the line for λ_G in Fig. 5.5), a population with $\lambda_G<1$ or $\bar{r}<0$ will be more likely to decline than to increase. Next we'll describe how to estimate the exponential population growth rate.

Quantifying population growth in a stochastic environment

So far I have emphasized that population growth for a time series cannot be estimated by a simple arithmetic mean of the λ values (λ_A), but rather should be based on the geometric mean population growth rate. What is the best way to estimate λ_G (or $\bar{r}$) and its associated estimate of uncertainty from field data (J.Y. Humbert & L.S. Mills, unpublished work)?

One way is described in Box 5.4. This straightforward approach of simply taking the geometric mean of all λ values in the time series (or the arithmetic mean of the r values)[6] works fine if estimates of abundance are available for each and every time interval. Unfortunately, for most studies holes creep into the time series because trucks

---

[6]The variance for $\hat{\bar{r}}$ across q estimates of r with no missing values is $\sigma^2 = \dfrac{1}{q-1}\sum_{i=1}^{q}\left(r_i-\bar{r}\right)^2.$

break down, field assistants get sick, weather keeps us from getting to the study site in a particular year, or funding fluctuates. If any breaks occur in the time series, so that some of the λ or r values refer to more than a single year (or time step), the values cannot be simply summed or averaged.

Because holes in the time series are an inescapable part of wildlife studies, how do we estimate the exponential growth rate ($\hat{r}$) and its variance (σ^2) in their presence? A simple and intuitive trend estimate, recommended by several papers and books (e.g. Caughley & Sinclair 1994, Caughley & Gunn 1996) is to plot the natural logarithm of all of the abundance estimates (ln N) against time; the slope of the line is an estimate of $\hat{r}$ (as in Fig. 5.2b). This method has several problems. First, there is no valid estimate of the variance in $\hat{r}$; the variance of the slope actually represents variance of the ln(abundance) values, not of the r_i, and will be less than the actual σ^2 of the population growth rate (Eberhardt & Simmons 1992). This is critical, because it means we can't test whether changes up or down are significant (and if you are trying to convince angry people that you have to change management based upon changes in population numbers, you'd like to have statistics on your side). Second, because successive abundances in a time series are not independent – when a particular abundance deviates from the mean regression line it will likely deviate in the next year too, causing what is known as autocorrelation – a basic assumption of linear regression is violated.

A much better way to estimate exponential trend ($\hat{r}$) and its variance (σ^2) is the **density-independent diffusion approximation method** (let's just call it the DA method) championed by Brian Dennis and colleagues (1991; see also Morris & Doak 2002). Essentially, you first transform time elapsed and population change for each interval to account for holes in the time series, then you do a regression of the transformed population changes against the transformed time intervals to estimate $\hat{r}$ and σ^2 (Box 5.5 gives a worked example). Specifically, the explanatory variable (x_i) is the transformed time interval between each successive pair of abundance estimates:

$$x_i = \sqrt{t_{i+1} - t_i} \tag{5.12}$$

The response variable (y_i) is the population change corrected for its time interval:

$$y_i = [\ln(N_{i+1}/N_i)]/x_i \tag{5.13}$$

By using population change (y_i) over the interval instead of the raw ln N values, the autocorrelation problem of the previous method is minimized. For ease in presentation I'll use a year as a time interval, but it could be any unit, such as months or days. Notice that if we had been able to obtain estimates of abundance each year, so that each interval of population change equals one ($x_i = 1$), then the response variable (y_i) would simply be the consecutive r_i estimates.

With the x_i and y_i calculated (easy to do in any spreadsheet program; try it with Box 5.5), you next perform a linear regression of the y values against the x values, forcing the regression intercept through 0. The slope of the regression is an estimate of $\hat{r}$. The estimated variance ($\hat{\sigma}^2$) for the estimated growth rate is the mean square error (called the residual mean square in some statistics packages) for the regression.

Box 5.5 How to calculate $\hat{r}$ and its variance ($\hat{\sigma}^2$) using the DA method

An isolated population of adders (a type of pit viper) in Sweden declined dramatically about 40 years ago, and has been studied intensively since 1981 by Thomas Madsen and colleagues (Madsen et al. 1996, 1999). The time series for adult male adders (1981–95) can be used to demonstrate how to calculate $\hat{r}$ and its variance ($\hat{\sigma}^2$) using the DA method. Because the snakes bask in the open in a small study area, so that virtually all adult males adders can be captured in each annual survey, sampling variance is nearly 0, so virtually all variance in the time series is process variance. Although the original data set is complete, I omitted 2 years (1987 and 1988) to show the common case of dealing with missing counts.

Year	Number of adult males	$\ln(N_{t+1}/N_t)$	$x_i = \sqrt{t_{i+1} - t_i}$	$y_i = \ln(N_{t+1}/N_t)/x_i$
1981	19	−0.0541	1	−0.0541
1982	18	0.201	1	0.201
1983	22	0.128	1	0.128
1984	25	−0.0834	1	−0.0834
1985	23	0.0000	1	0.0000
1986	23	−0.191 across three intervals	1.732	−0.110
1987	No count			
1988	No count			
1989	19	−0.111	1	−0.111
1990	17	−0.125	1	−0.125
1991	15	−0.511	1	−0.511
1992	9	−0.406	1	−0.406
1993	6	0.154	1	0.154
1994	7	−0.560	1	−0.560
1995	4			

Both the explanatory variable (time interval x_i) and the response variable (growth rate y_i) are transformed to account for the different number of years spanned by the growth rates. A linear regression of y_i against x_i, forcing the intercept through 0 (yes, it will look odd because so many of the x_i values are equal to 1), gives a slope of −0.11, a SE of the slope of 0.067, and a variance (error mean square) of 0.063. Therefore:

$$\hat{r} = -0.11 \text{ and } \hat{\sigma}^2 = 0.063$$

A confidence interval (CI) is calculated from the SE and the Student's t value. Specifically, a 90% CI is specified by $\hat{r} \pm [t_{0.10, \ q-1} * SE(\hat{r})]$, where q is the number of y_i in the regression (12 in this case). For these data the 90% CI of $\hat{r}$ is:

Box 5.5   Continued

upper: $-0.11+(1.796*0.067)=0.01$

lower: $-0.11-(1.796*0.067)=-0.23$

Although these count data imply a declining population (with 90% CI barely overlapping 0), the story has a happy ending: the inbreeding depression that contributed to the decline was reduced by translocations, with a considerable population increase after 1996 (see Chapter 9).

A confidence interval (CI) for $\hat{r}$ can be calculated from the standard error (SE), usually supplied by the statistics package and labeled as the SE of the slope; if not, you can hand-calculate SE from $\sqrt{\dfrac{\hat{\sigma}^2}{\text{Number of years spanned}}}$, where $\hat{\sigma}^2$ is the estimated variance and number of years spanned means the duration of the time series (unaffected by missing values in the time series). The confidence interval is: $\hat{r}\pm[t_{\alpha,q-1}*\text{SE}(\hat{r})]$ for the upper and lower intervals.

The $t_{\alpha,q-1}$ values are the critical values of the two-tailed Student's t distribution (available in any statistics book[7]) with a significance level of α (for example, $1-0.95=$ 0.05 for a 95% CI, or $1-0.90=0.1$ for a 90% CI) and q equaling the number of transitions used in the regression. As you recall, the 95% CI tells us that if we were to do this study several times using the same population and same sample size, 95% of the time the true growth rate would fall within the range of the estimated confidence interval. As I preached in Chapter 4 (on estimating vital rates), confidence intervals are essential because they help indicate whether an apparent trend is actually likely to be different from zero; whether or not our $\hat{r}$ confidence interval contains zero becomes especially important in management applications, for example in deciding whether to delist an endangered species that appears to be increasing[8].

The DA method does an excellent job of estimating density-independent trends, and their variance, in a variety of conditions (Humbert & Mills, in preparation)[9]. Other insights that can be gained from this regression method include confidence intervals for variance (σ^2), tests for outliers, tests for changes in trend in different segments of the time series, and tests for temporal autocorrelation in the population growth rate (i.e. whether a good or bad year for growth is correlated with whether previous or successive intervals were good or bad; see Morris & Doak 2002). The estimates of $\hat{r}$ and $\hat{\sigma}^2$ can also be used to predict probabilities of extinction from time-series data (see Chapter 12).

---

[7]You can also get the t statistic in Microsoft Excel by using =TINV (your probability, your $q-1$).

[8]A reminder: the variation in the time series should represent true process variation and not sample variance arising from estimating abundance (Chapter 2); if sample variance is not accounted for in the time series, the variance of the growth rate would be artificially large.

[9]Although the previous method, plotting ln N against time, performs well under some conditions the DA method always outperforms it; because the two methods can give qualitatively different results (one says increase while the other says decrease), you should always use the DA method.

How long a time series is needed to estimate trends reliably? The answer depends on many factors, of course, but a rule of thumb is at least 10–15 years or so (Swanson 1998, Holmes 2001, Morris & Doak 2002). However, here's a surprising but useful bit of mathematical trivia that can affect how you sample across the time series: the DA method for estimating $\hat{r}$ is absolutely equivalent to a form where only the first and last data points are used to estimate trend[10]:

$$\hat{r} = \frac{1}{\text{Total duration of survey}} * \ln\left(\frac{N_{last}}{N_{first}}\right)$$

Although this means you could quickly determine $\hat{r}$ with just this first-and-last-points method, it has no measure of variance so you'll still need to fall back to using all data points with the DA method to get a proper variance[11]. The fact that the DA method reduces to using just the first and last points to estimate trend – and that it estimates both $\hat{r}$ and σ^2 really well, even when up to half of the data points in a time series are missing (Humbert & Mills, in preparation) – suggests that if there is no reason to suspect cycles or other factors that would change the trend, the best strategy is to place more effort and money into fewer data points (with improved accuracy), even if it means lots of holes in the time series. Thus to estimate trend for a 15-year time series you might put more effort into only seven really good estimates of abundance rather than 15 mediocre ones.

Other approaches may also be used to estimate trends of wildlife populations over time. For example, Bayesian analyses (e.g. Taylor et al. 1996) are becoming more popular. Also, if mark–recapture data are available both λ and its variance can be calculated directly (see Nichols & Hines 2002 for nice overview and Cam et al. 2003 for an application based on marbled murrelets).

Summary

Population growth (and decline) is a multiplicative process, not an additive one, leading to exponential or geometric changes in the absence of density or individual-level effects. Although the bounce, or variation, that accompanies population changes over time can arise from deterministic factors coming and going, in part the variation in numbers will be an illusion arising from the fact that we typically have to sample (not census) a population; thus, nuisance sampling variation makes it look like the population varies more over time than it really does.

Even with these two sources of variation accounted for, however, there will be process variance in population growth due to demographic and environmental stochasticity. Stochastic variation in growth rates causes the long-term most-likely

---

[10]You've seen this before – this is mathematically identical to the estimate of λ in eqn. 5.4.

[11]Although some have suggested that the DA method wastes data because it reduces to only using two points in the time series, the alternative approach (least-squares regression of ln N against time) also weights more heavily the first and last data points in estimating the slope of the line.

abundance – based on the geometric growth rate – to be lower than that based on the arithmetic average λ. This leads to the counterintuitive finding that a population which has, on average, a positive growth rate may be most likely to decline. The best way to estimate population growth rate and its variance from a field-based time series is via a regression on transformed growth rate and time.

Although exponential and geometric growth are based on a number of assumptions, and no population can grow without limit for long, understanding this simple form will make it easier to understand more complicated models of population growth. Furthermore, we can learn a lot about mechanisms of population growth (increases or decreases) for real populations by comparing trajectories with what might be expected under simple geometric or exponential growth. The next two chapters will add complexity to our tools for assessing and predicting population change over time.

Further reading

Case, T. (2000) *An Illustrated Guide to Theoretical Ecology*. Oxford University Press, Oxford. An excellent overview of theory, clearly and cleverly explained.

McCallum, H. (2000) *Population Parameters: Estimation for Ecological Models*. Blackwell Science, Oxford. Provides a thorough background for estimating population growth rate.

Morris, W.F. and Doak, D.F. (2002) *Quantitative Conservation Biology: Theory and Practice of Population Viability Analysis*. Sinauer Associates, Sunderland, MA. The "go- to" source for time-series analysis, matrix modeling, and many other important topics.

6

Density-dependent population change

The elephant is reckoned the slowest breeder of all known animals, and I have taken some pains to estimate its probable minimum rate of natural increase; it will be safest to assume that it begins breeding when thirty years old, and goes on breeding till ninety years old, bringing forth six young in the interval, and surviving till one hundred years old; if this be so, after a period of from 740 to 750 years there would be nearly nineteen million elephants alive, descended from the first pair.

Charles Darwin (1859), *On the Origin of Species*

Introduction

So far we have learned how to describe and predict population change when additions and deletions to a population are unaffected by its own density. Of course, Darwin's quotation reminds us that it would be silly to expect populations to increase exponentially for long periods. Eventually, there will be feedback between the density of the population and its growth rate. **Density dependence** refers to the profound influence that a population's density has on the vital rates of individuals in the population; changes in vital rates, in turn, lead to changes in population growth rate[1].

This chapter will first cover classic negative density dependence, where high numbers lead to negative feedback on vital rates and population growth. We will then discuss positive density dependence, where survival or reproduction actually improves with higher density. Finally, we will explore some ways to incorporate density dependence in models that can be useful both for prediction and to show how surprisingly complex dynamics can emerge from simple processes.

---

[1]As in the last chapter, remember that population growth refers to increasing, decreasing, and constant populations.

Negative density dependence

Increases in density can exacerbate competition among members of a population and heighten susceptibility to predation, parasites, and disease. As long as humans have been observing other species (or their own species; see the quotation at the start of Chapter 5 from Malthus) we have noticed the occurrence and consequences of intraspecific competition among individuals of a species. Sometimes this competition occurs as direct **interference** or **contests** among individuals, as in fights for food, mates, or territories; winners get sufficient resources to survive and reproduce while losers may not. In other instances the competitive interaction is **exploitative**, or based on scrambles for resources, with simultaneous use of a common resource (often food) lowering the amount of that resource available to each individual without direct interference; this is like sharing a single pie with more and more people.

Obviously, the mechanisms and outcomes of competition can operate simultaneously, a fact nicely demonstrated with black-throated blue warblers that show clear negative density dependence (Fig. 6.1a). As abundance increases the warblers experience scramble competition, with reduced average fecundity across individuals. At the same time, contest competition occurs on a broad scale because increases in warbler abundance force additional individuals into areas of lower quality (Rodenhouse et al. 2003).

Competition is not the only mechanism that can lead to negative density dependence. For example, larger groups may be more conspicuous to predators, or have an increased probability of contracting parasites or contagious disease. In short, under negative density dependence births or survival – and the associated observed population growth rate – decrease as the population size increases (Fig. 6.1). Negative density dependence thus **regulates** population numbers within some equilibrium size range (carrying capacity) by decreasing population growth at high density and increasing it at low density.

Although regulation is a density-dependent process, **limiting factors** that determine the actual equilibrium population size range may be density-dependent or density-independent (Sinclair 1989, Caughley & Sinclair 1994, Hixon et al. 2002). As an example of both density dependence and environmental stochasticity (a density-independent factor) acting simultaneously as limiting factors, arctic ground squirrels experience strong density dependence on the proportion of females weaning a litter and the proportion of females surviving winter hibernation, but the weaning rate is also determined independently by weather (Karels & Boonstra 2000). Similarly, endangered San Joaquin kit foxes in California respond to both density (high density both caps territorial breeders and increases juvenile mortality) and rainfall (rain affects vegetation, which in turn affects small mammals, whose numbers can drive fox reproductive rate; White & Garrott 1999, Dennis & Otten 2000). For applied population biology the important point is to recognize that density can at times negatively affect vital rates, and the resulting population growth rate.

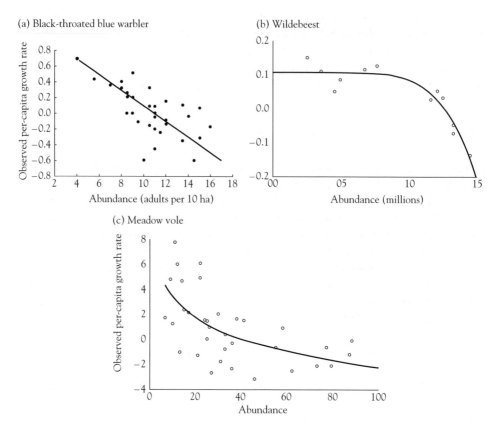

Fig. 6.1 Some examples of negative density dependence in wildlife populations. Observed per-capita population growth rate measured from time series of abundances $[\ln(N_t/N_{t-1})]$ is plotted against abundance. (a) Black-throated blue warblers (modified from Rodenhouse et al. 2003; reproduced by permission of The Royal Society); (b) wildebeest (from Turchin 1999 after Sinclair 1996; reproduced by permission of Oikos); (c) meadow vole (from Turchin 1999; reproduced by permission of Oikos). Notice that all sorts of linear and nonlinear responses of population growth to density are possible.

Positive density dependence

Of course, interactions among members of a population are not always negative. An increase in vital rates or population growth as density increases (or a decrease as density decreases) is referred to as positive density dependence[2]. The ecologist W.C. Allee was the first person to give comprehensive attention to the existence and

---

[2]Sometimes you will see positive density dependence referred to as **inverse density dependence**, where inverse is used because it is the opposite of traditional (negative) density dependence. In the fisheries literature this form of density dependence is often called **depensation**. I'll stick with positive density dependence: survival, reproduction, and population growth are positively related to density.

implications of positive density dependence (Allee 1931), so the phenomenon is often referred to as an **Allee effect** (for reviews see Courchamp et al. 1999, Stephens & Sunderland 1999). Box 6.1 shows a detailed example.

Some of the most important instances of positive density dependence occur when a population becomes very small, teetering on the brink of extinction[3]. In these cases,

---

Box 6.1 Multiple Allee effects in African wild dogs

There are fewer than 5000 African wild dogs remaining, and their persistence may be affected by Allee effects (Creel & Creel 2002). Larger pack sizes increase the likelihood of a successful kill in a given hunt, with heavier prey killed, shorter chases, and more meat per dog. Increasing pack size also confers benefits in predator defense. Larger packs can better defend their kills against spotted hyenas, and can more successfully counter attacks from other wild dog packs. Larger packs can also more easily defend their pups from predation.

The increased food acquisition and defense leads to a positive relationship between pack size and the number of offspring born and raised. Nonbreeders guard pups and feed them regurgitated meat. As pups grow older, nonbreeders continue to guard the carcass as pups eat. Through these mechanisms, packs with 10 or more adults raised three times as many yearlings compared to packs with nine or fewer adults (with 10.4 compared with 3.4 pups surviving 1 year, respectively). Packs with fewer than five adults are unable to raise pups (Creel & Creel 2002).

Of course, the benefits of increasing pack size evaporate at larger numbers as negative density dependence kicks in, with factors such as reproductive suppression at large pack size. Thus African wild dogs are an excellent example of positive density dependence at low numbers and negative density dependence at higher numbers.

A group of African wild dogs attacking a young wildebeest. Photograph by Scott Creel.

---

[3]Although some include as Allee effects both demographic stochasticity and inbreeding depression, I consider these elsewhere (see Chapters 5 and 9). These processes do depress population growth at low numbers, and therefore lead to the same outcome as Allee effects (Lande et al. 2003), but they can be modeled and managed separately from Allee effects (Chapter 12).

Table 6.1 Some possible mechanisms that lead to positive density dependence (Allee effects) in wild populations (Creel & Creel 2002).

Mechanism	Example
Minimizing predation	
Increased predator detection and defense	Survival rates in Dwarf mongoose in Kenya increase with group size because vigilance behavior by guards decreases predator attacks; the guards also defend against ground predators (Rasa 1989).
Greater confusion for predator	The confusion effect causes largemouth bass to take longer to capture silvery minnows as minnow school size increases (Landeau & Terborgh 1986).
Decreased probability of individual capture	Predator swamping: per-capita nest mortality is reduced in larger black brant colonies because predator numbers are overwhelmed (Raveling 1989).
Foraging advantages	
Access to foods	Small colonies of blind mole-rats are more likely to fail because they are unable to rapidly extend burrow systems to obtain food during the brief time period when the soil is moist and easily worked (Jarvis et al. 1998).
Increased resource detection	After being reduced in numbers by hunters and habitat alteration, a cause of further decline in passenger pigeons may have been that small flocks were compromised in their ability to find their patchy and sporadic food sources (e.g. acorns and nuts; Reed 1999).
Cooperative resource defense	Larger groups of coyotes are better able to defend carcasses against intruder coyotes (Bekoff & Wells 1986).
Mating and caring for young	
Finding mates	Glanville fritillary butterflies less likely to locate mates and successfully reproduce in small populations (Kuussaari et al. 1998).
Caring for young	The number of young fledged per nest increases with group size in white-fronted bee eaters because helpers reduce starvation of nestlings by provisioning food; helpers also assist in nest excavation, nest defense, and egg incubation (Emlen 1990).
Conditioning of environment	
Temperature tolerance	Bobwhite quail in large coveys standing in a circle with their tails towards the center of the circle are better able to survive extreme low temperatures (Allee et al. 1949).

extinction risk is exacerbated because individuals in small populations or groups are compromised for some or all of the following reasons (Table 6.1): (i) detection, confusion, or avoidance of predators, or the dilution of per-capita predation risk; (ii) foraging ability; (iii) mating and caring for young; or (iv) conditioning of the environment.

Positive density dependence can also affect dynamics in large populations. A spectacular illustration occurs in mormon crickets, which form migratory bands 16 km long and several kilometers wide packed with crickets; mortality within the group is nearly zero, while animals outside the group have high mortality (Sword et al. 2005). In invasive or pest populations, positive density dependence can affect how outbreaks occur and whether a pest will spread (Taylor & Hastings 2005). For example, after first being released in the New York City area about 1940, house finches spread relatively slowly at first, then abruptly accelerated their population growth and range expansion across eastern North America, consistent with crossing an Allee-effect threshold (Veit & Lewis 1996).

Overall, positive density dependence is probably widespread, especially in critically small wildlife populations (Liermann & Hilborn 2001). The most likely scenario for the majority of wildlife populations is that both positive and negative density dependence occur at different densities.

The logistic: one simple model of negative density-dependent population growth

Having considered some biological mechanisms that cause populations to exhibit negative and positive density dependence, we are ready to capture the effects of these processes on population growth. In the exponential growth model (Chapter 5), birth and death rates are unaffected by densities. Thus, for populations growing exponentially the change in numbers is described by

$$dN/dt = rN \qquad (6.1)$$

Exponential growth under this continuous time model indicates that at any time the population change equals the number of individuals (N) at that instant multiplied by r, the instantaneous growth rate per capita.

In contrast, with density dependence, population growth is not constant as population size changes over time. In particular, reproduction and/or survival changes with density, leading to potential changes in population growth. Figure 6.2 shows just a few ways that density dependence could change per-capita mortality ($1 -$ survival) and/or reproduction. The point at which per-capita mortality and reproduction are equal, so that the population just replaces itself and $\lambda = 1$ ($r = 0$), is called the **carrying capacity** (denoted by K). The carrying capacity is considered to be an **equilibrium** because if density is greater than K then mortality exceeds reproduction and the population will decrease to K; if it is less than K then reproduction exceeds mortality and the

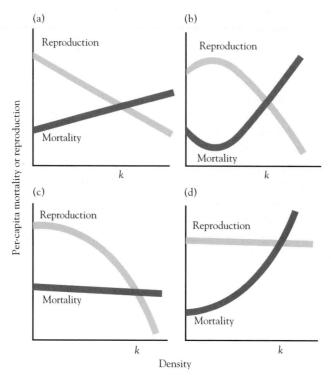

Fig. 6.2 Four possible ways that density dependence could affect per-capita mortality and reproduction rates. All graphs show only negative density dependence except (b), which at low density
shows a small range of positive density dependence. The bands embrace the uncertainty and
changes in the relationships. The region where the bands cross represents the carrying capacity
(*K*), where births and deaths are equal.

population increases towards K^4. Notice in Fig. 6.2 that the vital rates are drawn as
bands (not lines), so that the carrying capacity (where mortality balances reproduction) is a zone, not a single number. The bands reflect inevitable uncertainty in both
the density-dependence relationship and in vital rates and abundance over time, so
carrying capacity becomes a range of abundances, not some exact number. Another
important point conveyed by Fig. 6.2 is that one could draw an infinite number of
ways in which density dependence could change vital rates.

As per-capita reproduction and mortality change with density, the realized per-
capita growth of the population $[\ln(N_{t+1}/N_t)]$ becomes a function of both the instantaneous growth rate per capita of the exponential model (*r*; often called the **intrinsic
growth rate**) and of the way that the *r* value is affected by density. In a simple population model, the *r* gets multiplied by a term that dampens (or increases) it by an

---

[4]Don't fall into the trap of thinking of carrying capacity as the maximum population size observed.
For many reasons (including stochastic fluctuations) a population could, at times, exceed or be less
than *K*.

amount depending on the size of the population and the form that density dependence takes. The form of the relationship between density and realized per-capita growth rate $[\ln(N_t/N_{t-1})]$ could be linear or it could be a curve. The absolute simplest model is for negative density dependence to impose a linear decrease in realized per-capita growth rate as abundance increases. At very low abundance, realized per-capita growth rate is equal to r, the intrinsic rate under exponential growth. But as density increases, the realized per-capita growth rate declines in a steady, linear fashion. This is a **logistic growth model** (Fig. 6.3).

The logistic growth model, plotting abundance over time (Fig. 6.3b), has the characteristic S shape that we know and love from ecology classes. Notice too that the population size asymptotes at K, the carrying capacity. In other words, at K the population is stationary, neither increasing nor decreasing.

We can put the logistic curve of Fig. 6.3(b) into a continuous-time formula by simply modifying the exponential growth equation (eqn. 6.1) to dampen the expression of r as abundance increases:

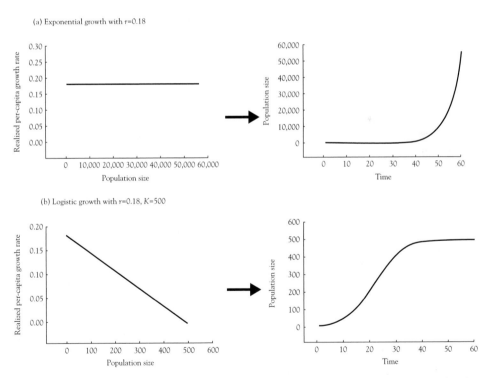

Fig. 6.3 The contrast between exponential and logistic growth. (a) Exponential growth. The realized per-capita population growth rate measured from a time series ($\ln[N_{t+1}/N_t]$) is equal to the intrinsic exponential growth rate (r), no matter what the population size is. This lack of density dependence leads to exponential growth of the population over time. (b) Logistic growth. With a linear decline in realized per-capita population growth rate as the population size increases, the population increases exponentially at first, then slows its growth as it approaches carrying capacity.

$$\frac{dN}{dt} = rN\left(1 - \frac{N}{K}\right) \tag{6.2}$$

To dissect this equation, notice that the rN part is just exponential growth (eqn. 6.1). The rest of the equation $(1-N/K)$ decrements population growth as density increases[5]: when abundance is very small relative to carrying capacity (small N compared to K), the part in parentheses is basically one and population change (dN/dt) is nearly exponential. When N equals K, the part in parentheses is zero so dN/dt is zero (no change in population). When N is greater than K the population declines.

From eqn. 6.2 we can derive another way of writing the per-capita growth rate:

$$\frac{dN}{dt} * \frac{1}{N} = r\left(1 - \frac{N}{K}\right) \tag{6.3}$$

The translation in words is again that per-capita population growth declines linearly as N goes from zero to K.

A topic of major interest in applied ecology, especially for thinking about harvest regulations, is the **recruitment** of new individuals or **yield** of the population. In practical terms, this is the net number of new individuals in a population over a unit of time; in terms of modeling it is characterized by dN/dt. For logistic growth, recruitment is maximized when the population is one-half the size of the carrying capacity (Fig. 6.4): at low abundance growth is essentially exponential, unhampered by negative density dependence, but the small numbers of breeders means that few individuals are born (dN/dt is small). Near K, dN/dt is also small, even though the population size is large, because negative density dependence is really dampening the per-capita growth rate of all the individuals. Another way to see that dN/dt is maximized at intermediate population sizes is to notice that in the S shape of the logistic growth curve (Fig. 6.4b) the tipping, or inflection, point − the steepest point of population increase − occurs at $0.5K$. In harvested populations, the high point of dN/dt is often described as the population size at which the **maximum sustained yield** could be taken; we'll put some serious cautions around this interpretation in Chapter 14.

The logistic equation can also be expressed in discrete time (May 1974) as:

$$N_{t+1} = N_t e^{\left(r\left[1-\left(\frac{N_t}{K}\right)\right]\right)} \tag{6.4}$$

The discrete time form of the equation introduces a time step, such that dynamics of the population at one time step are a function of abundance at the previous time (for example, the previous year). This time lag is realistic for many populations because reproduction or survival in one year often depends on conditions the previous year.

---

[5]The parenthetical part of eqn. 6.2 is derived from $\frac{K-N}{K} = \left(\frac{K}{K} - \frac{N}{K}\right) = \left(1 - \frac{N}{K}\right)$.

(a)

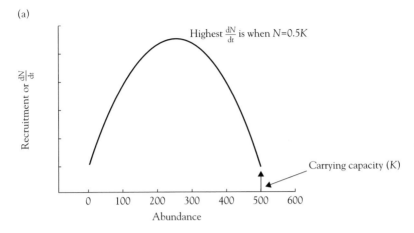

(b)

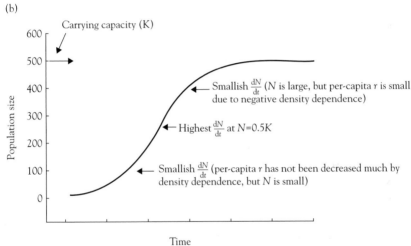

Fig. 6.4 (a) Population production or recruitment (dN/dt) for a population growing with logistic growth and a carrying capacity (K) of 500. Recruitment under logistic growth is maximized when the population is at 0.5K. (b) The basis for the recruitment in (a), demonstrating how recruitment changes as the population increases in numbers over time.

For example, winter starvation in feral Soay sheep and red grouse depends on population size the previous summer (Caughley & Sinclair 1994).

I'll say again that logistic growth is just one simple way of describing density dependence (actually, of negative density dependence only). Do not think of the logistic function as the only density-dependence function. The linear decline in per-capita growth rate with density as modeled by logistic growth (Fig. 6.3b) is not a biological law, or even necessarily general; in fact, a majority of species probably exhibit nonlinear density dependence, with their per-capita population growth rate declining relatively rapidly as population size increases from low density, then flattening out as carrying capacity is approached (Sibly et al. 2005); thus the relationship on the left-hand side of Fig. 6.3(b) would be concave. A variety of alternative equations exist to model concave (or even convex) relationships for density dependence, with the most widely used being the **theta-logistic**[6]. Also, simple **ceiling models** allow population growth to be exponential up to a ceiling that cannot be exceeded (applicable to cases where space is limited, as for birds that nest in tree hollows, or bison on the National Bison Range in western Montana where managers cull the herd each year to maintain a population below the expected carrying capacity). In Chapters 12 and 14 we will see how different forms of density dependence can affect extinction probability and harvest models. For now, the important point is that logistic growth shows us the consequences of density operating on populations, a useful contribution as long as we do not confuse the logistic with a law of population growth (Box 6.2).

Given the many biological reasons why small populations may exhibit positive density dependence, where per-capita growth rate increases with population density, it is useful to see how population growth would be affected if we add positive density dependence to the logistic equation[7]. If a population experienced this combination of

---

Box 6.2 The logistic equation captures one form of density dependence but is not a universal law

Although the use of an S-shaped logistic curve to capture negative density-dependent population growth goes back to at least 1838, Raymond Pearl became the hardcore marketing man for the logistic curve around 1920. Working with experimental populations of fruit flies, Pearl noticed that he could fit a very nice S-shaped logistic curve to describe the population growth trend of the flies over time. So far so good: he had a solid experimental demonstration of a form of density dependence. But the problem was that Pearl leapt from this observation to an intense campaign to argue that the logistic represented a law of population growth that all populations must follow. Unfortunately, Pearl's campaign haunts us still, as many people think only of logistic growth when they consider density dependence (see Kingsland 1985).

The strength of the logistic growth model is that it simplifies the messy dynamics of real populations; knowing how a population deviates from it is useful for refining initial assumptions so as to better understand how populations will change over time. However, while logistic growth is a mathematically simple and useful way to describe one form of negative density dependence, it really is just one way that density dependence may manifest itself in wild populations. Or, to state it more eloquently, we can turn to the words of Alfred Lotka, one of many who tried to put the brakes on Pearl's enthusiasm for logistic growth. Lotka argued that the logistic equation should merely serve as a catalyst for us to ask what mechanisms are affecting population dynamics: "An empirical formula is therefore not so much the solution of a problem as the challenge to such a solution. It is a point of interrogation, an animated question mark" (Lotka 1925 quoted by Kingsland 1985:85).

---

[6]The theta logistic model is a simple generalization of eqn. 6.4: $N_{t+1} = N_t e^{\left(r\left[1-\left(\frac{N_t}{K}\right)^\theta\right]\right)}$.

[7]For example, Courchamp et al. (1999) modified the logistic growth equation so that per-capita growth rate becomes negative below some threshold A: $\dfrac{dN}{dt} = rN\left(1-\dfrac{N}{K}\right)\left(\dfrac{N}{A}-1\right)$.

negative (logistic growth) and positive (Allee effect) density dependence, it would decline to extinction below the Allee-effect threshold A and increase if it were between A and K (Fig. 6.5). In fact, it may be very common in nature for populations to experience both positive and negative density dependence at different times and population sizes (Burgman et al. 1993). Whenever density dependence affects per-capita growth rate in any way other than that assumed by logistic growth, both the carrying capacity and the way the growth rate changes with density will vary. To emphasize this point, if you modeled a population using logistic growth, with a linear density-dependence function (Fig. 6.3b, left-hand panel), you would be assuming that the population would do well at very small population sizes. However, if the population actually experienced positive density dependence at low densities, it could decline to extinction below a threshold (as in Fig. 6.5, where populations of less than 25 decline toward extinction).

Density-dependent population growth will be influenced by stochasticity due to small numbers or environmental perturbations (see Chapter 5). As we saw for exponential growth, the expected population size under stochastic growth is lower than without stochasticity (Burgman et al. 1993:79–85). If the carrying capacity fluctuates, the average value of N will always be less than the average value of K,

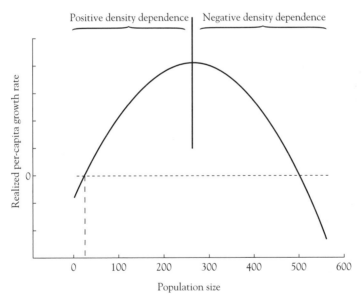

Fig. 6.5 The effect of positive and negative density dependence on the realized per-capita population growth rate $[\ln(N_t/N_{t-1})]$. Carrying capacity is 500 and r is 0.18, as in Fig. 6.3(b), but a positive density-dependence term (Allee effect) is added to the logistic model. When the population is below 25 it will decline to extinction, and positive density dependence dominates until the population size exceeds 250. When the population is above 500 it will decline until it reaches 500. Between 25 and 500 it will increase towards 500. Don't confuse this with Fig. 6.4(a); although they have similar-shaped curves they have different axes and cover different ideas.

because under logistic growth a population above K declines at a faster rate than a population increasing from a corresponding level below K (Gotelli 2001:38). For example, use eqn. 6.2 to calculate the change for a population with $r=0.1$, $K=500$, and $N=400$, compared to a population with the same r and K values but $N=600$.

Some counterintuitive dynamics: limit cycles and chaos

In 1974 Robert May reported a wild observation: populations growing according to the discrete logistic growth equation (eqn. 6.4), with absolutely no stochasticity, could show dynamics that bounce, or cycle, or become entirely unpredictable (Fig. 6.6). When r is less than 1.0, population growth follows a smooth, monotonic increase to carrying capacity, following the standard S-shaped curve discussed above. When r is between 1.0 and 2.0 logistic growth shows **damped oscillations**, so that the population overshoots K, then undershoots it by a lesser amount, and so on until it settles back down to the equilibrium. Between $2.0 < r < 2.69$, the population will show **stable limit cycles**: dynamics are still regular and predictable, with a repeated succession of valleys and peaks. So, things are getting interesting already: the logistic equation is leading to cycles created purely by the amount of population growth in discrete time (where population size next year depends on population size this year). If r increases more, so $r > 2.69$, you get **chaos** (Fig. 6.6e,f).

What is chaos? For practical purposes, chaos is distinguished by dynamics that are highly sensitive to initial conditions. Notice in Fig. 6.6(e,f) that the two populations are identical except for starting sizes of 10 animals in Fig. 6.6(e) and 11 animals in Fig. 6.6(f). The populations differ by only one individual initially, but by time step 40 they are vastly different (988 compared with 201 animals). Although the fluctuations generated under chaotic dynamics are unpredictable, they are different from random or stochastic fluctuations. One difference is that there is a structure to the unpredictable fluctuations under chaos (see Fig. 6.6e,f), with bursts of short-term cycles interspersed with stretches of irregular and erratic population sizes. Chaos is also distinguished from environmental stochasticity because if you start with exactly the same initial conditions, you will get exactly the same population trajectories every time (try it for yourself using eqn. 6.4 and an r value of, say, 2.8).

So, chaos is different from stochasticity. But the sensitivity to initial conditions reverberates into dynamics that over the long term are unpredictable and seemingly random. Thus if chaos is operating, you will not be able to predict future population dynamics, and a chaotic time series will appear as if it is fluctuating stochastically. Therefore, chaotic dynamics can drive apparently random fluctuations, even in an absolutely constant environment, and can exacerbate fluctuations when occurring in tandem with stochasticity (Hastings et al. 1993, Perry et al. 2000).

Nonlinear dynamics including stable limit cycles and chaos are profound in population biology first because they show that simple deterministic models can produce

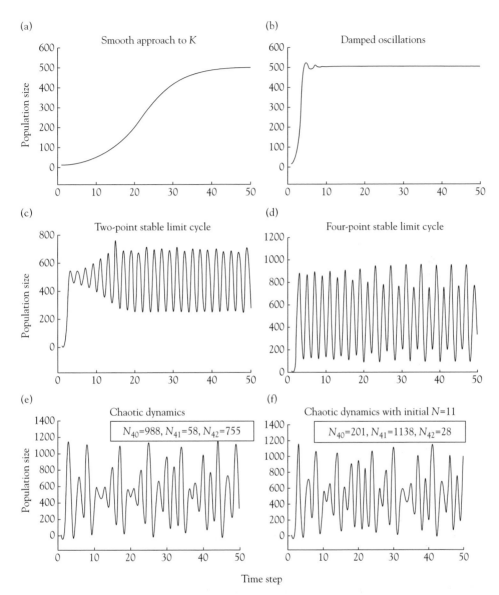

Fig. 6.6 Wild dynamics in population size over time that come from deterministic discrete logistic growth (eqn. 6.4) with a K value of 500 and an initial N value of 10. (a) $r=0.18$: a steady S-shaped curve flattens at K. (b) $r=1.5$: damped oscillations, overshooting then undershooting but settling down at K. (c) $r=2.2$: stable limit cycles of two points (two steps between each repeated high or low). (d) $r=2.6$: stable limit cycles of four points. (e) $r=2.9$: chaotic dynamics with starting $N=10$. (f) Chaotic dynamics with the same conditions as in (e) except initial $N=11$ to show the sensitivity to initial conditions. Although (e) and (f) may look similar at a quick glance, notice that abundances are actually vastly different.

what looks like environmentally induced cycles or random dynamics. Second, very subtle differences in starting conditions (say, in terms of initial population size) can have huge effects on the future population size. Third, chaotic dynamics abandon the concept of a traditional equilibrium density.

All of which brings us to the question of how often we need to know about this stuff in applied wildlife population ecology. First, for chaos to manifest through the single-species logistic growth equation, a time lag must be present. Time lags are quite common in most vertebrate populations. There can be time delays in the recovery of a resource (say, overbrowsed vegetation), in gestation time, or in behavior or physiology of individuals, producing a lagged response between change in resources and change in vital rates and population growth (e.g. Turchin 1999). An implicit time lag of one step is included in the discrete form of the logistic growth model, which is why May (1974) was able to use this simple equation to demonstrate the potential for chaos in biological populations.

Nevertheless, you may be disturbed by another necessary condition for chaos to occur via discrete logistic growth. Specifically, you may be astonished that I am taking time to worry about what happens when $r > 2.69$. After all, a wildlife population increasing with an r of 2.7 would have a $\lambda = e^r = 14.9$, meaning that it increases by 14-fold each time step!

In fact, if chaotic dynamics could only occur under the conditions specified by Robert May's groundbreaking work (discrete logistic growth with $r > 2.69$), such a complaint would be correct. But it turns out that many nonlinear processes affecting population growth could theoretically lead to chaotic dynamics, with or without discrete time (time lags) and with single or multiple species interacting (Hastings et al. 1993), at growth rates considerably lower than 2.69 in the discrete logistic model. Therefore, although almost no real wildlife population would have a growth rate high enough to drive chaotic dynamics using the simple May model, other ecological conditions could lead to nonlinear dynamics, and potentially chaos, in real populations.

The study of chaotic dynamics is blossoming in ecology, and current evidence indicates that chaotic dynamics could occur in real wildlife populations (Box 6.3). However, chaotic dynamics are certainly not lurking everywhere, and in fact traditional stochastic fluctuations probably explain much more of the noise in real ecological systems than do chaotic dynamics (Lande et al. 2003). Knowing how nonlinear dynamics and chaos might manifest in wildlife populations will help you understand the implications if future work demonstrates a wide occurrence of chaotic dynamics in field populations.

Summary

Density of conspecifics is a ubiquitous driver of vital rates for wildlife populations, with implications ranging from management of endangered species to harvest strategy assessment. Sometimes, density is really just a surrogate for proximate factors such as food availability, parasite burden, predation, or other variables that influence vital

Box 6.3 Chaotic dynamics in northern small mammals

Fluctuations in small mammal populations have been studied intensively for more than 100 years. In northern Europe, fluctuations in voles (*Clethrionomys* spp. and *Microtus* spp.) range from low-amplitude, noncyclic to high-amplitude, cyclic (periods of 4–5 years) and even chaotic dynamics, roughly following a south–north latitude gradient (Turchin & Hanski 1997; see figure). The dynamics of these species have been explained by an intensive linking of field data to quantitative population models. The models are based on logistic growth and include the effects of predator and prey on each other. Specialist predators (especially weasels) exhibit a delayed numerical response (see Chapter 8) to

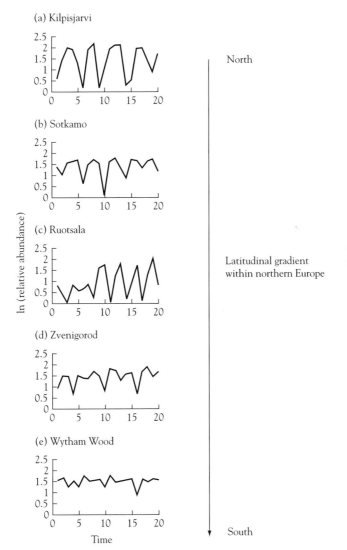

Population fluctuations of voles on a north–south gradient in northern Europe. Modified from Turchin and Hanski (1997). Reproduced by permission of the University of Chicago Press.

(*Continued*)

Box 6.3   Continued

changes in prey, creating a time-lagged density dependence in the voles. In the southern regions of northern Europe, generalist predators such as foxes and nomadic specialists such as short-eared owls dampen cycles because these predators are less dependent on local vole numbers. In contrast, a diminution of generalist predators in the northern region causes vole numbers to be much more closely coupled to the dynamics of the specialist predators. As a result, voles in the more northern regions show multi-annual, high-amplitude oscillations. Statistical analyses of the time series indicate that, in some cases, these oscillations are chaotic.

rates (Turchin 1999, Krebs 2003). Whether we consider density per se or the underlying outcomes of changing density, the effects on vital rates can change population growth rate for better or worse.

Negative density dependence increases vital rates (and population growth rate) as density decreases; thus declining small populations will become less likely to go extinct while growing larger populations will have reduced growth rates. On the other hand, positive density dependence – where smaller populations experience compromised predator avoidance, forage efficiency, or mating or caring for young – acts against small declining populations and in favor of large growing populations.

A common way to model negative density dependence is as a logistic growth function, whereby realized per-capita growth rate declines linearly with increasing density and becomes zero at the carrying capacity. As a distillation of complexity, logistic growth has been useful, for example making the strong graphic point that recruitment into a population is maximized when the population is held at half the carrying capacity, where a moderate number of individuals are breeding while incurring only moderate deleterious effects of density. The logistic model has also motivated study into how density dependence – potentially coupled with other factors that cause nonlinear relationships between density and population growth – can lead to stable limit cycles or even chaos, where dynamics become extraordinarily sensitive to initial conditions and fluctuations appear wildly random.

As useful as density dependence can be to understanding, modeling, and managing populations, we must be careful to embrace the uncertainty of how density dependence operates, acknowledging that it does not happen at all times and places and that the form and strength vary across time, space, and vital rates. Assuming only a simple logistic growth model for managing harvest or predicting extinction ignores the fact that other forms of negative density dependence, or positive density dependence, would lead to very different predictions. Ideally, we should use field data to quantify both the form and strength of density-dependence relationships for any particular population (Burgman et al. 1993, McCallum 2000, Turchin 2003). When data are insufficient to clearly demonstrate how density dependence acts, management implications can be explored using a number of different scenarios that might include exponential growth (no density dependence), and different forms of both negative and positive density dependence.

Further reading

Burgman, M.A., Ferson, S., and Akçakaya, H.R. (1993) *Risk Assessment in Conservation Biology.* Chapman and Hall, London. A practical and well-written book with useful applications of density-dependence concepts.

Gleick, J. (1987) *Chaos: Making a New Science.* Viking Penguin Books, New York. A compelling portrayal of chaotic dynamics in ecology and elsewhere.

7

Accounting for age- and sex-specific differences: population-projection models

For what is man? First, a child, soft-boned, unable to support itself on its rubbery legs, befouled with its excrement, that howls and laughs by turns, cries for the moon but hushes when it gets its mother's teat; a sleeper, eater, guzzler, howler, laugher, idiot, and a chewer of its toe; a little tender thing all blubbered with its spit, a reacher into fires, a beloved fool.

Thomas Wolfe (1942:432), You Can't Go Home Again

If you can't generalize from data there's nothing else you can do with it either. A science without generalization is no science at all. Imagine someone telling Einstein, 'You can't say "E=mc²." It's too general, too reductionist. We just want the facts of physics, not all this high-flown theory. Cuckoo.'

Robert Pirsig (1992), Lila

Introduction

Bull elk, cowbird eggs, frog larvae, mother wallabies, turtle hatchlings. Often wildlife ecologists care about particular parts of a population as much as they do the population as a whole. Whether the applied goal is to harvest, recover, reduce, or reintroduce wildlife populations, one cannot long avoid the dynamics of particular ages, stages, and sexes. The last two chapters have described a foundation for how to predict and describe changes in wildlife populations, but thus far dynamics have been described by a single term (λ or r) applied to the total population size (N). In this chapter I will explore how particular groups of individuals, and their birth and death rates, affect population growth and the likely numbers of individuals of different classes.

Here is a story to transition us, adding the wrinkles of age and sex structure (Coulson et al. 2001). Soay sheep studied on the island of Hirta, off the coast of Scotland, fluctuate dramatically, more than expected based on weather or density dependence alone. Why? In large part because survival of lambs and older males is heavily influenced by winter weather, whereas yearlings and prime-aged females are most affected by rainfall at the end of winter. Meanwhile, negative density dependence affects lambs and older females more than prime-aged adults or yearlings. In turn, the changing proportions of different sex and age classes caused by weather and density dependence cascades into effects on population growth (see also Box 9.3). Not all sheep are equal (Gaillard et al. 2001). Counting the sheep as equivalent, ignoring age and sex, would tell us very little about how ecological stresses affect population fluctuations or population growth.

For humans and a few other species, age can be tracked as a meaningful descriptor of an individual. However, vital rates in wild populations often depend on developmental, morphological, or even behavioral stages more than calendar age. Consider, for example, larval forms and adults in amphibians, fish of different sizes, or big trees and saplings. Furthermore, different stages can often be distinguished more easily in the field than age classes can, and management often centers more on recognizable stages than on ages (e.g. ungulate males of different sizes or antler-development stages). The predominance of stage structure means that throughout this chapter (and the book), I will refer to **stage** instead of repeating age or stage.

In basic ecology classes, age or stage structure is typically covered using life tables. Although life tables are important for basic ecological understanding, most of the applied things that life tables can tell us about wildlife population dynamics (e.g. estimates of λ, stage structure, and reproductive value) can be better estimated using matrix-projection models[1]. Thus I will skip life tables and focus on less well known yet more versatile and practical tools for understanding how structure affects wildlife population dynamics.

Specifically, the aim of this chapter is to describe the wonders of matrix-projection models for understanding wildlife populations. If the thought of matrix math makes you nervous, think of a population-projection matrix as merely a box to help keep straight the bookkeeping of birth and survival, a mathematical representation of biological processes. That's it, really. Lots of bells and whistles can be added to matrix projections, but at their heart they are less intimidating than they may look. So let's look at what a matrix is, then we will quickly come back to the surface to gulp the air of application to wildlife population biology.

Anatomy of a population-projection matrix

Throughout the chapter, I will use as a tangible example the common frog, a species found throughout much of Europe. The projection matrix, **M**, is a square of k columns and k rows, where k is the total number of stage classes (Fig. 7.1). Each element (or cell) of the matrix contains a value that is used to project stage-specific reproduction or survival forward one time step. A time step can be anything:

---

[1]For good overviews linking life tables to matrix models, see Noon and Sauer (1992) and Case (2000).

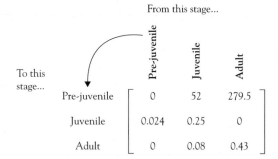

Fig. 7.1 Anatomy of a female-based projection matrix, using as an example the common frog (Biek et al. 2002; see also Box 4.8). This species has three stages: pre-juvenile (first year, consisting of the embryo, tadpole, and overwintering metamorph), juvenile (next 2 years), and adult. The projection interval, or time step, for this matrix is 1 year. The first row represents reproduction from each stage to the next year. The diagonal (e.g. $a_{2,2}=0.25$ and $a_{3,3}=0.43$; see text for an explanation of this notation) represents the proportion of individuals in a stage that will survive and still be in the same stage next year, while the subdiagonal (just below the diagonal; e.g. $a_{2,1}=0.024$ and $a_{3,2}=0.08$) represents the proportion surviving and advancing to the next stage next year.

for a yeast, the relevant time step of life and death might be an hour; for small mammals, it might be a month. For logistical and biological reasons, however, the most common time step for wildlife studies is a year, so throughout the book I will often use the terms year or annual as shorthand for the more general time step.

The elements of a matrix are described with subscripts that tell what row and column they are in (with the row first and the column second); for example, element $a_{2,1}$ is the element in row 2 and column 1. A handy way to decipher the biological meaning of any matrix element is to label the rows and columns of the matrix with the consecutive stages of your organism. Each element gives the transition – one time step later – from whatever column the number is in to whatever row the number is in. Another way to say the same thing is that $a_{i,j}$ represents the number of individuals contributed on average by each individual in class j at the current time step to class i at the next time step. For the common frog in Fig. 7.1, element $a_{2,1}$ is 0.024, meaning that on average 0.024 (or 2.4%) of the pre-juvenile frogs in the population survive to become juveniles the next year.

Notice in Fig. 7.1 that animals can remain in some stages for multiple time steps (for example 0.25 of the juveniles can remain as juveniles and 0.43 of the adults as adults). In a stage-based matrix, otherwise known as a **Lefkovitch matrix** (Lefkovitch 1965), transitions from any stage to any other stage can be accommodated. Stage-based matrices are more versatile than the original **Leslie matrix** (Leslie 1945), whereby vital rates depend on ages that are identifiable, and where the span of each age is the same as the length of the time step. In a Leslie matrix, an individual can only survive and transition to the next age, or die, so everything below the first row and not on the subdiagonal of the matrix must be zero. For practical purposes the distinction between Leslie and Lefkovitch matrices is only important to help you understand the terms in published papers.

With this brief lesson in projection-matrix anatomy, a few biological generalizations should become clear. First, each element of the top row of the matrix represents the reproductive contribution of each stage to the next time step. Second, the survival of individuals of any stage to the next time step (e.g. annual survival) can be determined from any matrix by adding up all the values for that column, excepting the first row[2]. For example, for the frogs in Fig. 7.1, annual survival of juveniles would equal 0.33, the sum of the proportion of juveniles that survive as juveniles (0.25) plus the proportion that survive and become adults (0.08). Third, the rates in the matrix must correspond to the stages you are interested in projecting. In particular, where the sexes have different survival rates, or where reproduction is known for females only, the vital rates are often female-based. In other cases, male-based models are most appropriate, as you will see for the red-cockaded woodpeckers in case study 1, below; two-sex matrix models are also possible, and you will see an example for ungulates in Chapter 14. The important thing is to be clear about which sexes are included in the projection, and how.

How timing of sampling affects the matrix

Because we are discussing applied population biology, let's think more about how to link the model to the field data, particularly to observable stages and to the timing of the surveys that produced counts of animals and estimates of vital rates (Fig. 7.2). Because each element of the top row contains the reproductive contribution of class j to the first stage class in the next time step, the top-row elements contain not only stage-specific fecundity (m_j), but also a term to advance the newborns to the next time step[3]. What does that mean? Well, newborns have to survive to be counted, or mothers counted last year have to survive to successfully bear their babies next year. So reproduction to the next time step depends on two terms: fecundity (m) and survival (P).

Exactly what we put into the elements of the top row depends on the kind of data collected. Suppose we were interested in projecting population growth for American bison, a species where most young are born at nearly the same time. For simplicity, assume all calves are born on May 31, and consider only the female portion of the population. If

---

[2]When building a matrix from field-collected vital rates, one of many decisions is how to partition annual survival into the appropriate elements in a stage-based model. Crouse et al. (1987) give a nice, simple approach to partitioning survival into matrix elements for a stage-based model.

[3]Remember, **fecundity** (or m_j) refers to the average number of offspring an individual in stage j produces in a year (see Chapter 4); if you are familiar with life tables, then m_j is the same as the m_x or b_x. I use **reproduction to next time step** for the top-row matrix elements that include both the fecundity and survival terms needed to project the fecundity to the next time step; this is often called **fertility** by human demographers, but fertility has different meanings in the ecological literature so I will avoid the term here.

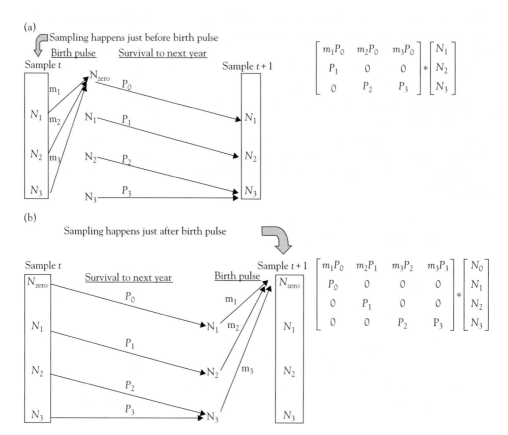

Fig. 7.2 General schematics of the birth and death processes captured when the sampling is either (a) before the birth pulse or (b) after the birth pulse. The animals sampled at times t and $t+1$ are boxed, with N_j representing number of individuals in each stage class j. This example assumes that animals stay in each stage for only one time step, except that those in the last stage can survive and remain in that stage for multiple time steps. Fecundity for each age class (m_j) represents the average number of offspring born to each individual of N_j. The probability of survival through one time step is represented by P_j. To the right of each schematic is the resulting projection matrix and population-size vector. In (a), note that newborns (N_0) are not seen until they have survived through their first year (P_0) to be counted as N_1 at the next sample interval; likewise, individuals in age class I (N_1) are just about to become 2 years old, and so on. The next batch of N_0 individuals are born just after sampling. In (b), note that there is an extra column and row in the post-birth-pulse matrix (compared to the case of the pre-birth pulse) because post-birth sampling occurs just after reproduction, making N_0 recognizable as its own class.

we sampled on May 30 (just before the birth pulse), then the youngest age class counted would be the calves born last May 31 that had lived to be counted just as they are about to become 1 year old. The reproductive contribution for each stage class, then, must include not only stage-specific fecundity (m_j; the average number of female calves born per year per female in stage j) but also the probability of newborns surviving to be counted at the end of their first year (call this P_0). Now, what if instead of sampling on

May 30 we sample on June 1, the day after the birth pulse? In this case we would sample newborns. We would know exactly how many female calves were born per female, but some of the mothers alive last year would have died during the year (remember again that the goal is to project the population forward through time). Thus the reproductive-contribution elements of the top row would include stage-specific fecundity (m_j) as well as survival of mothers in that stage to have the newborns (P_j).

People who spend a lot of time messing with population-projection matrices often denote as F_j each element of the top row, where each element is this composite of fecundity and survival of either the mothers or the newborns. Thus each element of the top row of the matrix represents the reproductive contribution to the next time step under either

- pre-birth-pulse sampling, $F_j = m_j P_0$, or
- post-birth-pulse sampling, $F_j = m_j P_j$.

So **post-birth** models have an extra stage class, because the newborns are recognizable as their own class (they were born just before sampling), whereas with **pre-birth**-pulse sampling we do not see newborns until they become class N_1 (as in Fig. 7.2a). I've been a little excruciating in detailing these two model types because it turns out to be a confusing topic in many ecology textbooks and published papers. If the accounting is kept straight, though, the two approaches give exactly the same population growth rate. And the strict pre- versus post-birth-pulse sampling can be relaxed to account for varying periods between the birth pulse and the sampling, or even to allow for continuous breeding (see Further reading). The development of the matrix for the common frog is shown in Fig. 7.3.

Projecting a matrix through time

How to project the matrix

Once the matrix model is filled with vital rates, it can be projected through time. The advantage of a matrix approach over the nonstructured models of the previous two chapters is that it keeps track of not just the total population size but also the numbers in each stage. In matrix terms, we will project through time the population-size vector. A vector is a skinny matrix of one column and k rows that contains the number of individuals in each of the k stages. To determine the population-size vector next year $[\mathbf{n}(t+1)]$, multiply the matrix $\mathbf{M}$ of vital rates by the vector of individuals at time t, $\mathbf{n}(t)$:

$$\mathbf{n}(t+1) = \mathbf{M} * \mathbf{n}(t) \tag{7.1}$$

By convention, matrices and vectors are shown in bold.

How do you multiply a matrix by a vector? Go across each row of the matrix, multiplying each element j by the same element of the vector (Fig. 7.4). Add up the products for one row to obtain the total number of individuals in that element of the vector. Again, if the math intimidates you, take a breath and realize that the projection

(a)

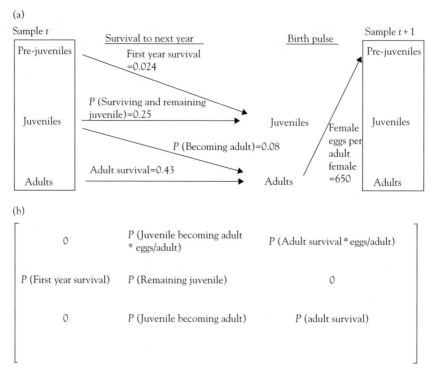

(b)

$$\begin{bmatrix} 0 & \text{P (Juvenile becoming adult} \atop * \text{ eggs/adult)} & \text{P (Adult survival} * \text{eggs/adult)} \\ \text{P (First year survival)} & \text{P (Remaining juvenile)} & 0 \\ 0 & \text{P (Juvenile becoming adult)} & \text{P (adult survival)} \end{bmatrix}$$

Fig. 7.3 A real-life example of a female-based post-birth-pulse matrix model for the common frog (Fig. 7.1). Female eggs per adult female refers to fecundity (see Box 4.8). (a) A diagramatic representation of the model; (b) the matrix (try plugging in the values and make sure you get the matrix in Fig. 7.1). Note that the matrix shows reproduction for juveniles (row 1, column 2) as well as adults (row 1, column 3) because a portion of the juveniles transition during the time step to become adults, at which point they reproduce. In general, for post-birth-pulse models for iteroparous species with n reproductive stages there should be $(n+1)$ non-zero elements in row 1.

of a matrix makes biological sense. The number of individuals in the first stage (newborns) next year comes from the reproductive contribution of each stage to the next time step (the top row of the matrix) multiplied by the number of individuals in each stage (the population vector). Likewise, the number of individuals advancing to a different stage or staying in the stage at the next time step are the product of survival (or other possible transitions below the first row) and the number of individuals in that stage.

Stable stage distribution and reproductive value

Matrix-projection methods can start with any number of individuals in different stages, and keep track of the relative number in each stage as well as population growth over time. This feature makes an important tool for many applications, ranging from tracking the possible growth of a translocated population to predicting what

$$
\begin{array}{ccccc}
\text{The matrix} & * & \begin{array}{c}\text{Population vector}\\ \text{in 2003}\end{array} & = & \begin{array}{c}\text{Population vector}\\ \text{in 2004}\end{array}
\end{array}
$$

$$
\begin{bmatrix} 0 & 52 & 279.5 \\ 0.024 & 0.25 & 0 \\ 0 & 0.08 & 0.43 \end{bmatrix} * \begin{bmatrix} 70 \\ 20 \\ 10 \end{bmatrix} = \begin{bmatrix} (0*70)+(52*20)+(279.5*10) \\ (0.024*70)+(0.25*20)+(0*10) \\ (0*70)+(0.08*20)+(0.43*10) \end{bmatrix} = \begin{bmatrix} 3835.00 \\ 6.68 \\ 5.90 \end{bmatrix}
$$

$$N_{2003}=100 \qquad\qquad\qquad\qquad\qquad\qquad N_{2004}=3848$$

Repeat multiplying the matrix by the current vector to get

$$
\begin{array}{cc}
\text{Population vector in 2005} & \text{Population vector in 2006}
\end{array}
$$

$$
\rightarrow \begin{bmatrix} 0+347.36+1649.05=1996.41 \\ 92.04+1.67+0=93.71 \\ 0+0.53+2.54=3.07 \end{bmatrix} \rightarrow \begin{bmatrix} 5731.38 \\ 71.34 \\ 8.82 \end{bmatrix}
$$

$$N_{2005}=2093 \qquad\qquad\qquad N_{2006}=5812$$

Fig. 7.4 An example of how to project a matrix through time. The sample matrix comes from the common frog (see Figs 7.1 and 7.3). A matrix of mean vital rates is projected for three time steps, beginning in the year 2003. Initially, our population has 70 pre-juveniles, 20 juveniles, and 10 adults. At the bottom of each vector is the total population size (N) for that year, rounded to the nearest whole female animal (as this is a female-based matrix).

might happen to certain stages during harvest. Although you can start with whatever number of individuals you want in each stage class, if vital rates stay relatively constant over time the population will converge on a population growth rate and stage distribution that is characteristic for that particular matrix. As a demonstration, Fig. 7.5 shows the projections for frogs from Fig. 7.4 for 14 years from the initial vector in 2003. Although growth rate fluctuated wildlife intially (e.g. $\lambda_{2003\text{-}04}=38.5$, $\lambda_{2004\text{-}05}=0.5$, $\lambda_{2005\text{-}06\text{-}}=2.8$; Fig. 7.4), by 2017 the population growth rate per year has become constant ($\lambda=1.46$; calculated by N_{t+1}/N_t). Also, the proportion of individuals in each class is constant, with about 98% of the population being pre-juveniles [e.g. $(306{,}931/313{,}490) * 100=98\%$], 1.9% juveniles, and 0.2% adults (for practice, calculate the age distribution for the year 2016 from the information in Fig. 7.5). This constant proportion of individuals in each stage class is known as the **stage age distribution** (SAD) or, more generally, the **stable stage distribution** (SSD). Nearly any population matrix – whether it represents a declining, increasing, or stationary population – will converge on a constant population growth and SSD if the vital rates making up the matrix stay relatively constant[4]. (The time to SSD depends on factors such as the initial age structure and the characteristics of the matrix itself, but should be achieved within 20 time steps or so for most vertebrate populations.) The population growth rate at SSD, and the SSD itself, are characteristic of the matrix, and are independent of the initial age distribution.

Although the SSD and λ_{SSD} are independent of initial stage distribution, the distribution of animals across stages influences both the time to reach SSD and the population abundance in the future. Consider population growth curves for our frogs again,

[4]There are matrices that will not converge to an SSD, including matrices that have only a single non-zero element in the first row, which can lead to stable oscillations (Leslie 1945, Caswell 2001).

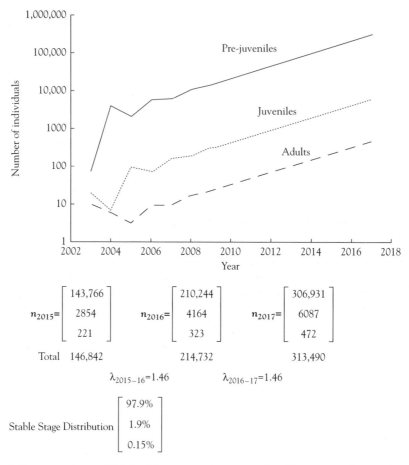

$$n_{2015}=\begin{bmatrix} 143{,}766 \\ 2854 \\ 221 \end{bmatrix} \qquad n_{2016}=\begin{bmatrix} 210{,}244 \\ 4164 \\ 323 \end{bmatrix} \qquad n_{2017}=\begin{bmatrix} 306{,}931 \\ 6087 \\ 472 \end{bmatrix}$$

Total   146,842   214,732   313,490

$$\lambda_{2015-16}=1.46 \qquad\qquad \lambda_{2016-17}=1.46$$

$$\text{Stable Stage Distribution} \begin{bmatrix} 97.9\% \\ 1.9\% \\ 0.15\% \end{bmatrix}$$

Fig. 7.5 Convergence to a SSD for the common frogs considered in previous figures. Population numbers over 14 years (from 2003 to 2017) are shown by stage class. The number of frogs is plotted on a logarithmic scale to accommodate the huge numbers of pre-juveniles, and because at SSD the trajectories become linear. Below the graph are the vectors (**n**), total population sizes, and geometric growth rates (λ) for the final 3 years. When the population reaches SSD, both the population growth rate (λ) and the proportion of individuals in each stage remain constant.

this time plotting total population size for 14 years with populations of 100 frogs that are seeded with all of one stage: 100 pre-juveniles, 100 juveniles, or 100 adults (Fig. 7.6). Even though all three populations start with the same initial number of frogs and the same constant set of vital rates, and after 14 years have all achieved the same growth rate at SSD, the numbers of frogs at time step 14 are vastly different for the three populations. Why? Because all stage classes are not created equal in their contribution to population growth, a phenomenon quantified by the concept of **reproductive value**. In Fig. 7.6 you can see that a population founded with 100 frogs, all adults, would reach a size of 2,125,660 frogs in 14 years, in contrast to the 477,171 individuals in the juvenile-seeded population and 7,843 in the population begun with 100 pre-juveniles. By convention, reproductive value is scaled relative to that of the first age

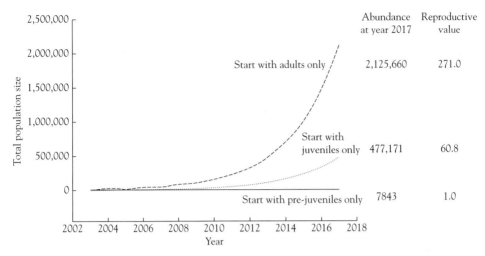

Fig. 7.6 A demonstration of reproductive value by projecting common frog population size beginning with 100 adults, 100 juveniles, or 100 pre-juveniles, with the constant vital-rate matrix from Fig. 7.1. Although the initial abundance, the projection matrix, and eventual population growth rate and SSD are identical in each case, the initial stage distribution causes bounce in population growth early on, and leads to drastic differences in abundance. Reproductive value is typically scaled relative to the first age class. The right side of the graph shows how reproductive value can be calculated based on relative abundances at SSD, dividing each abundance by that of the population begun with the first age class. (I used abundances in year 2017, after 14 years had passed, but you could use abundances any time after SSD was achieved.)

class. Therefore, for the frogs, adults have a reproductive value of 271.0, and juveniles of 60.8, compared to the reproductive value of 1.0 for pre-juveniles (Fig. 7.6).

Because the reproductive value quantifies how much each stage acts as a seed for future population growth (Caswell 1989:67), it has immense yet under-appreciated applications in wildlife population biology. Reproductive value is not a synonym for fecundity, or reproduction in the top row of the matrix. Rather, it takes into account reproductive output at that stage, as well as future reproduction, the likelihood to survive to those stages, and the population growth rate. In other words, reproductive value is a weighted average of present and future reproduction, accounting for population growth rate, that provides us with a practical way to assess contribution of different stages to future population growth (see Lanciani 1998, Case 2000).

So, here's what we've got so far on projecting matrices: Because all stages are not equal in their effects on population growth – that is, they have different reproductive values – the initial age distribution affects future abundance and the time required to reach SSD[5]. It also causes the population size to bounce around early on (as the age

---

[5]Look back at Fig. 1.3, showing how population momentum would cause the global human population size to increase even if women had only replacement numbers of children. The momentum is caused by an age structure leading to lots of babies even though modified vital rates would lead one to expect stationary population growth.

Box 7.1 How to calculate reproductive value, SSD, and the expected population growth at SSD

Because any constant population-projection matrix attains a constant SSD and λ, with each stage having a characteristic reproductive value, these are called **asymptotic matrix properties**. In the text I showed an approach to calculating each of the asymptotic matrix properties. For SSD and λ, project any initial population vector out by a number of times, say 100 time steps, and then calculate at time step 100 the proportion of individuals in each stage and the growth rate ($\lambda = N_{100}/N_{99}$). For reproductive value, you could use the seeding method (as in the frog example in Fig. 7.6).

Although these projection-based approaches are perfectly legitimate, intuitively transparent, and pretty easy to accomplish with simple multiplication that could be done, for example, in a Microsoft Excel spreadsheet, there are more elegant approaches to calculating the asymptotic matrix properties. For example, the **dominant eigenvalue** of the matrix, calculated using matrix math, equals λ, and its associated **right eigenvector** equals the SSD vector. Likewise, the **left eigenvector** of the dominant eigenvalue gives the vector of reproductive values.

distribution settles down to a SSD), and this is in the absence of any demographic or environmental stochasticity. However, the SSD and corresponding growth rate are a function of the matrix values and not the initial age distribution, so a population of any composition will eventually reach SSD and its associated λ as long as the matrix rates are relatively constant. Various ways to estimate reproductive value, SSD, and its associated λ value are described in Box 7.1.

For wildlife population management the implications of these population dynamics properties are profound. First, a set of vital rates represented as a projection matrix, coupled with a count of animals by stage class, provides insights into the inherent growth rate to be expected and the proportion of individuals eventually expected in each stage class over time. Second, the effect of age distribution means that a newly reintroduced population can be wildly erratic in its population growth – even without any stochasticity occurring – if the initial composition of the population is far from the expected SSD. Third, the reproductive value itself conveys the consequences of losing individuals of certain stages through harvest, or gaining them through translocations.

Before leaving this discussion on projecting matrix models through time, I should emphasize that I have only talked about density-independent matrix models. Density dependence can be added to matrix models (see the Further reading section of this chapter). I have also limited the discussion so far to the case where vital rates in the matrix are constant through time. Next I will briefly show how random variation (demographic and environmental stochasticity) can be incorporated into matrix projections.

Adding stochasticity to a matrix model

Although asymptotic properties such as the reproductive value, SSD, and population growth at SSD are useful, they are based on vital rates in the matrix being constant, or nearly so. But we know that vital rates are seldom constant for any length of time. And as we saw in Chapter 5, stochasticity has important implications, including the fact that it will decrease the likely future growth of a population compared with that expected from λ at the SSD[6].

Fortunately, computers make it quite easy to project a stage-structured model incorporating both environmental and demographic stochasticity (Chapter 5), assuming you have a specified starting population vector, and estimated means and variances for vital rates. To incorporate environmental stochasticity over time for a population, the computer builds a new matrix each time step, where each element in the matrix is chosen from a set of random numbers with a specified mean and process variance. The distribution of random numbers may be from a uniform distribution (all values equally likely to be chosen between a high and low value) or – more usually – from a distribution with central tendency, such as lognormal, normal, or beta (see Morris & Doak 2002). An alternate approach randomly picks one of several vital rates measured in the field (or even entire matrices of vital rates from field data; Bierzychudek 1982, Akçakaya 2000). Box 7.2 gives an example of environmental stochasticity in action for a population projection for the red-legged frog.

Demographic stochasticity in survival can be modeled to capture the real-world phenomenon whereby animals live or die as whole animals and not as fractions (Chapter 5). A common way to model demographic stochasticity in survival using the computer is to determine the fate of each individual in a stage based on the mean survival rate. Specifically, for each individual the computer picks a random number between 0 and 1; if the random number is less than the mean survival probability (also between 0 and 1) then the animal lives, if not, it dies. For example, suppose that your population vector has six yearlings, and the mean survival probability for yearlings is 0.8. Without demographic stochasticity, the expected number of subadults next time step is $(0.8*6=4.8)$. With demographic stochasticity, the computer might pick the following six random numbers: 0.32, 0.89, 0.51, 0.11, 0.94, 0.70. Thus, four of the animals would live, but two would die (the second and fifth). At small numbers the proportion of survivors can deviate greatly from that expected from the mean survival rate, just as a small number of coin tosses can lead to a big deviation from 50:50 heads/tails (see Chapter 5).

Sensitivity analysis

As we have seen, all stage classes are not created equal in their management importance, their effects on population growth, or in their relative abundance. Similarly, it

---

[6]Specifically, stochasticity will decrease population growth by an amount depending on which rate varies, how much it varies, and the sensitivity of λ to changes in that rate (Morris & Doak 2002:239). We will look at how to quantify sensitivity in the next section.

Box 7.2 An example of how to model environmental stochasticity, based on a population-projection matrix for red-legged frogs (*Rana aurora*)

Notice that this is a different frog species than the one discussed previously in this chapter. Data come from Biek et al. (2002).

Step 1: here is the matrix of vital rates.

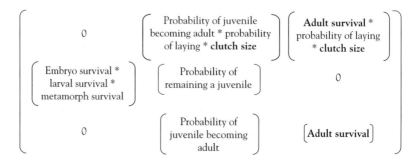

Step 2: environmental stochasticity for the two emboldened vital rates (clutch size and adult survival) is as follows.

	Clutch size	Adult survival
Mean	303	0.69
SD	95	0.13
Distribution for random numbers	Lognormal	Beta

Step 3: for five time steps, the vital rates chosen randomly from the specified distributions might, for example, be as follows.

Time step	Clutch size	Adult survival
1	287.6	0.66
2	326.8	0.71
3	252.0	0.93
4	382.9	0.55
5	251.9	0.60

Step 4: the distribution of vital rates chosen many times would look like the graphs below.

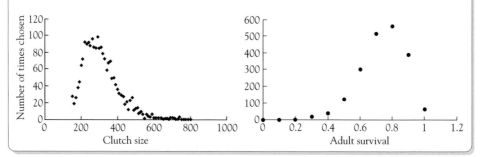

should not be surprising to learn that the vital rates themselves also vary widely in their effects on population growth and structure: all vital rates are not created equal. Intuition alone is insufficient to predict how changes in individual life-history components will affect population growth. Although one commonly hears conclusions like "forest fragmentation affects adult survival" or "acid rain affects clutch size," such statements do not indicate how the expected changes in vital rates affect population growth.

This simple demographic fact – that different vital rates do not have equal impacts on population growth rate – has been known for a long time (e.g. Cole 1954). But it was Hal Caswell's book on *Matrix Population Models* (1989), arriving on the heels of unprecedented access to desktop computers, that irrevocably convinced ecologists of the importance of a formal framework evaluating the effects of changes in vital rates. Sensitivity analysis provides that framework in the form of analytical and simulation-based tools to evaluate how past or future changes in life-history attributes or demographic vital rates affect population growth or persistence.

One of the earliest and most influential uses of sensitivity analysis in animal population biology targeted loggerhead sea turtles, which had been declining in the Atlantic by 3–5% per year for a long time. On the east coast of the USA enormous public sentiment built up concerning mortality of the eggs on the beach and the tiny (and adorable) hatchlings that were killed by predators, crushed by vehicles, and disoriented by lights as they tried to make their way from the nest to the ocean. Therefore, management focused on what seemed to be the obvious solution: increasing the survival of eggs and hatchlings. But in 1987 Deborah Crouse and colleagues published a sensitivity analysis that showed that even large increases in egg or hatchling survival would do little to reverse the population decline: the key was to increase survival of young adults in the ocean. It turned out that roughly 50% of loggerhead mortality in the Atlantic was due to young adult turtles becoming entangled in shrimp nets. The paper by Crouse et al. (1987) was key to the development of legislation requiring turtle-excluder devices to be installed by shrimpers (Crowder et al. 1994). In this case, sensitivity analysis of a matrix-projection model showed that intuition focusing on eggs and hatchlings alone to recover the species was wrong, and that a different management action would be much more beneficial and efficient for recovery.

Three main approaches are used by applied wildlife population ecologists to conduct sensitivity analyses (see Mills & Lindberg 2002 for more details). I will give a brief overview of these approaches, followed by a peek into the range of questions that can be answered with sensitivity analysis of matrix population models.

How to do a sensitivity analysis

Manual perturbation

The most basic approach to sensitivity analysis is to manually perturb, or change, the input to a population model and observe how the change affects the output[7].

---

[7]Output could be population growth rate, or it could be probabilities of quasi-extinction (see Chapter 14).

For example, one could ask how the deterministic population growth rate (λ) of a matrix model changes when adult survival is increased by 10% compared with an increase in fecundity. Management options can be explored by comparing the expected effects on population growth or persistence when each option changes certain vital rates by any pre-determined amount. The approach is infinitely flexible; sensitivity analysis via manual perturbations is not limited to investigating the importance of vital rates alone, but rather can explore a range of factors including density dependence, inbreeding depression, and movement among populations (Chapter 12). Also, manual-perturbation sensitivity analysis can incorporate different age or stage structures, quantifying the effect of age structure on population growth.

Analytical sensitivity and elasticity analysis

You have learned that different age classes may have very different relative abundances at SSD, and different effects on future population size (i.e. different reproductive values). Therefore, a vital rate for a certain age class will influence λ more if there are proportionately more of that age class (larger SSD) and if each individual of that age class has a larger impact (bigger reproductive value). Analytical **sensitivities** and **elasticities** elegantly combine the reproductive value of an age class with its expected SSD to evaluate how infinitesimal changes in individual vital rates will affect λ. Specifically, sensitivity for a vital rate that makes up matrix element $a_{i,j}$ (remember this is the matrix element in row i and column j) is a function of the reproductive value of the age class (v_i) and the SSD (w_j)[8]:

$$\text{Sensitivity of matrix element } a_{i,j} = \frac{\partial \lambda}{\partial a_{ij}} = \frac{v_i w_j}{\left(\overset{\text{Last stage class}}{\underset{k=1}{\sum}} v_k w_k \right)} \tag{7.2}$$

A larger reproductive value (v_i) or SSD (w_j) leads to a larger sensitivity. Notice that sensitivity is a partial derivative, defined as the infinitesimal absolute change in population growth rate given an infinitesimal absolute change in a vital rate or matrix element, while all other vital rates are held constant. As an alternative to the calculus in eqn. 7.2, you can also estimate sensitivity by making a tiny manual change to the vital rate of interest, and quantify how λ at SSD changes before and after the perturbation. For example, increase the element $a_{i,j}$ by 0.01, leaving everything else unchanged, and estimate λ before and after, as follows.

---

[8]If you prefer to think about equations graphically, consider that as a partial derivative, the sensitivity of matrix element $a_{i,j}$ equals the slope of the tangent to the curve relating population growth rate to the matrix element, evaluated at the mean element.

$$\text{Sensitivity of matrix element } a_{i,j} = \frac{\lambda_{a_{i,j}+0.02} - \lambda_{\text{original}}}{0.01} \qquad (7.3)$$

So, for the red-legged frogs in Box 7.2, analytical sensitivity would quantify how a tiny change in juvenile survival (say, from 0.69 to 0.70) would affect population growth compared with the same tiny change in another vital rate, such as clutch size (from 303 to 303.01). Although this may be useful for some applications, you can see immediately that from a practical perspective we have a scaling problem; the same absolute change of 0.01 in the mean of these two rates is very different for survival (a 1.4% change) compared with clutch size (a 0.003% change). That's where elasticity becomes useful. Elasticity is sensitivity's cousin, a metric that rescales the sensitivity to account for the magnitude of the vital rate. Thus elasticities are **proportional sensitivities** that describe the proportional change in λ given an infinitesimal one-at-a-time proportional change in a vital rate[9]:

$$\text{Elasticity of matrix element } a_{i,j} = (\text{sensitivity of } a_{i,j}) * \frac{a_{i,j}}{\lambda} \qquad (7.4)$$

When matrix elements are composed of more than one vital rate (e.g. where each element of the top row of a projection matrix contains both reproduction and survival components), or when a particular vital rate shows up in more than one matrix element, component sensitivities and elasticities can be calculated for each vital rate that appears in one or more matrix elements. Although the analytical formula for component sensitivities requires chain-rule differentiation for each $a_{i,j}$ that contains a particular vital rate x, in many cases the procedure is pretty simple[10].

  As a proportional measure of sensitivity, analytical elasticities are more widely used in applied population biology than sensitivities. Elasticities can be added together to predict the joint effect of changes in multiple rates (assuming the changes in vital rates and λ are linearly related). Elasticities of all matrix elements sum to one; elasticities of component vital rates do not add up to one but can still be ranked. Based on analysis of measured vital rates from hundreds of studies of different bird and mammal species, predictable patterns link life-history traits to the relative elasticities of different vital rates (Box 7.3).

---

[9]Analogous to sensitivity, elasticity is the slope of the tangent to the curve relating proportional changes in λ to proportional change in matrix elements.

[10]Assuming that each matrix element is a simple linear combination of different vital rates, here is a simple translation of the chain rule to calculate sensitivity for a vital rate x that is a component of more than one matrix element, where each element containing x has sensitivity $s_{i,j}$.

$$\sum_{1}^{\text{Elements containing } x} [(s_{i,j}) * (\text{product of components other t}$$

You can also use eqn. 7.3 to tweak a component vital rate and calculate sensitivity. Either way, the

$$\text{elasticity of vital rate } x = (\text{component sensitivity of vital rate } x) * \left(\frac{x}{\lambda}\right).$$

Box 7.3 How might we predict which vital rates will have highest elasticities for a wildlife species?

Although the best way to assess elasticity of a vital rate is to conduct analysis on a complete set of field-derived vital rates for a particular population, it is useful to know that some coarse generalizations can support general principles. For example, species with early maturation and large litters tend to have elasticities that are higher for reproduction (litter size and offspring survival) and lower for adult survival; conversely, in species with late maturation, fewer off-spring, and higher survival rates, population growth is affected more by adult survival than by reproduction (Heppell et al. 2000, Sæther & Bakke 2000, Oli & Dobson 2003). Survival of all stages will tend to have higher elasticities than reproductive output for most taxa with lifespans longer than a year (Crone 2001). The implication is that "in any sharp change of population growth rate for a long-lived species, one should first suspect a change in adult survival" (Lebreton & Clobert 1991:108).

Although these life-history general principles give us first-cut insights into which rates will have the highest elasticities, that does not mean that those rates are most important. Remember, vital rates with low elasticities but that change a large amount could actually affect the growth rate more than rates with high elasticities but that change little.

Analytical sensitivities and elasticities are easily applied, comparable across studies, and can be calculated from a single population matrix constructed from average or even best-guess vital rates. However, we should keep in mind their fundamental assumptions (Mills et al. 1999, 2001). First, they are asymptotic, relying on the population being at SSD (although this assumption can be relaxed; see Fox & Gurevitch 2000, Grant & Benton 2000, Caswell 2001). Second – and perhaps most importantly – analytical sensitivities and elasticities by themselves say nothing about how much vital rates change in nature or under management.

A classic case in point (Gaillard et al. 1998, Raithel et al. 2006) is that for ungulates in general, adult survival would be expected to have the highest elasticity by far. However, juvenile survival will be much more variable than adult survival, because juveniles are less buffered against density-dependent influences or environmental factors such as predation, bad weather, and so on. The fact that juvenile survival may easily vary from 0.1 to 0.7, whereas adult survival will tend to be much less variable, means that the rate with relatively low elasticity that changes a lot (e.g. juvenile survival) may affect population growth more than a rate with high elasticity that doesn't vary much (e.g. adult survival). Elasticities based on a mean matrix cannot capture how much a vital rate, and therefore population growth, can change in nature or under management.

An extension of analytical sensitivity and elasticity analysis, called **life-table-response experiments** (or LTREs for short), does explicitly account for variation with an analytical equation (Caswell 2001). For practical purposes, changes simulated on the computer are a more versatile way to the same end, so I will discuss such an approach next.

Life-stage simulation analysis

Wisdom and Mills (1997) developed a simulation-based approach to sensitivity analysis that might be considered a hybrid of the manual perturbation and analytical sensitivity/elasticity-based methods. The approach is called **life-stage simulation analysis** (LSA; Wisdom et al. 2000)[11] because it uses simulations to evaluate the impact of changes in different vital rates on elasticity rankings and λ. For the purposes of conservation decision-making, the user obtains (from the field if possible) both means and variances for vital rates. The variance should be based on process variance, uninflated by sample variance (Mills & Lindberg 2002). For projecting what might happen in the future, you can couple information from the past with specified changes in means and variances that are considered biologically, politically, and logistically possible under management in the future[12]. Correlations among vital rates are specified, if possible, as are the distribution functions for each vital rate (i.e. uniform, lognormal, beta, etc.). A computer program constructs many matrices with each rate in each matrix drawn from the specified distributions. Population growth rate is calculated for the matrix (usually asymptotic λ at SSD, although stochastic λ could also be calculated).

Output metrics in LSA include elasticity-based measures (e.g. the proportion of replicates where the vital rates shift rankings of elasticities, or the differences in elasticity values whenever the rankings of elasticities change across the replicates; Wisdom et al. 2000), as well as other metrics that avoid elasticity entirely. For example, one LSA output could be the percentage of replicates having positive population growth under different scenarios (Fig. 7.7).

Another way that LSA has commonly been used to evaluate the importance of vital rates for management is to regress λ on each vital rate as all rates change simultaneously (including the vital rate of interest) in 1000 or so simulated matrices (e.g. Wisdom & Mills 1997, Crooks et al. 1998, Cross & Beissinger 2001). The coefficient of determination (R^2) represents proportion of the variation in population growth rate that is explained by environmental variation in that vital rate, with all other vital rates varying simultaneously. When all main effects and interactions are included, the R^2 values sum to one. When λ is a linear function of the vital rates, the slope of the line equals the analytical sensitivity and R^2 is a function of both the slope (i.e. analytical sensitivity) and the proportionate variation in that vital rate, adjusted for covariance among vital rates. The same relationships hold for elasticity if the regression is done on log-transformed data (Brault & Caswell 1993, Horvitz et al. 1997). Therefore, the

---

[11]Morris & Doak (2002:344–8) refer to this approach as a "simulation-based sensitivity analysis."

[12]I avoid use of the terms prospective and retrospective sensitivity analysis (Caswell 2001) because these terms have been used to imply that the inclusion of variation in a sensitivity analysis prohibits one from asking what might occur under future management. When conducting a sensitivity analysis of potential management scenarios, it seems more constructive to simply make explicit whether or not variation is included, the origin of the estimates of both variation and mean rates, and the rationale for potential future changes in vital rates (Mills et al. 2001, Wisdom et al. 2000, Mills & Lindberg 2002).

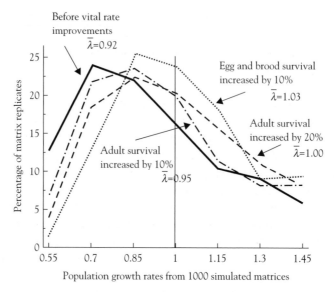

Fig. 7.7 An example of one form of LSA output for greater prairie-chickens, a species whose populations are declining, scattered, and potentially vulnerable to extirpation (see Wisdom & Mills 1997). The potential consequences to expected population growth (λ) for different potential management strategies are derived from simulations of mean (and variances) of vital rates using a LSA. The expected distribution of λ in 1000 simulated matrix projections before vital-rate improvements is shown by the solid line; dashed and dotted lines show how the distribution of expected growth rates change as particular vital rates are improved by increasing their means by specified amounts (and decreasing their variation by 20%). $\bar{\lambda}$ gives the mean of the λ for that distribution. Increasing by 10% the average survival during part of the first year (egg and brood survival) gives the biggest increase in the expected population growth (dotted line). Increasing by 10% the average adult survival gives much less improvement (dotted/dashed line). To get about the same increase in population growth as the 10% increase in first-year survival, average adult survival would need to be increased by 20% (dashed line). Modified from Wisdom et al. (2000). Reproduced with permission of the ESA.

simulation-based LSA R^2 can be compared with analytical life-table-response experiment approaches, in that both account for infinitesimal effects (e.g. elasticity) as well as the range in variation of different rates (Wisdom et al. 2000). However, LSA is more flexible than life-table-response experiments because any sort of change can be simulated, and a variety of output metrics are possible.

Case studies

To end the chapter, I will consider four case studies that used the application of matrix projections and sensitivity analysis to inform management, often in nonintuitive ways.

Case study 1: what are the best management actions to recover an endangered species?

Red-cockaded woodpeckers are an endangered species endemic to mature pine forests of the southeastern USA. They are cooperative breeders, with males staying on natal territories as nonbreeding helpers to the breeding pair for up to 11 years before inheriting natal territories or dispersing. How should managers decide which management strategies are most likely to increase the population growth of this endangered species? Detailed field studies provided critical insights into vital rates, behavior, and potential effects of management actions (Walters 1991), which could then be extended with population models exploring how population growth and persistence could be most efficiently increased through management.

Selina Heppell and colleagues (1994) used an innovative matrix-modeling approach based on male red-cockaded woodpeckers and using behavioral transitions associated with helping in addition to the usual size or age transitions (Fig. 7.8a). For example, fecundity in the top row was defined as the number of fledglings produced by individuals that survived the year and were helpers or breeders in that time step. Four management techniques with specific predicted effects on one or more vital rates were evaluated using the manual-perturbation approach coupled with elasticity analysis. Each of these actions targeted specific vital rates, thereby affecting one or more elements $a_{i,j}$ of the matrix in Fig. 7.8(a). The management options were as follows.

1 Remove invaders: remove cavity invaders such as flying squirrels and other woodpecker species that inhabit red-cockaded woodpecker nest cavities, increasing woodpecker fecundity (all elements of top row).

2 Female translocation: capture and relocate female red-cockaded woodpecker fledglings to solitary male territories, causing more solitary males to become breeders (increase $a_{6,4}$ and $a_{1,4}$).

3 Cavities in occupied territories: drill cavities in existing territories, increasing the fecundity of breeders (increasing all of the top row) and the probability that fledglings become helpers ($a_{2,1}$) while decreasing the chance that fledglings become breeders ($a_{5,1}$).

4 Cavities in unoccupied territories: increase new territories by drilling artificial cavities in unused yet suitable habitat and by reducing hardwood understory. This action should increase both fledgling-to-breeder and helper-to-breeder transitions (top row, and $a_{5,1}$ and $a_{6,2}$) while decreasing fledgling mortality (all of first column).

What would be the predicted relative effect of these management actions? Creating new territories through cavity construction in unoccupied territories (management option 4) had the biggest benefit (Fig. 7.8b) through its effects on two vital rates with relatively high elasticity (fledgling-to-breeder and helper-to-breeder transitions). Removing cavity invaders and placing more cavities in unoccupied territories were also potentially reasonable approaches, whereas female translocation (option 2) seems a waste of money. Importantly, Fig. 7.8 also underscores a recurring theme: the best strategy ultimately depends on how much rates could be changed. For example, a 25% change by removing invaders (option 1) would increase red-cockaded woodpeckers more than a smaller (10%) change via the best strategy of installing cavities in unoccupied territories.

(a)

Transition	From fledglings	From helpers	From floaters	From solitary males	From 1-year-old breeders	From ≥2-year-old breeders
Fledglings produced	0.080	0.266	0.324	0.275	0.486	0.522
To helpers	0.294	0.494	0.000	0.000	0.000	0.000
To floaters	0.031	0.000	0.000	0.000	0.000	0.000
To solitary males	0.043	0.020	0.172	0.216	0.000	0.000
To 1-year-old breeders	0.074	0.000	0.000	0.000	0.000	0.000
To ≥2-year-old breeders	0.000	0.257	0.483	0.410	0.725	0.800

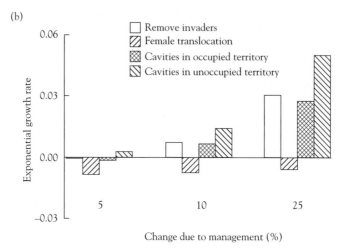

Fig. 7.8 Male-based population-projection modeling to inform management of the endangered red-cockaded woodpecker (from Heppell et al. 1994; reproduced by permission of The Wildlife Society). (a) The stage-based matrix model; (b) expected growth rates expected following implementation of the four management alternatives, each of which was expected to influence different vital rates.

A critical part of red-cockaded woodpecker life history not incorporated into this model is the spatial structure among multiple groups. Because helper males do not disperse far to fill a breeding vacancy, the new territories established with cavities need to be close to currently occupied territories. Although matrix models can be built to incorporate multiple populations with specified movement among them, individual-based models provide an alternative framework to matrix models for population projections. Using an individual-based model that was also spatially explicit, Walters et al. (2002) extended the findings of Heppell et al. (1994) to conclude that new artificial cavity sites need to be aggregated, or clumped near current territories, emphasizing not only density but also spatial distribution of the managed cavities; even quite small local populations could be supported if the territory groups were aggregated. Importantly, these recommendations about cavity establishment – born of a union of excellent field work coupled with thoughtful population modeling – have been incorporated into the new federal recovery plan for red-cockaded woodpeckers (US Fish and Wildlife Service 2003).

Case study 2: what are the most efficient management actions to reduce a pest population?

Brown-headed cowbirds are a nest parasite native to the short-grass Great Plains of North America. Although scattered populations may have been present historically throughout much of North America (Morrison & Hahn 2002), their numbers and distribution increased with landscape fragmentation associated with European settlement and agriculture during the 19th and 20th centuries. Because cowbirds exist at high densities in many agricultural areas, and each female can lay up to 40 eggs per season in the nests of other species, cowbirds can reduce nest success of their passerine hosts. To reduce the effects of cowbirds on native and threatened species, land managers have implemented control efforts since at least the early 1970s. For example, trapping efforts in Texas have removed 3000–5000 female cowbirds per year, and in Michigan control programs to protect Kirtland's warblers have removed 3000 or more cowbirds and eggs each year (Kelly & DeCapita 1982). Given limited funds, is the most efficient way to decrease cowbird population growth to remove eggs, to remove adults on the breeding or wintering range, or some other action?

Citta and Mills (1999) used both analytical sensitivity analysis and LSA to examine the consequences of cowbird control efforts. The LSA scatterplots of the variation in λ explained by each vital rate in 1000 simulated cowbird matrices indicate that egg survival alone explains a preponderance (61%) of the variation in λ (Fig. 7.9a). By contrast, fecundity and survival of other stages all explain less than 15% of variation in λ (Fig. 7.9b–f). Why? It turns out that the proportional infinitesimal effect of each rate (i.e. analytical elasticity) is equal for egg, nestling, and yearling survival, but egg survival probably varies a lot more (see the range on the x axis in Fig. 7.9) and so has a greater opportunity to affect λ. Thus LSA captured the fact that both elasticity – the infinitesimal effect – and the variation in a rate determine the merit of altering certain vital rates.

So, the next question is how easy it would be to change egg survival, which is normally highly variable. Killing certain stages (adults or eggs) may be more or less palatable to the public, and more or less feasible under field conditions, both of which affect how much a rate can be changed. At SSD, 92% of the cowbirds would be eggs, so that in a population of 5000 cowbirds you would have to destroy more than 400 eggs to change survival enough to cause λ to be less than 1.0; to remove that many eggs would involve intense effort, making it not only expensive but also prone to disrupting the host species that are innocently caring for the cowbird's eggs (Citta & Mills 1999).

The conclusion of the modeling was that management-induced changes in neither fecundity nor egg survival alone would easily affect population growth. Likewise, adult survival on the breeding grounds would have to be reduced by a lot to cause λ to decline, an option limited by the compensation that occurs via replacement by floaters and immigrants; reducing survival on wintering grounds is logistically daunting. Thus, although local cowbird reductions can successfully protect sensitive host species on a local scale (Rothstein & Cook 2000), the sensitivity analysis clarified that easy fixes in the form of removals of one stage would not reduce long-term population growth of the parasitic pest. Instead, multiple vital rates would have to be hammered simultaneously, probably by managing land use across a landscape.

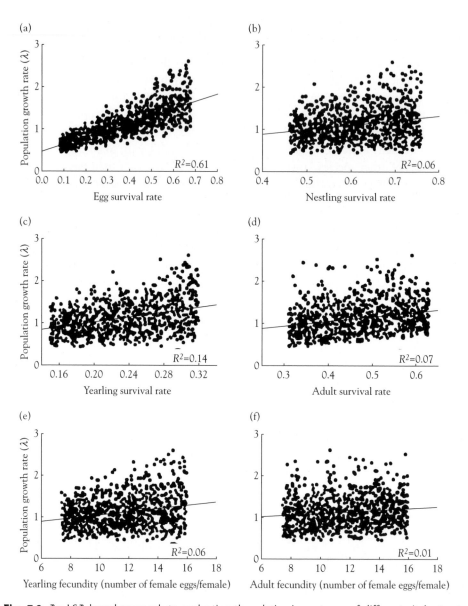

Fig. 7.9 An LSA-based approach to evaluating the relative importance of different vital rates to population growth in brown-headed cowbirds. The R^2 value describes the proportion of variation in λ explained by variation in a vital rate, based on 1000 simulated matrices where vital rates were chosen from the range of variation determined from published studies. Notice that egg survival alone accounts for 61% of the variation in λ. From Citta & Mills (1999)

Case study 3: how should a harvested species be managed?

Migratory waterfowl have been intensively studied and managed, to both protect populations and provide compatible hunting opportunities. In the USA about $50 million in migratory bird conservation funds (primarily funded by duck stamps bought by hunters) are dispersed each year to protect and enhance wetlands and grasslands for waterfowl habitat. Traditionally, less than 40% of these funds has been apportioned to breeding areas. An ongoing debate has centered on how much effort (and money) should be dedicated to protection and enhancement of habitats on breeding areas (especially wetlands and nesting habitat) compared with non-breeding areas (especially wetlands for migratory and wintering waterfowl).

Hoekman et al. (2002) used LSA to assess the effects of management and environmental variation on population growth of the North American mid-continent mallard population, considering both the infinitesimal effect of each vital rate, and the observed variation in each. The LSA indicated that vital rates on breeding grounds (hen survival during the breeding season, clutch size and nest success, and survival of ducklings) collectively explained about 84% of the variation in λ, compared to only 9% explained by nonbreeding survival on migration and wintering areas[13] (Fig. 7.10).

The finding that the contribution to duck population growth of nonbreeding survival is dwarfed by vital rates on breeding grounds has profoundly influenced waterfowl management. An expert panel assembled in 2004 by the US Fish and Wildlife Service came to a striking science-based conclusion based largely on the sensitivity analysis: given that variation in vital rates on breeding grounds explains the vast majority of the variation in λ for mid-continent mallards, and given the general absence of strong differences in the ability to change vital rates on breeding compared with nonbreeding areas through habitat management, the panel recommended that approximately 90% of the waterfowl conservation funds should go to breeding areas (Cox et al. 2004). This suggestion has been elevated to the top levels of the US Fish and Wildlife Service, and although politics will certainly play a role, it appears likely that proportionately more management funding will shift to the breeding grounds. Simultaneously, current adaptive harvest management for waterfowl (see Chapter 14) is recognizing the need to incorporate breeding-area processes to optimize harvest management. Thus, a matrix population model has distilled a nonintuitive insight (that breeding-ground dynamics drive population growth) that is changing the trajectory of waterfowl funding and management. These results have also been useful in reassessing research priorities, with increased funding directed toward sources of variation in nest and duckling survival.

Case study 4: what research is needed to understand global amphibian declines?

Beginning around 1990, researchers from around the world sounded an alarm that amphibian numbers seemed to be declining. The call to action came from monitoring studies, and for the last decade or so the question has been how the declines would best be reversed. Work on a variety of species has shown how various vital rates might be affected by ultraviolet radiation, pH, disease, habitat destruction, or other factors. But there has been a missing link between the data showing declines and the data

---

[13]The final 7% can be thought of as statistical noise, accounted for by interactions among rates and the nonlinear response of λ to the changes in vital rates.

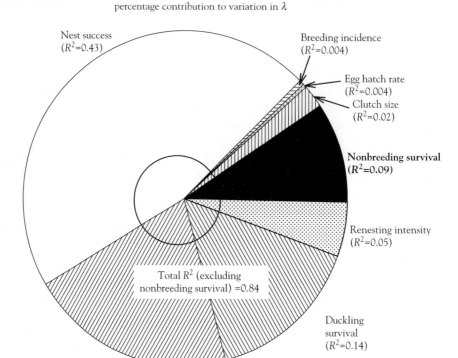

Mid-continent mallard population:
percentage contribution to variation in λ

Nest success
(R^2=0.43)

Breeding incidence
(R^2=0.004)

Egg hatch rate
(R^2=0.004)
Clutch size
(R^2=0.02)

**Nonbreeding survival
(R^2=0.09)**

Renesting intensity
(R^2=0.05)

Total R^2 (excluding
nonbreeding survival) =0.84

Duckling
survival
(R^2=0.14)

Breeding season
survival
(R^2=0.19)

Fig. 7.10 Results of an LSA analysis for female mid-continent mallards in North America (Hoekman et al. 2002). Each pie slice shows the proportion of variance in population growth rate explained by that vital rate in 1000 simulations of vital rates drawn from field studies. In other words, the plot shows the R^2 from regression plots of vital rates against population growth determined as in the cowbird example (Fig. 7.9). Approximately 84% of the variation in population growth rate is expected to arise from breeding-ground vital rates. The 7% not accounted for in the pie can be thought of as statistical noise, accounted for by interactions among the rates and the nonlinear responses of population growth to the changes in vital rates.

showing that vital rates have changed by certain amounts: would those changes in vital rates be likely to cause the observed declines?

Biek et al. (2002) conducted sensitivity analysis for three potentially declining amphibian species for which there were reasonable vital-rate estimates and purported mechanisms driving reduced vital rates: western toads, red-legged frogs, and common frogs (the latter two species formed the basis for the matrix examples in this chapter). In all three species, post-metamorph survival (juvenile or adult) had the highest elasticity, indicating that λ was most likely to be decreased by a given reduction in these rates compared with others (such as embryonic or larval survival) that have been the target of most experimental studies. Manual perturbation and LSA enriched the conclusions by reinforcing (as in other case studies) that if rates with low elasticities vary a lot, then they can affect λ even more than rates with higher elasticities.

Summary

Understanding the effects of age or stage structure on population processes is critical for wildlife population ecologists, both because different stages are important to management decision-making (e.g. bull elk compared with fawns, or turtle eggs compared with ocean-going juveniles) and because structure affects population growth. Matrix population models are certainly not the only way to account for population structure, but they are popular due to their relative simplicity and straightforward links to vital rates measured in the field.

The fact that different stages are not equal in their influence on population growth means that a saavy population ecologist will quantify the reproductive values of each stage to help inform translocations, harvest strategies, or control of pest species. If vital rates remain relatively constant over time, a population will achieve a SSD and a population growth rate characteristic of those vital rates.

Because both reproductive values and proportions in the population will differ between stages, survival and reproduction in different stages do not have the same effects on population growth. In other words, just as different stages are not equal in their effects on population growth, neither are vital rates equal. The broad and important field of sensitivity analysis seeks to quantify the relative importance of different vital rates and the expected efficacy of different management actions on population growth or persistence. Analytical sensitivity and elasticity show how much an infinitesimal change in each vital rate might affect population growth. As useful as this insight is, remember that the amount that a rate can change in nature or under management will also affect how important a vital rate is to population growth.

The two sensitivity analysis methods that do the best job of specifically and intuitively incorporating a specified range of variation in vital rates include manual perturbations and LSA. Manual perturbations of vital rates can contrast specific predictions from management actions that are expected to have specific effect, an approach that identified useful steps for managers to take in the recovery of red-cockaded woodpeckers. LSA can simulate many possible matrices from user-specified means and variances. Its output can be variable, including assessment of the stability of elasticity rankings across variation in vital rates as well as direct insight into how changes in certain rates are expected to affect population growth. Management of pest species (e.g. cowbirds), harvested species (e.g. mallards), and other species of concern (e.g. declining amphibians) would be more efficiently directed using LSA to evaluate the effects of management scenarios on expected population growth rate.

I will end with an apt metaphor borrowed from Ron Reynolds of the US Fish and Wildlife Service. A good general always goes into battle with a thoughtful focus on achieving the most with the resources at hand; troops are not scattered randomly across the battlefield or positioned according to political whims. Likewise, a good wildlife manager should use the insights from population-projection models and sensitivity analysis to see how proposed actions could ripple through to affect population growth in ways that are not obvious, revealing which actions will be a waste of time and money and which would be cost-effective. Population-projection models frame the biological context to help win the management battle.

Further reading

Caswell, H. (1989/2001) *Matrix Population Models: Construction, Analysis, and Interpretation*, 1st and 2nd edns. Sinauer Associates, Sunderland, MA. The single most important reference for the mathematics of matrix models.

Noon, B.R. and Sauer, J.R. (1992) Population models for passerine birds: structure, parameterization, and analysis. In: *Wildlife 2001: Populations* (eds. D.C. McCullough and R.H. Barrett), pp. 441–64. Elsevier Applied Science, London. Although specifically focused on birds, this is one of the most readable and practical discussions of building and analyzing matrix populations.

8

Predation and wildlife populations

If there are any marks at all of special design in creation, one of the things most evidently designed is that a large proportion of all animals should pass their existence in tormenting and devouring other animals.

J.S. Mill (1874; cited by Taylor 1984:1)

The large, ferocious gray or buffalo wolf, the sneaking, snarling coyote, and a species apparently between the two, of a dark-brown or black color, were once exceedingly numerous in all portions of the Park, but the value of their hides and their easy slaughter with strychnine-poisoned carcasses of animals have nearly led to their extermination.

P. Norris (1881), Second Superintendent of Yellowstone National Park

Introduction

With its gore, its excitement, and its brutal finality, predation has always fascinated humans. Biologists have built on the core intrigue of predator–prey dynamics, co-opting the term **arms race** – widely used to refer to how human armies inevitably esca-late technology to keep up with each other – to describe the evolutionary changes in both predators and prey (Dawkins & Krebs 1979). For wildlife prey we see speed, poisons, coloration, armor, alertness, and deception, matched on the field of battle by similar traits in the predator.

As wildlife population biologists, an understanding of predation is important because the public is vocal and curious about what happens to predators and prey, and because predation plays such an important role in population dynamics. Some of the most controversial issues in wildlife and conservation biology hinge on the extent to which predators affect prey numbers. As one recent example of a theme that has played out all over the world for centuries, on January 12, 2006 the *Idaho Statesman* newspa-per reported that the Idaho Department of Fish and Game "plans to kill up to 75% of the wolves in the Lolo elk zone to bolster struggling elk herds there."

Do wolves and other predators control or adversely affect their prey, so that killing the predators will in fact bolster the abundance of their prey? Similarly, in the context of invasive species (Chapter 11), would a predator biocontrol agent successfully reduce the numbers of a pest or invasive species? Might native species be driven inadvertently toward extinction by the introduced biocontrol predator?

To help shed light on these questions, this chapter focuses on the effect of predators on prey dynamics and, to a lesser extent, the effect of prey on predator numbers. I will emphasize concepts, avoiding a plunge into the sea of predator–prey models, including the famous Lotka–Volterra predator–prey equations of heuristic value to general ecology, but of less practical value in an applied population biology text.

Finally, before jumping in we must define two key terms. First, the concept of predators **controlling** prey. Taylor (1984) has noted that the word control has been used in predator–prey discussion to mean almost anything and, therefore, nothing. When I refer to predators controlling prey I will specify particular outcomes, such as predation regulating prey numbers and affecting fluctuations around an equilibrium, or acting to limit prey at low numbers, including extinction[1].

The second term to define is what is meant by *predation*. Do herbivores prey on plants, do decomposers prey on dead animals, or granivores on seeds? Is a parasite or disease a predator? Is it predation if an animal kills one of its own species in a fight? For the purposes of this chapter I will focus primarily on animals killing and consuming animals, recognizing that even in this narrow definition there can be surprises: the main killer (and consumer) of pre-weaning snowshoe hares are not big-fanged carnivores, but rather red squirrels and ground squirrels (O'Donoghue 1994).

Does predation affect prey numbers?

As you might expect, the best short answer to this question is sometimes or that it depends (Box 8.1). We don't have to look very far to see examples where predators limit prey population size – potentially all the way to extinction – or cause oscillations in prey abundance to be either exacerbated or dampened. Some of the most spectacular examples of control by predators are with recently introduced predators, both when they arrive and after they are removed. Cats, rats, brown tree snakes, and foxes have caused devastating extinctions around the world when they arrive in a new area. Indeed, 40% of the extinctions of birds on islands have been caused by predation by introduced animals (Estes et al. 2001).

One reason why native prey – particularly on islands – can be so badly affected by introduced predators is that the prey are a big step behind in the arms race, lacking the adaptations necessary to escape or even to fear the predators. The loss of anti-predator behaviors, leading to **ecological naiveté** of prey on islands, could arise either

---

[1]Recall from Chapter 6 that **regulation** refers to maintaining numbers within some equilibrium range through density-dependent processes while limiting factors determine the actual equilibrium numbers and may be density-dependent or -independent.

Box 8.1 Do predators in New Zealand affect two species of shearwaters?

This chapter is all about why predators have demographic effects on some prey populations but not others. An instructive case involves two species of shearwater, burrowing petrels of conservation concern in New Zealand, where the management concern is whether control of exotic predators would be a more efficient path to recovery than reducing browsing damage by introduced mammals or establishing new breeding sites.

Predators of shearwaters in New Zealand include, most prominently, stoats (a type of weasel otherwise known as ermine) introduced to New Zealand in the 1880s, as well as other introduced mammalian predators such as rats and cats. The main factors that affect how the Hutton's and sooty shearwaters are affected by predators include the following.

- The location of colonies affects the suite of predators. Hutton's shearwaters nest above the snowline, and stoats are their only substantial predator. By contrast, sooty shearwaters nest close to sea level and must contend with a suite of introduced predators including not only stoats but also cats and rats.
- Size of existing colonies affects the impact of predation. The two remaining colonies of Hutton's shearwaters contain about 110,000 and 10,000 breeding pairs. Because predator (stoat) numbers are limited by a lack of prey over the winter (when shearwaters and many other species are gone), the predation rate is fairly dilute. On the other hand, sooty shearwater colonies are perhaps 90–9% smaller, so kills can have a much larger impact.

Therefore, predation on sooty shearwaters has led to low and highly variable breeding success and adult survival. To increase sooty shearwater abundance, the only real management solution is aggressive reduction of the whole suite of predator species, including not only stoats but also cats and rats. By contrast, the relatively low predation rates on Hutton's shearwater indicate that even if all stoats could be killed, population growth for this prey species would be marginally affected, indicating that the best management strategy for Hutton's shearwaters would be to control destructive browsing by introduced mammals and to establish alternative breeding sites (Cuthbert et al. 2001, Cuthbert & Davis 2002, Jones 2002).

---

from the chance loss of key traits when an island is founded by a few individuals or from relaxed selection on anti-predator behaviors that are potentially expensive to maintain (Blumstein & Daniel 2005). Quammen (1996:205–6) gives examples:

> Loss of wariness is sometimes manifest as ingenuous nesting behavior: In the Galapagos, the blue-footed booby puts its eggs onto a bare patch of ground, unprotected, unconcealed, not even cushioned by a cradle of vegetation. Another form of ingenuous nesting involves building a nest in plain view on a tree limb, where it can easily be raided by a climbing predator. The Mariana crow practices that sort of reckless behavior on the island of Guam. A more cautious bird might at least conceal the nest, or place it beyond reach at the end of a thin branch, or suspend it in an elaborate woven pouch, as the tropical oropendolas

do. But oropendolas are mainland species, surrounded by predators and obliged to be more cautious. Boobies can be boobies ... These animals aren't imbecilic. Evolution has merely prepared them for life in a little world that is simpler and more innocent than the big world.

When they have evolved together, predators and prey interact on more equal footing, but still prey density or fluctuations can be affected by predation. The classic cycles of snowshoe hare in North America are driven at least in part by predation (Krebs et al. 1995), as are the regular, widespread cycles of northern small mammals (Korpimäki & Norrdahl 1998). More surprisingly, mammalian carnivores often kill other carnivores (**intraguild predation**), accounting for up to 68% of known mortalities in some species and at times limiting numbers (Palomares & Caro 1999).

A more subtle, but potentially pervasive line of evidence for effects of predators comes from changes more than one trophic level removed from a top predator. **Mesopredator release** (Soulé et al. 1988) occurs when mesopredators (mid-level predators) are regulated by top predators through either predation or competition. If the top predator is removed, a top-down **trophic cascade** can occur, whereby the mesopredators increase in number and in turn decrease abundance of their prey. A classic example has been documented in southern California, where intensive urbanization has destroyed most of the native sage-scrub habitat (Crooks & Soulé 1999). With the decline or absence of coyotes from this system, both native mesopredators (striped skunk, raccoon, and grey fox) and exotic meso-predators (especially domestic cats) were released from predation and competition from coyotes (the cat response also occurred because without coyotes around owners tended to let their cats outside more often). The resulting high numbers of mesopredators cascaded into both higher prey mortality (cats around a single moderately sized canyon killed more than 500 birds, nearly 1000 rodents, and over 600 lizards per year), and reduced abundances of scrub-breeding birds. Trophic cascades initiated by vertebrate predators in terrestrial systems are fairly common in nature (Schmitz et al. 2000).

Despite the range of examples where predators do reduce numbers of their prey, we also see plenty of places in the wild where prey continue to persist and even flourish with predators in their midst. To foreshadow a theme of the chapter, prey are active participants in the life and death process, evolving and behaving to reduce their chances of being killed. Even predator-naïve animals can harbor innate reactions of caution that can reduce vulnerability to novel predators. For example, the last population of the rufous hare-wallaby on the Australian mainland was destroyed by a fire and foxes in 1991 so the species persisted only on two islands off the coast; however, captive-breeding efforts showed that hare-wallabies can be trained to avoid cat and fox predators, which they will be exposed to when reintroduction programs begin (McLean et al. 1996). At a population level, the death of prey individuals, no matter how massive or macabre it may seem to us, does not necessarily result in a smaller prey population; consider that roughly one-third to one-half of all bird nests are destroyed by predators, but the decline of bird populations following such predation is certainly not inevitable (Côté & Sutherland 1997).

In short, predators and prey are entwined in a dance of evolution and population response. The best generalization we can make on population response is that predation can certainly regulate and help limit numbers of prey, but is unlikely to drive prey populations to extinction unless introduced species are involved or the prey population is small and fragmented or otherwise affected by other recent perturbations (Macdonald et al. 1999). To extend this generalization, we will closely examine three main factors that determine whether a predator will limit or regulate prey in any particular case: the predation rate of the predator on the prey (in turn a function of predator and prey numbers, and the number of prey killed per predator), the degree to which the predation can be compensated for by the prey, and which individuals are killed. Considering these factors will move us away from the relatively empty question of whether predation affects prey numbers and toward the more interesting and useful question of whether predators in a particular setting are likely to affect the dynamics of their prey.

Factors affecting how predation impacts prey numbers

Percentage of the prey population killed

Prey face a world that is "red in tooth and claw" (as Lord Tennyson put it), populated by predators that can respond to an increasing number of prey by increasing their own numbers and by killing more per predator per unit time. Therefore, the total number of prey killed will be a product of both the number of predators (the predator numerical response) and how many prey each individual predator kills (the predator functional response). The **predation rate**, or percentage of the prey population killed per unit time, is

$$\text{Predation rate} = \frac{\text{Number of prey killed}}{\text{Prey abundance}} 100 \tag{8.1}$$

Thus, in this section I first discuss the predator numerical and functional responses, which collectively determine the number of prey killed, then merge those with prey abundance to explore the implications for predation rates.

Numerical responses of predators

The **numerical response** reflects the change in number of predators as prey abundance changes; more precisely, it is the equilibrium numbers of predators present at a given prey density (there could be a time lag between current prey numbers and the eventual equilibrium predator number). Within a population, the numerical response will be a function of the predator's birth and death rates, which we know can be captured as λ or r.

In addition to the numerical response mounted from within the predator population, more rapid numerical increases in a predator's population can be driven by an

aggregative response, whereby predators converge from elsewhere to consume prey. Aggregative responses are of special interest in the agricultural-pest arena, because the numerical response of, say, an avian predator to an outbreaking insect pest will be too slow to be effective, whereas an aggregative response may lead to a very rapid increase in local predator density. To cite one such case, Carolina chickadees rapidly congregate in woodlands with greater densities of leaf-mining moths, aiding in suppression of the moth (Connor et al. 1999).

Predicting and interpreting numerical responses becomes more complicated when there is more than one predator or prey. With several predator species, reducing the abundance of one predator (say through predator control) could actually increase the numerical response of other predators due to trophic cascades or relaxed competition. This seems to be what happened in New Zealand when attempts to remove stoats to protect nesting birds (Box 8.1) increased introduced rat numbers (one prey of stoats), which in turn increased predation on sooty shearwaters (Lyver et al. 2000).

Multiple prey species can strongly affect the numerical response of predators on prey. Because some prey species are better able to increase or sustain their numbers in the presence of predation, they may facilitate a numerical response in the predators that results in a decrease of other prey species. Thus, what seems like competition between alternate prey species may actually be **enemy-mediated apparent competition** (Chaneton & Bonsall 2000), where prey species affect each other's abundances through their effects on the numerical response of a shared predator. For example, woodland caribou in Canada are exposed to multiple native predators (especially wolves, cougars, and bears) that in turn are supported by multiple prey (especially moose and deer) that do quite well in the human-modified landscape. The incidental take of caribou by the abundant subsidized predators reduces caribou population growth (Wittmer et al. 2005).

A special form of enemy-mediated apparent competition, termed **hyperpredation** (Smith & Quin 1996), occurs with the introduction of both a predator and an introduced prey that is able to sustain or increase its numbers in the face of predation. The introduced species might seem to be merely a competitor with natives – say, rabbits introduced to an area with native rodents, lizards, or birds – but through hyperpredation it could also drastically impact the native prey by sustaining much higher numbers of the shared predator than could be supported by the native species alone (Box 8.2). Even if the native prey is only a by-catch of secondary importance to the predator, the predator's numerical response – subsidized by the introduced prey – can devastate the native prey.

Interestingly, with cats and cat food humans are introducing to native systems both a hyperpredator and its introduced prey. Cat food can maintain both domestic and semi-feral farm cats at densities far higher than native carnivores (Woods et al. 2003, Kays & DeWan 2004). In Great Britain the cat population of approximately 9 million is about 20 times that of stoats and weasels and more than 30 times that of foxes. Cat numbers in the USA total perhaps 80–100 million owned, stray, or feral cats. Given that each cat kills anywhere from tens to hundreds of wild birds and mammals each year, allowing cats to roam free unleashes the fury of their numerical effects on native wildlife.

Box 8.2 Introduced rabbits lead to hyperpredation by cats on native species

Rabbits have been introduced – usually intentionally – to hundreds of islands worldwide. They adapt well to most conditions, eat a variety of plants, and have exceptionally high population growth rates. Rabbits certainly have well-known direct effects on both the vegetation and on other grazing species that are competitively inferior. Less well appreciated and probably more insidious, however, are the indirect effects they can have on native wildlife via apparent competition and hyperpredation (Smith & Quin 1996, Courchamp et al. 1999, 2000, Norbury 2001). Consider the response of feral and domestic cats (*Felis silvestris catus*) to a bountiful food source of rabbits, captured in the figure.

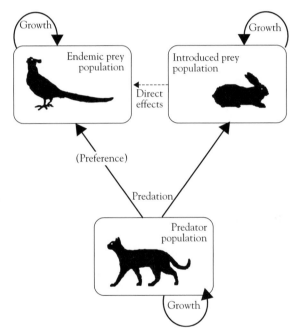

Introduced rabbits affect other herbivores both directly and indirectly by sustaining predators. From Courchamp et al. (1999).

Cats have caused local extinctions of native birds and mammals throughout the world (Mack et al. 2000, Risbey et al. 2000). Often feral cat numbers are limited by seasonal lows in prey abundance, and they may be limited in space by intervening areas with few native prey. However, when rabbits arrive and increase in numbers across the landscape, cats can prey on them whenever and wherever native prey are sparse, initiating a vicious numerical response. Here are just two of many documented examples.

- On the sub-antarctic island of Macquarie, introduced cats persisted with parakeets for more then 60 years. However, within 20 years of rabbit introduction the parakeet was extinct because rabbits increased cat numbers, even as the only indigenous land birds present during winter (parakeets, rails, and a teal species) declined.

(Continued)

Box 8.2 Continued

- In New Zealand, both cats and introduced stoat populations are supported by rabbits, and highly endangered native grand and Otago skinks suffer elevated predation as a result. The effects are worst when rabbit density fluctuates, because the sustained predator community switches to skinks most ferociously when rabbit numbers temporarily decrease.

The moral of the story is to deal with not only the predators, but also with the rabbits. Control of rabbits needs to be sustained, because if it is tentative, allowing rabbits to bounce back in repeated pulses, the predator suite could switch to native fauna during rabbit lows and cause even more damage.

Awareness of enemy-mediated apparent competition can lead to better management decisions that may not be obvious (Box 8.2). If an introduced predator is demolishing the native fauna, and its abundance is subsidized via hyperpredation on an introduced prey species, removing the predator would be easier if the introduced prey were simultaneously removed. Likewise, removing only an introduced competitive prey species as a means to increase a native species could cause more harm than good if an associated hyperpredator is not simultaneously removed. A classic case involved proposals to remove feral pigs from the California Channel Islands (USA), both because the pigs have badly damaged the islands' native vegetation, and have supported through hyperpredation increased numbers of introduced golden eagles, which in turn have caused the precipitous decline of endemic and endangered island foxes. Although the pig removal would seem to be a straightforward and sensible plan, eradicating pigs without also reducing the eagles could actually trigger fox extinction because eagles will likely kill more foxes as pigs decline (Courchamp et al. 2003).

Functional responses of predators

The functional response, or **kill rate**, describes the number of prey killed per predator per unit time. As the prey numbers increase, kill rate could respond (or not) in many different ways. Although predator–prey theorists have categorized a variety of functional responses (Jeschke et al. 2002), we will focus on the two most likely for wildlife predators, named by Holling (1959) as **Type 2 and Type 3 functional responses**[2]. These are shown on the two panels on the left of Fig. 8.1; Type 2 has a hyperbolic curve whereas Type 3 shows a sigmoidal increase.

For any functional response, the kill rate must always flatten at a maximum because there is limited time available for hunting and killing. In particular, **search time** is required to locate prey, and **handling time** to pursue, kill, and eat the prey. The functional response can also be limited by **satiation** where a full stomach takes away motivation to eat more. However, the functional response can exceed what would be

---

[2]We're ignoring Type 1, a straight-line relationship between prey density and functional response, for which it is hard to come up with biologically realistic mechanisms.

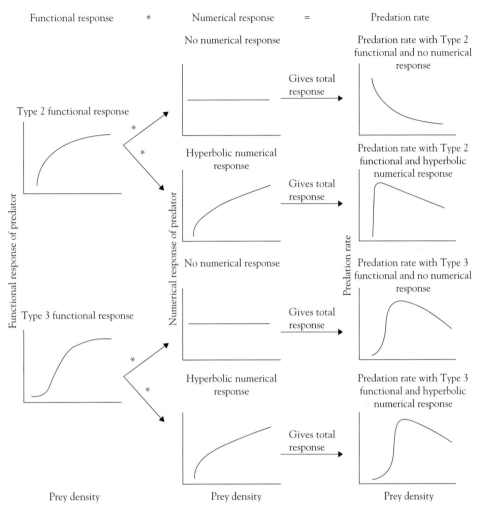

Fig. 8.1 How a predator's functional and numerical response can combine to affect predation rate on a prey population, using as examples curves derived from analyses on wolves and moose (Messier 1994, 1995). The left-hand panels are functional response curves (moose kills per wolf per 100 days) that are either Type 2 or Type 3. Each functional response is multiplied by a numerical response (middle panels; wolves per 1000 km²) that is either constant (unresponsive to density) or a hyperbolic increase resulting from local births and/or an aggregative response by the predator. The right-hand panels show the predation rate (percentage of moose population killed by wolves per year).

expected based on the energy needs of the predator (Kruuk 1972, Short et al. 2001): four or fewer red foxes killed up to 230 adult black-headed gulls in one night, eating fewer than 3% of them; in two separate instances in Australia a single introduced fox killed 11 wallabies and 74 penguins over several days, eating almost none of the victims; up to 19 spotted hyenas killed 82 Thomson's gazelle and badly injured 27 more in one night, eating only 16% of the kill.

Such seemingly heinous acts by predators raise intense emotions in humans because the gratuitous killing can seem to be an immoral waste of life. Why do predators do it? In some cases, when predators encounter easily accessible domestic prey or arrive in a system with naïve wild prey they initiate **surplus killing**, whereby animals are killed but not eaten. The **henhouse syndrome**, leading to surplus killing, is an almost inevitable result of a high-performance predator confronted with an easy target (see Box 8.3). A textbook case of **partial prey consumption** that comes with surplus killing can be found in brown and black bears eating salmon, where bears consume less of each fish when more fish are available. Furthermore, all fish and fish parts are not equal: unspawned fish – with higher muscle quality – are eaten more than spawned-out fish, and high-energy parts like brains and eggs preferentially consumed (Gende et al. 2001).

In addition to surplus killing via the henhouse syndrome, **excessive killing** beyond immediate energetic needs may be an adaptive strategy for foraging over a longer time period. One striking illustration can be found with least weasels (Jedrzejewska & Jedrzejewska 1989) that killed and consumed bank vole prey approximately in proportion to their energetic needs each day during the summer and fall, but killed (and cached in their nests) more than they needed as the Polish winter cold descended; when temperatures got really cold weasels stopped hunting and instead ate out of their cache, a highly adaptive trait facilitating survival through cold winters (not unlike the nuts in a squirrel's hoard). Similarly, coyotes in the Yukon of Canada cache entire carcasses of nearly half of their snowshoe hare kills in early winter, and return to eat most of

---

Box 8.3 The henhouse syndrome: surplus killing by predators

The behavioral programming of the act of predation can lead to the killing of far more prey than necessary to fulfill energetic demands. Often called the **henhouse syndrome** because it can happen when a predator gets into a chicken coop, such surplus killing arises from the ethology of predation. Each of the four behaviorally distinct behaviors involved in predation (search, pursue, kill, and consume) are independently reinforced (Kruuk 1972). That is, the animal is rewarded not just by completing the whole predation act – eating the prey – but also by successfully carrying out each of the four behavioral components independently. (Think about why this must be true: for a young predator to learn its craft, where most early attempts fail to culminate in a prey in the belly, there must be positive reinforcement, or psychological encouragement, for performing each stage on the way to consuming the prey item.) A decrease in time spent performing any one or more of these behaviors will elevate the functional response. Normally, each step is time-consuming because the arms race adaptations of most prey challenge predators at each step of the search–pursue–kill–consume process. But if the predator is presented with an unusual case where search and pursuit are made ridiculously easy – say the prey is penned or ecologically naïve – the predator can simply perform the act of killing again and again and again. The predators are not morally bereft, nor are such killers a case of problem or rogue individuals. For a predator faced with available prey, trivial costs of killing, and little risk of injury, there simply is no adaptive reason why it should stop killing, regardless of whether the prey are eaten.

them over the next few months of deep winter even when they are covered by half a meter of snow (O'Donoghue et al. 1998). Of course, a carcass killed but not eaten by a predator is not wasted in an ecological sense because scavengers and decomposers will consume it. In fact, some scavengers depend on excess kill, as when common ravens treat gunshots as a dinner bell and fly towards the sound with the expectation of finding a 70-kg elk gut pile to scavenge from a successful hunter (White 2005).

Even with surplus or excessive killing, the required search and handling time will still determine an upper limit to the functional response. Likewise, the overall shape of the functional response curve is also determined by search and handling time, along with satiation. For example, the tapering of kill rate (see the whole Type 2 curve and the right-hand side of the Type 3 curve in the left-hand panels of Fig. 8.1) can arise from a decreasing motivation to hunt as the belly becomes full, or from less time available as a higher and higher proportion of the predator's time is taken up by handling. The left-hand side of the Type 3 curve, showing an increasing kill rate with increasing prey density, can be driven by additional factors. As prey increase from very low numbers, predators increasingly learn how to recognize, subdue, and consume the prey, developing a **search image** to increase kill rate. A newly acquired search image can cause the predator to switch to a prey as it becomes more numerous. Another mechanism leading to the increasing kill rate in the Type 3 curve is prey behavior: if prey use camouflage, or safe hiding places that are limited in number, both strategies will result in a larger proportion of prey taken as their numbers increase.

How do functional responses affect predation rate on the prey, independent of the predator's numeric response? For a fixed predator density (see the flat horizontal lines showing no numerical response in the middle panels of Fig. 8.1), a Type 2 kill rate (functional response) creates positive density dependence in prey survival; as prey numbers increase so too does prey survival because predation rate decreases. The positive relationship between prey survival and prey numbers tells us that at low or declining prey density, a Type 2 functional response can create strong Allee effects (Sinclair et al. 1998, Gascoigne & Lipcius 2004; see Chapter 6). On the other hand, a Type 3 functional response will tend to relieve declining prey from predation pressure at very low prey densities by facilitating higher survival (lower predation rate). If, however, a large prey population under predation with a Type 3 functional response collapses due to poor environmental conditions (perhaps a drought or heavy snow), its ability to increase may be compromised by the negative density dependence in prey survival at low to medium densities (i.e. as prey density increases, predation rate increases and prey survival declines; Fig. 8.1). Thus the prey could be stuck in a **predator pit**, a low density from which they cannot recover unless predation rates dramatically decline.

The functional response curves distill predator responses as a function of prey density (and so are often called **prey-dependent models**). A countercurrent to functional response curves plotted against prey density has emphasized that the kill rate depends on lots of things other than prey numbers. Certainly this is true; as we have seen, kill rate can be affected by context – evolutionary background and age structure as well as habitat and weather conditions – and also by other species including alternate prey and predators. One alternative way of capturing some of these other influences on functional response has been to plot kill rate not against prey number but rather against the ratio of prey to predator population sizes (Abrams & Ginzburg 2000). Such **ratio-dependent predator–prey models** can be useful complements to the traditional ones based only on prey density, helping us to understand the factors other than prey numbers that affect kill rates in wild populations (Vucetich et al. 2002).

Ultimately, the shape of the functional response curve for any predator–prey system is a manifestation of how well the predator and the prey are doing in the arms race[3]. Prey strive to minimize the functional response by defense and camouflage, while predators improve their search image and decrease travel and processing time between kills. Of the hundreds of examples that could be given of behavioral and morphological responses by prey to reduce functional response in the arms race with predators, one of my favorites is that the black tips on the relatively long tails of weasels confuse aerial predators and deflect attacks away from vital parts of the weasel (Powell 1982). In short, the concept of functional response is a useful heuristic component of interpreting predation rate, and improved methods allow it to be estimated, with variance, from field data (Hebblewhite et al. 2003, Joly & Patterson 2003).

Total predation rate

Next let's explore the combined effect of functional and numerical responses to determine the total number of prey killed at different prey densities, or overall predation rate (eqn. 8.1). Of the many possible combinations of numerical and functional responses that lead to different predation rates, I chose for Fig. 8.1 a few examples that could be reasonable for a wolf/moose predator–prey system (Messier 1994, 1995). Although I've avoided units to make Fig. 8.1 less busy, let's work through an example of how the predation rate (shown by the right-hand side of Fig. 8.1) is calculated from predator functional and numerical responses (the left-hand and middle panels of Fig. 8.1), and prey number. Suppose:

- moose density$=2$ moose/km^2
- wolf functional response$=2.73$ moose killed/wolf per 100 days
- wolf numerical response$=41.9$ wolves/1000 km^2 $=0.0419$ wolves/km^2
- total kill$=2.73*0.0419=0.114$ moose killed/km^2 per 100 days.

Thus, the **annual predation rate** would be total killed per 100 days$*3.65/2$ moose$=0.21$, or 21% of the moose in the area killed by wolves per year. (You should try it with different numbers to convince yourself that you could draw the predation-rate curves in Fig. 8.1 if you were given the moose density and the wolf functional and numerical response values.)

As already discussed, a Type 2 functional response without a numerical response can cause the predation rate to increase as prey numbers decrease, leading to an Allee effect, often called **destabilizing** because the predation rate gets worse and worse as

---

[3]Although, as a generality, functional responses similar to Type 2 are probably most common for wildlife populations (Gascoigne & Lipcius 2004).

prey populations get smaller and smaller (Fig. 8.1, upper right-hand panel). When a Type 2 functional response is accompanied by a positive predator numerical response (Fig. 8.1, right-hand panels, second row), the total predation rate can be low at very low prey numbers (leading to high prey survival) but can be destabilizing at higher prey numbers. Importantly, if the hyperbolic numerical response is moved upward, as would be expected in a multi-prey system where the predator could sustain itself at reasonably high numbers independently of the prey being considered, then the total predation rate curve becomes destabilizing at all prey densities (Messier 1995). Thus, potentially severe Allee effects (destabilizing positive density dependence) due to predation are likely when the functional response is Type 2, and when the predator numbers are limited by factors other than the prey in question (Sinclair et al. 1998, Gascoigne & Lipcius 2004).

For an endangered prey population, the theory just discussed implies that hyper-predation and multiple native prey can allow predators to persist at high numbers even when the endangered prey is nearly extirpated, thereby initiating further decline of the prey. For example, the apparent competition on woodland caribou described above means that the incidental take of caribou by predators whose numbers are subsidized by other prey will cause the small caribou populations to suffer proportionately worse predation mortality (Wittmer et al. 2005). Similarly, smaller populations of native skinks in New Zealand are less able to sustain the losses from a suite of introduced predators sustained by rabbits (Norbury 2001).

So predators can kill a lot of prey through numerical and functional responses, and that can lead to a high predation rate. But, perhaps counterintuitively, a high predation rate does not necessarily mean that predators will limit prey population growth. Why not? There are two reasons. First, mortality due to predation may be compensated for. Second, which age or stage class gets killed matters for prey population growth. We'll explore each of these next.

Compensation of predation rate

When Paul Errington started observing predation on muskrats and bobwhite quail in the mid-1940s, the theory of predation in wildlife biology was simple: predators kill prey, so the removal of predators should mean more prey. Errington (1946) challenged that dogma. Behaviors such as territoriality may limit population size for many prey, making certain individuals (e.g. social subordinates) vulnerable to dying from disease or starvation if they are not killed by predators. Errington (1956) called these individuals the "doomed surplus," surely one of the most compelling phrases of ecological jargon of all time. Taylor (1984:28) notes that "by reducing predators to the ecological equivalent of garbage collectors, Errington undoubtedly served to forestall the conscious eradication of a number of carnivorous birds and mammals from North America."

Although it may be disconcerting to think about a doomed surplus in a population, the phrase makes it easy to realize that mortality due to predation may be at least partly **compensatory**. The mortality arising from predators killing the doomed surplus will be compensated for with lower mortality from other sources, say due to weather. Thus predation merely replaces other forms of mortality, leading to no net loss in prey numbers. In symbols, the annual survival rate under predation at some time t (S_t) is the same as the survival rate in the absence of predation (S_0). In a classic example, red grouse in Scotland that do not obtain territories in the autumn absorb nearly all of the mortality for the population. When a territory holder does die, a nonterritorial bird quickly takes its place, keeping density steady even when predators remove a large number of grouse (Jenkins et al. 1964). Compensation in survival can only go so far, because predation mortality can only be fully compensatory if it does not exceed other nonpredation-related mortality sources.

So with compensatory mortality, realized annual survival (S_t) is unaffected by predation rate. By contrast, if predation operates as an **additive** form of mortality, survival becomes a product of both not being killed by predators ($1-M_P$) and surviving everything else (S_0):

$$\text{Realized survival under additive predation} = S_t = S_0(1-M_P) = S_0 - S_0 M_P \qquad (8.2)$$

Errington was insightful enough to realize that mortality due to predation could be compensated for not only by other forms of mortality, but also by increases in other vital rates such as reproduction or immigration into a depredated population. Some of the most obvious examples of increasing reproduction to compensate for predation come from multiple clutches in birds. Mallard ducks are a notable example, as they rarely double brood (produce a second clutch after hatching ducklings), but if their nest is depredated they typically renest, and can do so up to five times in one season if nests are preyed upon repeatedly (Hoekman et al. 2005). Compensation for predation also occurs by immigration. For instance, despite humans killing more than 50% of an introduced red fox population each year as part of an effort to protect endangered birds in California the foxes persisted, in part because up to half of the population was immigrants coming in from neighboring populations (Harding et al. 2001). Because compensation of predator mortality can occur not only through survival (when the doomed surplus are taken) but also through increased reproduction and immigration, populations with compensation can sustain high predation rates.

The extent of compensation of predator mortality becomes of intense management interest when evaluating the efficacy of predator control, because compensation for predator mortality undercuts the utility of predator control (Côté & Sutherland 1997, Banks 1999). However, the greatest interest in compensation of predator-caused mortality centers on harvest of wildlife by humans as predators. Therefore, I will wait until Chapter 14 to explore compensatory mortality further, with lots more examples. For now, I'll leave you with the general understanding that predation rate alone cannot predict whether predators will reduce the numbers or dynamics of a prey population; fully compensatory predation will not affect prey at all, even if predation rate is high, while fully additive mortality from predation will decrease survival rates. Predation will rarely be fully additive or compensatory, but rather occurs on a continuum.

Having established two of the factors determining the effect of predators on their prey – the predation rate and compensation – we will next explore the third main factor, the age or stage of the prey killed.

Who gets killed

For predicting effects of predators on prey populations, which age class gets killed becomes important for three reasons. First, all age classes are not equally killable, so available age classes can affect the functional response. For example, American prong-horn on the National Bison Range of Montana currently face a single substantial predator, the coyote, which kills approximately 90% of fawns in their first year but cannot kill adults (Byers 1997)[4]. Second, age classes differ in the extent to which the predation mortality can be compensated; for instance, hatchling mortality in birds might be relatively easily compensated for by multiple additional clutches, whereas there may be less latitude to compensate for adult mortality.

Finally, as we have seen, all age classes and vital rates are not created equal in their effects on prey population growth. We can assess whether a given predation rate is likely to affect the prey λ value by calculating reproductive values and performing sensitivity analyses, as in the last chapter. Of course, we would keep in mind that a large mortality for an age class with a small effect on prey λ value could affect the prey as much as or more than a smaller change in a rate with a large effect on prey population growth. But the bottom line is that, depending on the age class killed, a high proportion of prey killed will not necessarily affect population growth even if the prey is unable to compensate for the predation mortality.

In short, sound estimates of vital rates can be married to projection-matrix models to gain management insights into effects of predation on particular stages. A good example expands the shearwater case study (Box 8.1). Using a matrix-projection model and an LSA-style approach incorporating uncertainty to explore how the λ value of Hutton's shearwater would vary across management changes, Richard Cuthbert and colleagues (2001, 2002) found that small changes in adult survival affect population growth more than even fairly large changes in chick or fledgling survival. Because stoats prey on chicks more than adults, and the highest mortality risk for adults occurs away from the breeding ground where stoats are, the management recommendation was to divert attention away from stoat predation on chicks and instead focus on minimizing the smaller level of stoat predation on adults and on other adult mortality sources such as by-catch of shearwaters from ocean fishing. Thus the sensitivity analysis ties back to the argument that reducing numbers of introduced stoats will be a relatively inefficient management option for the conservation of this shearwater species (Box 8.1).

Other examples abound where the effects of predators have been elucidated by formal analysis of which age or stage of prey is being killed. Although cheetah cubs are heavily preyed upon by lions and hyenas, an LSA sensitivity analysis incorporating both mean vital rates and their likely changes under management found

---

[4]Byers (1997) makes a compelling case that the remarkable adaptations of adult pronghorn for speed (approaching 100 km/h) are a "ghost of predation past," when Pleistocene predators including chee-tahs and hyenas would have preyed on adults. The return of wolves to the pronghorn range may once again impose predation on adult pronghorn.

that management focusing solely on reducing predation on cubs would be less effective than actions to increase – even slightly – survival of adults (Crooks et al. 1998). Likewise, the short-necked turtle in Australia is beginning to endure high predation from introduced red foxes; although the effects of fox predation on nests appear horrific, with rates exceeding 95% in some areas, the turtles would actually be better served by management to reduce adult mortality, which is much lower than nest predation but contributes more to turtle population growth (Spencer & Thompson 2005). Finally, you may recall from Chapter 7 that breeding-ground vital rates for mallards, which are often driven by predation, influence population growth more than do vital rates in the nonbreeding season, which includes harvest by hunters.

Summary

The question of whether predators control prey is huge in applied wildlife population biology, with implications ranging from whether predator reduction will protect introduced endangered prey or increase ungulate prey for hunters, to whether introduced predators are likely to decimate their prey. To answer the question with a broad yes or no is ecologically naïve. Rather, we can answer the question for any particular case by assessing three primary details.

First, we need to know the predation rate, or percentage of the prey population killed by predators. The predation rate is the number of prey killed divided by prey abundance; the number of prey killed is the product of the numerical and functional response. The numerical response describes the number of predators as prey numbers change. Multiple predator species can complicate the numerical response because reduction of one predator could increase the numerical response of other predators due to competitive release or trophic cascades. Multiple prey also complicate the predator numerical response through apparent competition or hyperpredation, where one prey sustains high numbers of a predator which in turn affects another prey species (as a special form of apparent competition, hyperpredation tends to involve an introduced predator and prey affecting native prey).

The other component affecting the number of prey killed is the functional response, or kill rate. Defined as the number of prey killed per predator per unit time, the kill rate is limited by satiation and limits in time available to search for and handle prey. The kill rate may well exceed immediate energetic requirements, however, if surplus killing occurs or if kills are cached to be used over longer time periods. Complex behaviors and feedbacks between predator and prey determine the shape of the functional response curve, with predator learning and prey escape behavior playing roles. A Type 2 functional response curve could create an Allee effect in small prey populations, decreasing survival as prey numbers decrease; by contrast a Type 3 response would tend to stabilize small prey numbers. Ratio-dependent models are an alternative to functional response plotted against prey density.

Even for a certain predation rate, two other details must be known to determine whether predation will affect a prey's population dynamics. First, we must know whether the predation mortality is compensated for. Compensation occurs via lower mortality in other parts of the year, lower mortality in other life stages, and/or by increased reproduction or immigration. If predation mortality is compensated for, then predation is unlikely to affect prey density or fluctuations, whereas additive predator mortality is more likely to affect prey numbers.

Finally, the effect of predators on a prey population will depend on who gets killed. Because all age or stage classes are not equal in their vulnerability to predation, in their ability to compensate for mortality, or in their effect on population growth rate, massive predation can occur on certain age classes with very little impact on population growth. Alternatively, small additive mortality rates from predation imposed on age classes with high reproductive value and/or making up a large proportion of the population can substantially lower population growth.

Predation is awe-inspiring, bone-chilling, and a major driver of population dynamics for many wildlife species. The predation rate by age class and the extent to which mortality due to predation can be compensated will vary over time and space, affected by weather, habitat changes, parasites and diseases, and other factors. By measuring these factors over space and time, the effect of a predator on a prey can be resolved.

Further reading

Errington, P.L. (1946) Predation and vertebrate populations. *Quarterly Review of Biology* **21**, 144–77; 221–45. A true classic, filled with insights that continue to be timely even now.

Taylor, R.J. (1984) *Predation.* Chapman and Hall, New York. This slim volume rings with an engaging style that packs in an enormous amount of theory, math, and applied thoughts on predation.

9

Genetic variation and fitness in wildlife populations

> *That any evil directly follows from the closest interbreeding has been denied by many persons; but rarely by any practical breeder; and never, as far as I know, by one who has largely bred animals which propagate their kind quickly. Many physiologists attribute the evil exclusively to the combination and consequent increase of morbid tendencies common to both parents: and that this is an active source of mischief there can be no doubt.*
>
> **Charles Darwin (1896:94; from Allendorf & Luikart 2006),**
> ***The Variation of Animals and Plants under Domestication***

Introduction

Deformed sperm in Florida panthers, lower survival for song sparrows with lower genetic variation, and compromised ability of inbred red-cockaded woodpeckers to adapt to global warming. In all of these cases, genetic variation intersects with population dynamics. Just as density dependence, predation, and interspecific competition affect vital rates – in turn affecting population dynamics – so too can levels of genetic variation feed into population processes.

Genetic variation plays a fundamental role in affecting both the long- and short-term dynamics of wildlife populations. Most of this chapter will be spent considering when, how, and why genetic variation can be lost, and how genetic variation interacts with environmental and deterministic factors to affect population persistence over the short term. To begin with, however, we will step back to consider the importance of genetic variation over a longer time scale (see Soulé & Wilcox 1980, Frankel & Soulé 1981).

Long-term benefits of genetic variation

Genetic variation allows long-term adaptation

Genetic variation is the stuff of diversification, of adaptation, of speciation, of evolution; without it, a population or species lacks the raw material to evolve in response to changing conditions. Indeed, the profoundly influential Fundamental Theorem of Natural Selection, developed by Sir Ronald Fisher in 1930, is based on this idea that the rate of evolutionary change in a population depends on the amount of genetic diversity available. In an applied sense, low genetic variation may compromise the long-term ability of populations to adapt to toxins, disease, or global warming (Soulé 1980, Schiegg et al. 2002, Reed et al. 2003).

Consider how lack of genotypic variation can limit a population's ability to respond to disease. The vertebrate body has an incredible defense system, whose genetic basis is encoded, in part, at the major histocompatibility complex (MHC). The MHC is made up of many variable nuclear genes that work together to build a well-stocked arsenal to recognize and destroy intruder genomes (Aguilar et al. 2004), so a catastrophic loss in MHC variation may compromise the ability of individuals in a population to mount an immune response to a novel disease. For example, wild pocket gophers from small isolated populations have very low MHC variation and suffer more severe and long-lasting infections when experimentally infected with hepatitis B (Sanjayan et al. 1996, Zegers 2000); interestingly, these animals were also able to accept reciprocal skin grafts from each other, indicating that genetic variation was so low that the MHC genes were unable to distinguish self from nonself[1]. Similarly, stranded California sea lions with lower heterozygosity were more likely to harbor infectious diseases and parasites (bacterial and helminth) and take longer to recover; not only do these animals cost more to treat and rehabilitate, they also could act as reservoirs of infectious disease (Acevedo-Whitehouse et al. 2003). Such studies on wild vertebrates support concerns that endangered species with very low genetic variation may be unable to adapt to new diseases, climate shifts, or other changes (e.g. Meagher 1999, Hedrick 2003).

Genetic variation provides blueprints

Another long-term concern related to the loss of genetic diversity is that as species, or locally adapted populations, are lost so too are the blueprints of life. Aside from the ethical and philosophical concerns of losing distinct forms of life from the planet, there is a utilitiarian drawback: we lose the raw material that we may need for our own ends. We use wild organisms for food, recreation, medicines, clothing, shelter, sources for spiritual inspiration and scientific understanding, and services ranging from crop pollination to pollution removal (Hunter 2002). The medicinal uses can be particularly striking: for example, in the USA nearly half of all medicines contain active ingredients obtained directly from plants, microorganisms, or animals. These active ingredients are often from plants that might be considered to be of little value – who would have guessed that extracts of willow bark (*Salix* spp), used by ancient Greeks and Native North Americans, would provide the material for isolating salicylic acid, the painkilling ingredient in modern aspirin!

---

[1] In human transplants, a battery of medicines must be used to override destruction of the nonself tissue.

As another example, consider the search for a variety of rice that would be resistant to a crippling disease called grassy stunt virus. After screening over 6000 rice varieties, scientists found one resistant variety that had recently gone extinct in the wild due to dam construction. Luckily, the variety was represented in a museum collection, so the disease-resistant strain of rice could be developed and is now grown extensively in Asia (Hunter 2002). Varieties of organisms arise from genetic variation; if the variation is lost we lose the templates for the rich tapestry of life that we value, enjoy, and use.

What determines levels of genetic variation in populations?

The big four: mutation, gene flow, natural selection, and genetic drift

Genetic variation is maintained in populations via a dance among four processes: mutation, gene flow, natural selection, and genetic drift. Although the ultimate source of variation is mutation, phenotypic changes due to mutation tend to occur slowly. By contrast, gene flow from one population to another can maintain and substantially increase variation in local populations (Chapter 10). Population geneticists refer to gene flow as **migration**, not to be confused with the ecological meaning of migration as the seasonal movements of animals. The effect of natural selection – the third factor affecting genetic variation – is complex. Selection can favor heterozygotes, or could decrease variation if individuals at just one end of the phenotypic spectrum are favored. In any case, selection is the only mechanism that produces adaptive evolutionary change among the individuals that make up a population.

Of the four factors affecting genetic variation, the one that inexorably acts to decrease variation within populations is genetic drift. Genetic drift occurs when allele frequencies change randomly, or drift, from one generation to the next just because some alleles get passed on from parents to offspring while others do not (Box 9.1)[2]. Genetic drift causes allele frequencies to change, leading both to the loss of alleles as certain alleles are randomly **fixed** (achieve a frequency of 1.0) and to a decrease in heterozygosity (or random increase in homozygosity). The random changes in allele frequencies under genetic drift are in strong contrast to natural selection, where the most suitable alleles are transmitted, leading to local adaptation. The strength of genetic drift is inversely related to population size.

Of course, all four of these population genetic processes – mutation, gene flow, natural selection, and drift – interact. One set of interactions of intense interest to wildlife population biology is the relationship between selection and genetic drift. In large populations, heterozygosity and allelic diversity change slowly in response to natural selection and an underlying mutation rate. However, when the population is small, natural selection becomes overwhelmed by genetic drift unless selection is very strong. Collapsing a small mountain of population genetics theory into a brief rule of

---

[2]If it helps, think of genetic drift as the genetic cousin of demographic stochasticity (Chapter 5), with sampling variation having greater and greater impact as abundance is reduced.

Box 9.1 Loss of variation due to drift

Here is a simple example showing how genetic drift is a random process whereby alleles are lost by chance in a small, randomly mating population, leading to increased homozygosity. Consider six unrelated animals founding a population. Each year, male–female couples are formed (fathers are shown with solid lines, mothers with dashed lines) and give birth to two offspring (one male, one female) before dying. We follow the fate of just two alleles (A, a) at one locus carried by each individual. Although one or the other allele will inevitably become fixed (achieve a frequency of 1.0), there are many possibilities over three generations; here I show just one possible outcome, with the frequency of allele A in the right-hand column.

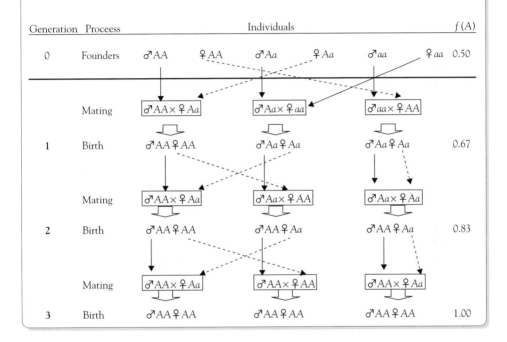

Generation Proceess   Individuals $f(A)$

thumb (Allendorf & Luikart 2006), allele frequencies will be driven primarily by genetic drift and not selection when the product of the genetically effective population size (N_e; to be described in more detail below) and the selection coefficient (s; the proportionate decrease in fitness of homozygotes under selection) is less than one ($N_e s < 1$). We can restate this by saying that a population with N_e smaller than $1/s$ will be ruled by random genetic drift and its inexorable march towards lower heterozygosity[3]. For example, the Chatham Island black robin in New Zealand persisted at less

---

[3]So, for example, a deleterious allele that reduces fitness by, say 1%, would have its frequencies driven by chance (drift) and not by selection if N_e is less than 100 (1/0.01 = 100). Selection coefficients on individual loci are typically less than about 5% (Frankham et al. 2002:215, Miller & Lambert 2004).

than 30 birds for approximately 100 years, reaching a low of five individuals in 1980; although they fortunately have increased to approximately 250 individuals now, they show a lack of both neutral DNA variation and a lack of variation at MHC loci, which are normally under strong selection to be variable (Miller & Lambert 2004)[4]. The fact that drift can overwhelm selection in small populations has profound implications for management, because compromised ability to locally adapt makes them unable to deal with novel stressors such as diseases.

Genetic changes due to fragmentation

Now, let's assemble this background on population genetic processes and focus on the response following population fragmentation, where a large population becomes small or a small connected population becomes isolated (Mills & Tallmon 1999). In Fig. 9.1, population C becomes small after being severed from the large population A, with changes in allele frequencies and loss of heterozygosity and allelic diversity over subsequent generations. Similarly, small population B initially has similar allele frequencies and heterozygosities to A because of gene flow, but after fragmentation loses genetic variation via genetic drift. By contrast, little genetic drift occurs in large population A, and heterozygosity and allelic diversity remain more constant over the relatively short term. Notice that even though random genetic drift is dominating small isolated populations B and C, leading to loss of heterozygosity and allelic diversity within the populations, allelic diversity and polymorphism across the group of populations may still be retained because different alleles are lost randomly within individual populations but are still present in the collection.

Are these processes relevant in the real world? Absolutely. At the extremes of isolation, genetic variation tends to be lower in populations on islands than for those on mainlands (Frankham 1997). The same is often true for translocated wildlife populations, whose frequent loss of genetic variation due to small founding population sizes exemplifies the **founder effect** on genetic diversity: for example, 10–20 years after reintroduction, four populations of bighorn sheep founded with between eight and 69 individuals tended to have lower levels of genetic variation than did the source population (Fitzsimmons et al. 1997). Similarly, allele frequencies in eight populations of alpine ibex populations founded by translocation (between five and 34 animals) diverged from the source population at a rate consistent with expectations under genetic drift (Scribner & Stüwe 1994). In general, genetic variation tends to decrease as population size decreases across a wide range of wildlife species (Frankham 1996).

So there is really very little debate that genetic variation can be lost in nature when isolation is complete and/or population sizes are small. But are these processes

---

[4]An interesting contrast was found for San Nicolas Island foxes, isolated for perhaps 1000 years and experiencing bottlenecks perhaps as low as 10 foxes, with DNA fingerprints and microsatellites showing no variation but with MHC diversity maintained (Aguilar et al. 2004). The differences between the bottlenecked robin and fox populations are not entirely explained, but may be due to stronger selection on MHC diversity in the foxes.

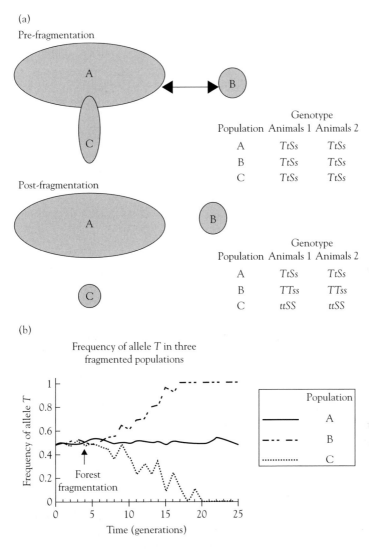

Fig. 9.1 Schematic of potential genetic consequences that can occur when population fragmentation decreases abundance and connectivity among wildlife populations. (a) Circled areas represent suitable habitat and contain different wildlife populations. Arrows represent gene flow across a semi-hospitable habitat. The genotypes of only two individuals and two genes (loci) are shown per population, although in real studies many genes and many individuals would be sampled. Genotypes of animals in different populations are identical before fragmentation, but diverge for the isolated populations after fragmentation. (b) Example of changes in the frequencies of the *T* and *t* alleles in the three populations. The fragmentation event occurs in the third generation. Thereafter, genetic drift causes allele frequencies in the reduced-size (population C) and reduced-connectivity (population B) populations to diverge from each other and from population A, fixing the *T* allele in population B and the *t* allele in population C. The time for divergence could range from just a few to hundreds of generations depending on the severity of the bottleneck and the initial frequencies of the respective alleles. Allele frequencies in population A remain relatively stable due to its large size. Modified from Mills and Tallmon (1999). Reproduced by permission of Brill Academic Publishers.

relevant to human-caused fragmentation? Outcomes will vary widely depending on the specifics, but reductions in genetic variation certainly can occur. For example, reduction in geographic range and population size due to habitat fragmentation, disease, and/or overharvest has decreased genetic variation in species ranging from greater prairie-chickens (Bouzat et al. 1998a,b), to koalas (Houlden et al. 1996), wild turkeys (Leberg 1991), common frogs (Hitchings & Beebee 1997), geckos (Sarre 1995), and wombats (Taylor et al. 1994). Clearly, human-caused changes can decrease genetic variation over ecologically relevant time scales (Spielman et al. 2004).

Obviously, as with any process in nature, complexities and exceptions will arise from both ecological processes and limitations of measurement techniques. So we should not be surprised in cases where genetic variation is not decreased following population fragmentation. For example, Leung et al. (1993) found a reduction in heterozygosity on a true island compared to a control population for Australian fawn-footed mosaic-tailed rat but no similar loss on three forest fragments created about 70 years ago. Similarly, clearcuts surrounding forest fragments in Oregon contained very few California red-backed voles, implying ecological isolation, but heterozygosity or allelic diversity was not lower in 13 small remnants compared to five large control populations (Tallmon et al. 2002).

At least three reasons, often acting together, explain why a loss of genetic variation may not be detected in wildlife populations that have been fragmented. First, fragmentation may not lead to genetic isolation, as only a small level of gene flow is needed to maintain genetic variation (as we shall explore shortly). Second, sufficient time must pass to lead to a detectable change in genetic composition. After a drastic change in population size or connectivity, several to as many as hundreds of generations must elapse before measures of variation reflect those changes; consequently, recently-formed habitat fragments are less likely to show decreased levels of genetic variation compared to fragments or islands isolated for hundreds or thousands of years. Finally, fragmentation may not lead to detectable changes in genetic variation because populations are often naturally fragmented across a landscape. Historical barriers to dispersal, such as rivers and mountain ranges, contribute to statistical noise that can obscure the signal from recent changes due to habitat fragmentation. For example, Cunningham and Moritz (1998) found that the effects of recent forest clearing had less effect on genetic variation in the prickly skink than did historical effects of glacial retreat and expansion. For all of these reasons, a signal may not be detected from genetic analysis, even if demographically important changes are occurring due to isolation and decreased abundance (Gaines et al. 1997).

Quantifying the loss of heterozygosity: the inbreeding coefficient

Theory predicts and empirical evidence supports that genetic variation within a population can be lost due to genetic drift following fragmentation, at a rate inversely related to the population size. Now is a good time to quantify this loss with the often-used term, **inbreeding**.

Defining inbreeding

Inbreeding refers to the loss in heterozygosity arising from mating of individuals related by ancestry. For humans, what comes to mind is incest (or consanguineous unions) among close relatives. Although social taboos tend to limit such preferential mating among humans, striking exceptions exist, as for example perhaps one in six marriages in Roman Egypt were between full siblings, with the female spouse described as "my wife and sister of the same father and the same mother" (Aoki 2005:14). Although preferential mating between relatives is the form of inbreeding that can happen in large populations, we have seen that in small populations heterozygosity can also be lost due to genetic drift. In the case of genetic drift, mating is random but because all individuals are related in a small population, each mating inevitably occurs between relatives.

Wright (1969) developed **inbreeding coefficient** terminology to describe loss of heterozygosity arising from both nonrandom mating and genetic drift. The inbreeding coefficient (F) is subscripted with I (individuals), S (subpopulation), or T (total population). F_{IS} relates to inbreeding in individuals relative to the subpopulation to which they belong, quantifying the reduction in heterozygosity of individuals due to preferential (nonrandom) mating with relatives. F_{ST} quantifies an inbreeding effect arising in subpopulations relative to the total population of which they are a part, the reduction in heterozygosity of a subpopulation due to genetic drift with random mating in a finite population.

Putting these two forms of inbreeding together, the overall inbreeding coefficient of an individual (denoted F_{IT}) includes the contributions from both F_{IS} and F_{ST}. Mathematically, the probability that an individual is not inbred relative to the total population ($1-F_{IT}$) is a product of the probability that it is not inbred due to nonrandom mating of individuals within subpopulations ($1-F_{IS}$) and the probability that the subpopulation has not lost heterozygosity due to genetic drift ($1-F_{ST}$; Wright 1969:486):

$$(1-F_{IT})=(1-F_{IS})*(1-F_{ST}) \tag{9.1a}$$

which can be rearranged to

$$F_{IT}=F_{IS}+F_{ST}-(F_{IS}*F_{ST}) \tag{9.1b}$$

For studies of effects of fragmentation on wildlife populations, F_{IS} is typically assumed to be zero, because preferential mating between relatives within subpopulations is generally avoided in wild populations (Ralls et al. 1986, Hoogland 1995:354)[5]. As just one example, skinks in Australia actively avoided mating with close kin, even after habitat

---

[5] In molecular studies of wild populations the F_{IS} is estimated as the deviation from Hardy–Weinberg equilibrium. If it is greater than zero it may indicate preferential mating among relatives (inbreeding), while if less than zero it may indicate heterozygote advantage or avoidance of inbreeding.

fragmentation caused higher relatedness between potential mates (Stow & Sunnucks 2004).

Putting zeros into eqn. 9.1(b) for F_{IS} gives $F_{IT}=F_{ST}$, making the overall inbreeding coefficient (F_{IT}) tantamount to the loss of diversity due to drift (F_{ST}). Therefore, in most free-ranging wildlife populations, the inbreeding coefficient (F_{ST}) is often called the **fixation index** (remember, fixation is when drift causes one allele to attain a frequency of 1.0 while all others at that locus are lost), emphasizing that in fragmented populations the loss of genetic variation comes from drift, not from preferential mating among relatives. When F_{ST} is near zero, populations have similar allele frequencies, implying little differentiation among populations due to drift. As F_{ST} increases toward 1.0, genetic drift causes heterozygotes to be lost within populations and different populations become fixed for different alleles.

Estimating the inbreeding coefficient in wildlife populations

The inbreeding coefficient can be calculated in three primary ways. One way commonly used by breeders and zoo managers is to track the matings of an individual's ancestors, thereby determining the extent to which an individual shares identical genes inherited from different ancestors (e.g. brothers and sisters share on average half of their genes, as do parents and offspring, so mating between relatives will quickly result in offspring that have identical alleles at many loci). In this case, the overall inbreeding coefficient (F_{IT}) is estimated directly. Although it can be difficult to determine pedigrees under field conditions, this approach can sometimes be successful in wild populations (Haig & Ballou 2002).

If F_{ST} approximates F_{IT} in wild populations, another way to measure the inbreeding coefficient (fixation index) is to determine the proportional reduction in heterozygosity in subpopulations due to genetic drift (see Raybould et al. 2001). Specifically, H_S is the average expected heterozygosity within subpopulations at Hardy–Weinberg equilibrium (Chapter 3) and H_T is the expected proportion of heterozygotes if the subpopulations were pooled and mated randomly[6]. Then F_{ST} is

$$F_{ST}=1-(H_S/H_T) \tag{9.2}$$

Think of this as the reduction in heterozygosity relative to a large, outbred population. For example, if a target population (occupying a fragment of interest) has a heterozygosity of 0.4 ($H_S=0.4$) while the expected heterozygosity in a large unfragmented population is 0.6 ($H_T=0.6$), then there is a 33% reduction in heterozygosity ($F_{ST}=0.33$).

The third way to estimate the inbreeding coefficient or fixation index (F_{ST}) is based on the loss of heterozygosity due to genetic drift when subpopulations are small. F_{ST} is expected to increase (and heterozygosity to decrease) each generation by an amount

---

[6]H_T is calculated from expected total heterozygosity at Hardy–Weinberg equilibrium based on mean allele frequencies of all populations.

equal to $1/(2N_e)$, where N_e is the effective population size. After t generations of genetic drift, then F_{ST} (or simply F) at time t is expected to be

$$F_t = 1 - \left(1 - \frac{1}{2N_e}\right)^t \qquad (9.3)$$

The effective population size

But what is this N_e, the effective population size? In essence, N_e formalizes the fact that individuals do not contribute equally to the gene pool[7]. The effective population size is the size of an ideal population that would lose heterozygosity due to drift (or increase its F_{ST}) at the same rate as the real population in question. An ideal population for gene transmission has constant size, and discrete generations in which individual reproductive success is random. Such beasts as the **ideal population** and N_e are needed because different species have different mating systems, making them vary widely in how efficiently and fairly they pass down their genes. How else could we compare genetic effective size in a monogamous species, where almost all parents contribute genes to the next generation, to something highly polygynous like elephant seals, where huge males weighing up to 4 tonnes exclude other males from mating in harems of over 100 females (Hoelzel et al. 1993)? N_e provides a standardized baseline against which we can compare the expected loss of genetic variation from real wildlife populations.

Obviously, real populations are not ideal gene transmitters, so N_e will virtually always be less than N, the total number of individuals in the population (see Box 9.2 for an example). The three main factors causing the N_e/N ratio to be less than 1.0 are uneven breeding sex ratio, fluctuations in population size over time, and variance in family size (or reproductive success) due to factors including age or social structure. As a rule of thumb, the ratio of N_e to N for wildlife populations is roughly 0.2–0.3 (Frankham 1995, Kalinowski & Waples 2002, Waples 2002).

The N_e is sometimes estimated by back-calculating from an estimated abundance (N) and assuming an N_e/N ratio of 0.2 or 0.3. Effective population size can also be estimated directly using demographic equations and simulations (e.g. Harris & Allendorf 1989, Kelly 2001), or with genetic measures (see Schwartz et al. 1998). For example, grizzly bears in Yellowstone National Park were estimated to have an N_e of about 80 across the 20th century and an approximate N_e of 100 currently, with an N_e/N ratio of about 0.3 (Miller & Waits 2003).

When does inbreeding lead to inbreeding depression?

So far I have discussed genetic variation – how it is lost and how to measure its loss – and mentioned some long-term consequences of that loss to a population's ability to

---

[7]Once again, a nondemocratic phenomenon in an ecological process!

Box 9.2 Estimating N_e and the N_e/N ratio

Consider just two of the factors that can depress the ratio of effective population size (N_e) to head count (N). First, a skewed sex ratio will depress N_e:

$$N_e = 4N_{ef} * N_{em}/(N_{ef} + N_{em})$$

where N_{ef} and N_{em} are the effective number of females and males (approximated by the numbers of breeders). As an exercise for yourself, try plugging into this formula 100 animals with an increasingly skewed sex ratio (from 50:50 to, say, 75:25 and then 90:10): the N_e will decline.

The second real-world factor considered in this example is the depression of the N_e/N ratio due to fluctuations in population size over time. The average N_e over time is not the familiar arithmetic mean, but rather the harmonic mean of the effective population sizes over t generations:

$$N_e = \frac{t}{\sum_{i=1}^{t}(1/N_{ei})}$$

An important property of the harmonic mean is that it is dominated by small numbers; the N_e over time will be close to the smallest N_e during the time series.

As an example, consider data from a small population of snakes (adder, *Vipera berus*) that has been isolated for at least a century due to the expansion of agricultural activities in southern Sweden (Madsen et al. 1996, 1999). The following table lists, for 7 consecutive years, the head count (N) and effective population size (N_e).

	Adult head count (N)			**Effective population size (N_e)**		
Year	**Female**	**Male**	**Total N**	**N_{ef}**	**N_{em}**	**Total N_e**
1984	13	25	38	9	13	21.27
1985	17	23	40	1	2	2.67
1986	11	23	34	5	13	14.44
1987	22	20	42	14	16	29.87
1988	17	20	37	6	12	16.00
1989	22	19	41	15	18	32.73
1990	17	17	34	4	10	11.43
Harmonic mean			–			9.9
Arithmetic mean			38			–

The N_e/N ratio incorporating the combined effects of both sex ratio and fluctuations over time is (Kalinowski & Waples 2002) (harmonic mean N_e)/(arithmetic mean N). For this example, this N_e/N ratio is 9.9/38.0=0.24. Hence, the genetically effective population size is about one-quarter that of the total number of individuals. Therefore, drift can have more dramatic impacts on this population's genetic variability than the head count alone might suggest.

adapt and survive in a dynamic world. But I have not yet fully answered the "so what?" question: does inbreeding due to population fragmentation and isolation lead to demographic consequences that affect wildlife populations over the short term? In other words, does inbreeding (loss of heterozygosity) lead to **inbreeding depression** manifested as a decrease in demographic vital rates?

Inbreeding depression for domestic animals and wild animals in zoos

Humans have long known that inbreeding in domestic animals can have negative demographic consequences whereby demographic rates are depressed (see the quote at the start of the chapter). Establishing the effects of inbreeding depression in non-domesticated animals has been a challenge, because both levels of inbreeding and changes in vital rates must be measured. The pioneering work of Kathy Ralls and colleagues (Ralls et al. 1988) showed that wildlife species in captivity often suffer inbreeding depression, with an average 33% reduction in juvenile survival across 38 normally outbred mammal species inbred in zoos to a level equivalent to parent–offspring or full-sibling matings.

Inbreeding depression for wild populations

But what about free-living wildlife populations? Compelling evidence has accumulated in recent years to demonstrate that inbreeding due to genetic drift in fragmented populations can lead to inbreeding depression in the wild (Hedrick & Kalinowski 2000, Keller & Waller 2002). Based on 169 estimates of fitness and inbreeding from 35 species in the wild, approximately 55% of the data-sets showed detectable inbreeding depression, with an average reduction in fitness of 27% for inbred birds and mammals compared to outbred ones (Crnokrak & Roff 1999; see also Reed & Frankham 2003). Table 9.1 summarizes three recent studies testing for inbreeding depression in wild terrestrial vertebrates.

   Importantly, inbreeding depression will tend to be greater in the wild – perhaps six times greater (Crnokrak & Roff 1999) – than measured in most captive studies, because subtle decrements in fitness due to inbreeding are more likely to be expressed in more stressful environments (see Bijlsma et al. 2000). To evaluate how inbreeding depression changed in a typical laboratory compared with more stressful environments, Meagher et al. (2000) trapped wild house mice and created lines that were either inbred (to $F=0.25$) or outbred. In the relatively stress-free conditions of the laboratory, inbreeding did not affect survival and had only small effects on reproductive success. By contrast, in seminatural enclosures that experimentally mimicked wild stressful conditions, inbred males exhibited lower survival than outbred males and had lower reproductive success because they tended to be defeated or die in intense mating-territory fights (Fig. 9.2). Similarly, Keller et al. (2002) found that for cactus finches inbreeding depression of adult survival was five times worse under poor environmental conditions. Although exceptions certainly exist, you can expect that inbreeding depression is apt to be worse as populations face increasingly stressful conditions.

Table 9.1 Three recent studies that detected inbreeding depression in wild terrestrial vertebrates. Other examples are cited in the text, and in Crnokrak and Roff (1999) and Keller and Waller (2002).

Species	Approach	Inbreeding depression	Reference
Golden lion tamarins	Monitoring was done in a forest fragment for 13 years; the study compared inbred and non-inbred survival.	Overall survival rate during the first 7 days of life was 95% (411/434) for non-inbred offspring compared with 75% (35/47) for inbred offspring.	Dietz et al. (2000)
Song sparrows	The study compared inbreeding levels of survivors and mortalities following severe winter kill; it also compared inbreeding levels with lifetime fitness for approximately 20 years of field data.	Survivors were consistently less inbred than those killed in catastrophic storms. Also, survival rate and lifetime reproductive success were reduced for inbred individuals.	Keller et al. (1994), Keller (1998)
Red-cockaded woodpeckers	In one of largest remaining populations in wild (more than 500 individuals), inbreeding based on pedigrees was analyzed across four generations (16 years).	Inbreeding increased nest failures and decreased mean annual number of yearlings produced per pair (44% fewer when $F \geq 0.125$); also, inbred birds were less able to adjust laying dates to changes in climate.	Daniels and Walters (2000), Schiegg et al. (2002)

    Most studies of inbreeding in both captivity and in the wild underestimate the total effect of inbreeding depression because they evaluate only one vital rate (juvenile survival is the most common). For example, Keller (1998) found little to no inbreeding depression on juvenile survival of song sparrows on Mandarte Island, British Columbia. However, after consolidating the effects of inbreeding depression on different component vital rates (survival from egg to breeding, adult survival, reproductive success, and maternal effects of inbreeding on hatching success), Keller (1998) estimated considerable costs of inbreeding: an egg with an inbreeding coefficient of 0.25 would experience a total loss of fitness of 79% over its lifetime (see also Meagher et al. 2000 for mice and Saccheri et al. 1998 for butterflies).

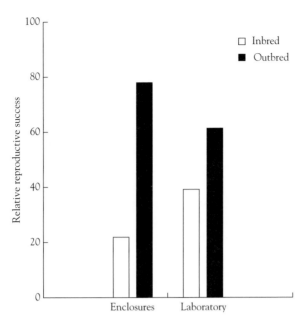

Fig. 9.2 Inbreeding depression in stressful seminatural enclosures is greater than that in the laboratory for male house mice (Meagher et al. 2000). In both enclosures and in the laboratory, inbred mice had lower relative reproductive success than outbred mice. However, the difference between fitness of inbred and outbred male mice (cost of inbreeding) is much less in the laboratory than it is in the semi-natural enclosures, where mice compete intensely for territories and mates.

Can wild populations adapt to inbreeding through purging?

The level of inbreeding depression in any particular population can be affected by its historical exposure to small population sizes and the extent to which it is adapted to inbreeding. Over the long term, natural selection will tend to remove harmful (deleterious) alleles in a process called **purging**, whereby animals carrying those alleles have lower reproduction or survival and so are less likely to transmit the alleles (Charlesworth & Charlesworth 1999, Bijlsma et. al. 2000, Keller & Waller 2002). Can we then expect purging to decrease inbreeding depression over time? Purging will be most effective in reducing inbreeding depression if the cause of the decreased vital rate is the expression of a few highly deleterious recessive alleles; that is, alleles that express their harmful phenotype when they become homozygous, and whose negative effects on fitness are bad enough that selection against them is strong (remember that selection overpowers drift only when $N_e * s$ is greater than 1). Because the primary mechanism of inbreeding depression is probably not highly deleterious alleles, but rather the cumulative expression of many homozygous alleles of small effect, purging often will be ineffective. For example, Ballou (1997) found that although purging could be detected for most mammalian species, it was very small in effect (see also Byers &

Waller 1999). Similarly, Bijlsma et al. (2000) found persistent inbreeding depression affecting extinction of experimental fruit fly populations even after 45 generations, and Eldridge et al. (1999) documented inbreeding depression in black-footed rock-wallabies after 1600 generations of isolation on an island. Purging may occur, however, when inbreeding is relatively slow, as found with fruit flies (Swindell & Bouzat 2006).

From a wildlife population perspective, even if purging does occur it carries a substantial price. First, the reduction of genetic variation could compromise future fitness via decreased adaptive potential. Second, the isolated population bears the cost of lowered demographic rates during the period that purging occurs; in other words, while purging sounds so pure and desirable, remember that bad genes can only be purged when animals die or fail to reproduce! In short, although we should expect inbreeding depression to be worse in recently isolated or contracted populations compared to historically small ones, a long period of being small or isolated is not necessarily an antidote to inbreeding depression.

Another genetic mechanism that could reduce vital rates: mutations in mtDNA

So far I have focused on how loss of nuclear variation due to genetic drift could decrease vital rates. However, mitochondrial mutations may also affect individual fitness and population viability (Gemmell & Allendorf 2001). Because mtDNA is transmitted maternally (Chapter 3), harmful mutations that affect only males will not be subject to natural selection. Furthermore, because mitochondria are haploid and derive from only one parent, the effective population size of the mitochondrial genome is only about one-quarter that of the nuclear genome, so that harmful alleles could more readily be fixed due to genetic drift. Mitochondrial mutations are known to reduce sperm function and male fertility, which could affect population growth rate directly and decrease N_e by increasing variability in male reproductive success (Rand 2001).

Inbreeding depression meets other concerns in fragmented populations

Although inbreeding depression can occur in the wild, its impact will vary widely across species and circumstances (Pray et al. 1994, Keller & Waller 2002). Also, a decrease in one vital rate due to inbreeding sometimes can be compensated for by an increase in another rate, as where egg mortality due to inbreeding is accompanied by higher survival of the inbred offspring (Van Noordwijk & Scharloo 1981).

Furthermore, small populations subject to inbreeding can simultaneously be adversely affected by weather, Allee effect, disease, or predation, so that in many cases inbreeding depression may be of less concern than other threats. As we will see in Chapter 12, population models provide a framework to incorporate the interaction between genetics and other factors to influence the persistence of wildlife populations. Instead of dichotomizing genetic versus nongenetic factors acting upon wildlife population persistence, or arguing that one or the other is universally more or less important, we should ask which vital rates (if any) are affected by inbreeding depression, how important those vital rates are to population growth (remember from Chapter 7 that all vital rates are not equal), and how the demographic costs due to inbreeding interact with other factors (Mills & Smouse 1994, Lacy 1997). Box 9.3 describes an integrated, long-term, and comprehensive case study of a natural population, incorporating analysis of genetic factors with the influence of demography, weather, and disease.

Box 9.3 The Soay sheep story

Soay sheep, a primitive domestic breed introduced to the island of Soay, off the west of Scotland, about 4000 years ago, were subsequently introduced to nearby Hirta Island in 1932. They have been studied continuously since 1985 (Paterson et al. 1998, Coltman et al. 1999, O'Brien 2000). Field data indicate that weather, density dependence, sex and age structure, parasites, and genetic variation intertwine to drive dynamics in this population. The nematode parasite load varies among individuals, but does not greatly affect fitness except during severe winters (such as 1989, 1992, and 1995; see figure panel a), when sheep with a high parasite load tend to starve and die.

But what drives individual parasite load? Lambs and yearlings are more susceptible, as are males. Also, adults in more dense populations have nearly double the parasite load than the same populations at low density following a winter with high mortality. The interesting story is made even more fascinating by the link to genetic variation. More homozygous individuals tend to have a greater parasite load, contributing to the lower survival in harsh winters (figure panel b). Clearly, it is folly to treat genetic, demographic, or environmental factors in isolation if we want to understand and manage populations.

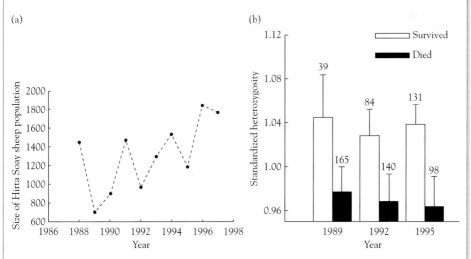

Soay sheep on the island of Hirta. (a) Population size over time. (b) Mean individual heterozygosity is higher for sheep that survived each of three severe winters (white columns) compared with those that died (black columns). Heterozygosity exceeds 1.0 because it has been standardized. The numbers above the SE bars are sample sizes. Modified from Coltman et al. (1999). Reproduced by the permission of the Society for the Study of Evolution.

What to do when faced with inbreeding depression?

If we accept the weight of evidence that inbreeding depression could be a concern in wild populations, then we must consider what should be done in those cases. I will not consider management of genetic variation in zoo or other captive populations, where matings can be manipulated (Oyler-McCance & Leberg 2005). In wild populations, the obvious best strategy when confronted with inbreeding depression is to facilitate an increase in population size while minimizing the impact of outside factors that could make the population decline (e.g. predation, climatic factors, or disease). Unfortunately, such a solution is often just a platitude, an empty suggestion that is true enough but impossible to implement.

This brings us to the only practical alternative to dealing with inbreeding depression, at least in the short-term until the deterministic factors causing the decline can be reversed: break the inbreeding by bringing new individuals into the population. Increased connectivity or gene flow can be achieved by approaches ranging from managing the matrix habitat, to construction of movement corridors, to physical movement of animals from place to place. Given the potential benefit of gene flow to small isolated populations, we must first consider potential disadvantages of imposing immigrants on a population suffering from inbreeding depression, and then decide how to balance the pros and cons.

Outbreeding depression and other potential disadvantages of gene flow

From a nongenetic point of view, the downsides to imposing gene flow include the possibility of spreading contagious diseases (Cunningham 1996, Miller et al. 1999) and the disruption of behaviors or social structure (Frankel & Soulé 1981). From a genetic standpoint, **outbreeding depression** is a potential concern if fitness is reduced following matings of individuals from different populations (see Dudash & Fenster 2000). Two related mechanisms can cause outbreeding depression. First, populations may be locally adapted to different environmental conditions, so that the offspring of parents from different populations are not well adapted to either location. Second, if different populations evolve different **coadapted gene complexes**, or sets of genes that occur and interact well together, then outbreeding can disrupt these gene complexes and decrease fitness (Edmands 1999).

Examples of outbreeding depression in mammals and birds are relatively few, especially when compared to the opposing force of inbreeding depression (Frankham et al. 2002). One of the best examples of outbreeding depression in vertebrates comes from two subspecies of largemouth bass in the USA: northern and Florida (Hallerman 2003). Experimental transplants of Florida largemouth bass into the northern range resulted in lower overwinter survival and reproductive success for the Florida bass compared to the pure northern bass, with hybrids having intermediate vital rates. One mechanism for the outbreeding depression appears to be inability of the Florida and hybrid bass to shunt energy from somatic growth to storage reserves for use during cold northern winters.

Appropriate levels of connectivity

Balanced against the potential demographic and genetic pitfalls of imposing immigrants are the compelling genetic benefits to outbreeding when a population is suffering from inbreeding depression (we'll also discuss the demographic benefits of connectivity in Chapter 10). From a genetic perspective the question becomes whether immigrants will lead to **genetic rescue** where population fitness increases by more than can be attributed to the demographic contribution of the immigrants (Tallmon et al. 2004). No universal answer exists to that question, nor should there be; this is a case where the biologist realizes that appropriate levels of connectivity will be situation-specific, depending on the relative importance of all of these genetic factors, coupled with other demographic factors (such as reproductive values of immigrants), the behavioral context of mating and survival, and the environment.

That said, one general rule does account for the pros and cons of connectivity on genetic variation. We have already mentioned the interplay between genetic drift and gene flow in determining the expected patterns of genetic divergence among a set of subpopulations. Too little gene flow will lead to a loss of heterozygosity within small populations and increase in among-population divergence due to inbreeding and genetic drift. Too much gene flow will homogenize gene frequencies among populations, swamping the ability of populations to adapt to local environmental conditions. Following the initial work of Wright (1931), the **one-migrant-per-generation rule** (the OMPG rule) has emerged as the level of gene flow sufficient to prevent the loss of alleles and minimize loss of heterozygosity within subpopulations (Fig. 9.3), while still allowing divergence in allele frequencies to occur among subpopulations. So one breeding individual (migrant) entering a population each generation achieves the balance between genetic tradeoffs, regardless of the population size[8]. A battery of assumptions underlie the OMPG rule, but it has stood up surprisingly well under simulation and empirical studies (Spielman & Frankham 1992, Hedrick 1995). Taking into account some of the real-world considerations for applying the OMPG rule to wild populations, and noting that many other factors could warrant higher or lower levels of connectivity, a minimum of one and a maximum of 10 migrants per generation would be an appropriate level of connectivity for genetic purposes, although more migrants may be necessary if populations fluctuate greatly in size over time (Mills & Allendorf 1996, Vucetich & Waite 2000).

Four case studies

In closing we will visit four applied projects that have initiated genetic rescue as a practical management solution to inbreeding depression.

---

[8]Here is the intuitive reason for why the OMPG rule is independent of population size: small populations lose variation rapidly due to genetic drift, but a single migrant counteracts drift because it makes up a larger proportion of the population. A single migrant makes a smaller proportional contribution to a larger population, but such a population also loses variation more slowly.

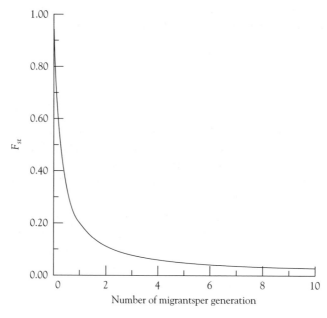

Fig. 9.3 Relationship between the inbreeding coefficient due to drift (F_{ST}) and the number of migrants (breeding individuals from outside a local subpopulation) per generation. One migrant per generation leads to a minimal loss of heterozygosity or alleles within subpopulations, while still allowing for local adaptation among subpopulations. Accounting for real-world complications (e.g. due to mating potential, social structure, relatedness of migrants, stability of population trend, and so on), one ideal migrant could translate to up to 10 or more actual individuals. Modified from Mills and Allendorf (1996). Reproduced by permission of Blackwell Publishing Ltd.

Greater prairie-chicken

As native grasslands were increasingly fragmented, the Illinois population of the greater prairie-chicken became isolated from other populations and declined from about 2000 individuals in the early 1960s to fewer than 50 birds by the early 1990s. Three observations linked the decline to inbreeding depression due to genetic drift. First, the decline occurred despite intense and somewhat successful efforts to control predators and increase the quality and quantity of habitat. Second, the decline was accompanied by a decrease in genetic variation for the persisting Illinois birds compared to both the still-large populations in neighboring states and to historical samples collected from the Illinois area before the demographic contraction. Third, translocations of prairie-chickens from the neighboring states since 1992 increased the low egg fertility and hatching success. The prospect that inbreeding depressed hatching success, and gene flow restored it, is made more striking by the fact that demographic sensitivity analysis for this species has shown hatching success to be the life stage that surpasses all others in its impact on population growth rate. Thus, this example not only points to gene flow as an appropriate tool to reverse inbreeding depression, it also underscores the importance of using our knowledge that not all vital rates are equal to indicate whether inbreeding depression for a certain vital rate is likely to have a substantial population-level effect (from Bouzat et al. 1998a,b, Soulé & Mills 1998, Westemeier et al. 1998).

Adder

Along the Swedish south coast, a population of snakes has been isolated from other populations for at least a century. Approximately 35 years ago, the population was greatly reduced (fewer than 45 adults) due to human development and the destruction of hibernation sites. Compared to other nonisolated Swedish populations, the isolated population showed a high proportion of deformed or stillborn offspring and very low genetic variation. In 1992, 20 adult males from another population were brought into the population, which by then was reduced to only 10 males. Not only did genetic variation increase, but the proportion of stillborn offspring fell. Importantly, after enough time had passed for outbred offspring to reproduce, population growth responded strongly, with a rapid and dramatic increase in the number of adders (Fig. 9.4; from Madsen et al. 1996, 1999).

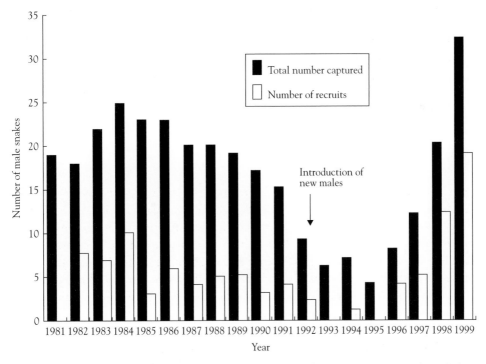

Fig. 9.4 The introduction of new male snakes led to not only an increase in genetic variation, but also a reversal of population decline in adders. Bars show the annual total number of males (females remain in burrows and so are hard to count, while virtually all males can be counted, making this a true census of males) and the number of newly recruited males (males introduced in 1992 are not included). Modified from Madsen et al. (1999). Reproduced by permission of Nature Publishing Group.

Grey wolf

Following the human persecution typical for wolves worldwide, wolves on the Scandianvian peninsula were extinct by the 1960s. Fortunately and surprisingly, in 1983 a breeding pack was discovered more than 900 km from the nearest extant wolves in Finland and Russia.

Historical genetic variation could be evaluated based on tooth samples of 30 museum specimens obtained over the 100-year period before the extinction of the 1960s. The historical samples, coupled with continuous monitoring and the nearly complete sampling of the newly founded population using various genetic measures (maternally inherited mtDNA, paternally inherited Y-chromosome markers, and microsatellites) facilitated two important insights. First, the Scandinavian population was founded by one male and one female who came from the eastern (Finland/Russia) population; they were not survivors from the exterminated Scandinavian population. Second, the heterozygosity of the re-established population was greatly increased in 1991 by a single male immigrant from the eastern population (heterozygosity went from a mean of 0.49 in 1985–90 to 0.62 in 1991–5, a level close to the heterozygosity of both the historical population and the extant eastern population). Not only did the heterozygosity increase (and inbreeding coefficient decrease) with the arrival of the immigrant, population growth (Fig. 9.5) increased dramatically from fewer than 10 wolves through the 1980s to 90–100 individuals by around 2000 (an annual growth rate of $\lambda = 1.29$ following the male's arrival). Thus, a population founded by two wolves lost variation and limped along with fewer than 10 individuals for about a decade,

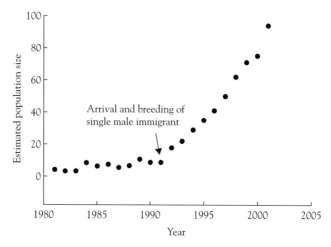

Fig. 9.5 Introduction of genes from a single male wolf is associated with a drastic increase in a small isolated wolf population (data from Vilà et al. 2003; reproduced by permission of The Royal Society). The population was isolated until a single male immigrant arrived in 1991; his genes were incorporated rapidly into the population and numbers increased drastically.

until the infusion of new genetic material from a single immigrant led to rapid population increase (from Vilà et al. 2003)[9].

Florida panther

As a subspecies of the widespread mountain lion (cougar, puma), Florida panthers were severely impacted by habitat fragmentation and unregulated killing until the mid-1960s. By the 1980s and 1990s, they had declined to about 60 or fewer individuals and were suffering from inbreeding effects on fitness, mainly in males (reduced sperm viability, increased male sterility, and undescended testicles); they also showed increased frequencies of heart defects and kinked tails. Their genetic variation was much lower than other North American mountain lions, and much lower than that found in historical samples from around 1900. Because of concerns over inbreeding depression, a carefully considered program initiated in 1995 led to the release of eight females from the closest natural population (*Felis concolor stanleyana* in Texas), with which the Florida panthers probably interbred historically. As of 2001, the Texas females and their offspring had successfully outcrossed with Florida panthers, with approximately half of the living population related to the Texas transplants. Fitness attributes have not yet been assessed. Maehr and Lacy (2002) note two concerns of the translocation that are generally relevant to the use of gene flow to reverse inbreeding depression. First, the genetic contributions from Texas animals may swamp the adaptive Florida panther alleles. Second, managers should not lose sight of the fact that a genetic restoration plan should only be one step of restoration, with emphasis on conservation and habitat restoration to facilitate population expansion (from Hedrick & Kalinowski 2000, Land & Lacy 2000, Maehr & Lacy 2002).

General rules

We have covered a lot of material at the interface of genetics and demography in this chapter. In closing I will extract a handful of general rules for considering and acting on the loss of genetic variation for a wildlife population.

- Focus more on the loss of genetic variation (heterozygosity and allelic diversity) than on the current levels. Some evidence links standing levels of heterozygosity to fitness (Allendorf & Leary 1986, Rhodes & Smith 1992, Reed & Frankham 2003), but it is not consistent (Britten 1996). Instead, the reduction in genetic variation should be of primary concern.
- Be more concerned about loss of genetic variation if it happens relatively quickly. Although the ability of populations to adapt to inbreeding through purging is uncer-

---

[9]The failure to breed prior to the male wolf's arrival may have been behavioral avoidance of inbreeding rather than inbreeding depression per se. From a practical perspective, the effect on population growth is the same: even if the male initiated **behavioral rescue** and not **genetic rescue**, heterozygosity of the population increased and the population size grew.

tain, we would expect the worst cases of inbreeding depression to occur in large populations that become small relatively quickly.

- Populations with low growth rates will tend to be more susceptible to effects of inbreeding depression. Populations with high growth rates will be better able to rebound to larger sizes after crashes or translocations, accumulating less inbreeding and restoring genetic variation via mutation.

- Remember, all vital rates are not created equal. For example, if inbreeding depression affects a vital rate with little effect on population growth rate, then you should worry less than if the depression affects a rate with high impact.

- If inbreeding depression occurs in an isolated population, consider ways of implementing gene flow to break the inbreeding. Although many factors must be considered, between one and 10 immigrants per generation is a reasonable starting point to balance genetic considerations over the short term while working to reverse the causes of population decline.

Summary

Genetic variation, and its effects on wildlife populations, is as real and important to understand as other factors such as predation or introduced species. In recent years, our ability to measure and interpret changes in genetic variation has blossomed. Loss of genetic variation can compromise the long-term ability of individuals to adapt to change. Also, inbreeding due to genetic drift in a small population can lead to a reduction in demographic vital rates and population growth. Because inbreeding and other genetic factors will interact with other factors to affect the short and long-term persistence of wildlife populations, an applied wildlife population biologist must understand the basic forces that affect genetic variation: genetic drift, gene flow, natural selection, and mutation. Armed with this knowledge, we can predict when we should be most concerned about the loss of genetic variation, and what to do about it.

Further reading

Allendorf, F.W. and Luikart, G. (2006) *Conservation and the Genetics of Populations*. Blackwell Publishing, Cambridge, MA. A detailed description of concept, tools, and application of conservation genetics.

Frankel, O.H. and Soulé, M.E. (1981) *Conservation and Evolution*. Cambridge University Press, Cambridge. One of the key early works articulating a link between genetics and the dynamics of plant and animal populations.

Frankham, R., Ballou, J.D., and Briscoe, D.A. (2002) *Introduction to Conservation Genetics*. Cambridge University Press, Cambridge. A comprehensive textbook with many excellent examples.

Schonewald-Cox, C.M., Chambers, S.M., MacBryde, B., and Thomas, W.L. (1983) *Genetics and Conservation: a Reference for Managing Wild Animal and Plant Populations*. Benjamin/Cummings, Menlo Park, CA. An early classic of papers assembled to focus genetic insights on practical management problems.

10

Dynamics of multiple populations

If it's not the Concho water snake, it's the muriqui. If it's not the muriqui, it's the Florida panther. If it's not the Florida panther, it's the eastern barred bandicoot in Australia, or the tiger in Asia, or the cheetah in Africa, or the indri in Madagascar, or the northern spotted owl in the Pacific Northwest, or the black-footed ferret in Wyoming, or the Bay checkerspot butterfly in California. Or it's the grizzly bear, which in the contiguous United States is now confined to a half-dozen islands of montane forest . . . The pattern is widespread. All over the planet, the distributional maps of imperiled species are patchy. The patches are winking . . .

David Quammen (1996:602), *The Song of the Dodo*

Introduction

Juvenile Columbia spotted frogs in Montana don't sit idly in ponds: up to 62% of them move among ponds each year, cruising up to 5 km and gaining 750 m in elevation on their mountain jaunts (Fig. 10.1). White-tailed deer, never known for being shy about their movements, affect disease spread by dispersing further when there is less forest cover (Long et al. 2005). Recovery of threatened Canada lynx in the USA may depend as much on maintaining movement from Canada into the USA as it does on managing existing populations (Schwartz et al. 2002). Across the world, in Australia, the loss of native vegetation affects dispersal and persistence of blue-breasted fairy-wrens (Brooker & Brooker 2002, Smith & Hellman 2002).

So, it appears that animals often move between populations and that this movement is important. Having spent a large part of the book focusing mostly on single populations, we will now delve into multiple populations across a landscape. Certainly, in some cases only a single population exists because the species is naturally endemic or close to the brink of extinction. But in most cases wildlife species occur across the landscape in multiple populations, and connectivity among them becomes as important as dynamics within the individual populations. Here are just a few of the many applications underscoring the importance of the population biology of multiple populations.

Fig. 10.1 Movements of juvenile Columbia spotted frogs from low-elevation ponds to a high-elevation lake in Montana. The inset shows a juvenile Columbia spotted frog (approximately 25 mm in total length). The dotted lines show movements of multiple frogs: (a) elevation gain of 770 m over a horizontal distance of 4240 m (18° mean incline); (b) elevation gain of 760 m over 4620 m (16° incline); (c) elevation gain of 700 m over 1930 m (36° mean incline). From Funk et al. (2005). Reproduced by permission of The Royal Society.

- If a wildlife population goes from being continuous to being broken into several units with little to no connectivity, what might happen?
- For a growing population, is the increase driven by inherently good conditions, or merely due to immigrant subsidies from other populations?
- Alternatively, is a population decreasing due to poor local conditions, or because it is a remarkable exporter, thereby supporting other populations in the landscape?
- If you are reintroducing a species to the wild, are you better off establishing one or several populations; if several, how many, and how far apart?

All of these questions require us to explicitly consider dynamics among populations as well as processes within populations.

In this chapter I will first define what connectivity means for wildlife population dynamics, and how to measure it. Next I'll review a range of ways that populations may respond to changes in connectivity, including the creation of multiple isolated populations, metapopulations, and sources or sinks. Finally, I'll discuss the primary ways of maintaining or restoring connectivity, including corridors, managing the intervening matrix, and translocations.

Connectivity among populations

What is connectivity?

So far I've used **connectivity** as a broad and vague term, but more precise terminology is needed to describe specific consequences for wildlife populations (Mills et al. 2003). The act of permanent movement of an individual away from the home area to another population is called **dispersal**. **Emigration** refers to dispersal out of, and **immigration** dispersal into, a target population. We care about emigration and immigration because they are key players in that little equation introduced in Chapter 4 (eqn. 4.1) that describes the processes affecting abundance in the next time step (N_{t+1}):

$$N_{t+1} = N_t + B + I - D - E \qquad (10.1)$$

where B and I are the number of animals that arrive due to birth or immigration, and D and E are those that leave by dying or emigrating. Of course, an animal could disperse to an area not currently occupied by the species, resulting in **colonization** if the species has never occupied the site and **recolonization** if it has.

The population genetics world uses the term **migration** (abbreviated to m) to refer to the proportion of individuals that move between populations, establish residence, and breed (Chapter 9). Migration in this sense is thus the same as **gene flow**, and is not to be confused with migration in the ecological context of seasonal movements across elevation or latitude. Notice that dispersal per se has immediate demographic effects in that an individual leaves one population and, if it survives, joins another, but unless it breeds (constituting gene flow or migration) it will not affect genetic structure or directly increase long-term population size.

Consequences of connectivity for wildlife populations

Here are four overlapping areas where connectivity matters for wildlife populations.

Persistence and fluctuations of populations

Jim Brown and Astrid Kodric-Brown (1977) coined the term **rescue effect** to describe the fact that genetic and demographic contributions of immigrants tend to increase abundance and fitness of extant populations. The genetic rescue comes from the breaking of inbreeding depression (remember the OMPG rule from Chapter 9). Demographic rescue refers to immigrants pushing the population size away from the zone where demographic stochasticity drives extinction probability (Chapter 5); dispersal among populations can also synchronize their dynamics and alter persistence by dampening fluctuations within populations. Alternatively, if connectivity for one species facilitates movements of its predators or diseases, the abundance or persistence of the target species may actually be decreased.

Colonization and recolonization of empty sites

Connectivity allows new sites never occupied to be colonized, permitting a response to changing environmental conditions such as global climate change. Also, sites occupied previously but now extinct can be recolonized, increasing the persistence of the entire suite of populations. Even for the relatively common and highly vagile scarlet tanager, a neotropical migrant scattered about eastern North America, increased distances between forest patches reduce the probability of recolonization of an empty patch (Hames et al. 2001).

Abundance of populations providing dispersers

If animals only dispersed when a population reached carrying capacity, then abundance might be unaffected as dispersers would merely be escaping negative density dependence. But dispersal often happens when the population is well below carrying capacity (**pre-saturation dispersal**; Lidicker 1975), so emigration can drain the population. We'll return to these ideas more formally later when we talk about sources and sinks. The concern about emigration being a drain on a population becomes particularly acute with translocations, where the cost to the donor population must be balanced against the benefit for the recipient.

Taxonomic designation

Because gene flow affects genetic differentiation and adaptive differences among populations across the landscape, some have argued that the degree of gene flow should be a fundamental criterion in deciding when populations should be considered distinct species or subject to special taxonomic designations or management actions (Crandall et al. 2000).

Measuring connectivity among wildlife populations

We can make some solid generalizations about dispersal for wildlife species (Van Vuren 1998, Clobert et al. 2001, Goudet et al. 2002). First, it occurs to some extent for virtually all species. There can be advantages to staying at home and not dispersing (**philopatry**), such as avoiding the terrible conditions, hostile residents, or lack of mates that may be encountered in the unfamiliar terrain of dispersal[1]. But the disadvantages of philopatry – including inbreeding, variability in the

---

[1]Dispersal is not always such a dramatic leap into the unknown, as dispersers often have a good knowledge of their travel routes and where they will settle thanks to previous exploratory forays and complex behavioral assessments (Van Vuren 1998). Over time, dispersal paths may become hard-wired.

environment, and competition for resources – can be even greater. Furthermore, mortality during dispersal may be compensated for by higher reproduction or survival once dispersers establish (e.g. yellow-bellied marmots; Van Vuren & Armitage 1994). At times the benefits of dispersal can be affected by how humans influence the dispersal landscape, such as when snowshoe hares show little mortality due to dispersal (Gillis & Krebs 2000) unless they disperse through open-canopy logged patches (Griffin 2004). In short, benefits of dispersal outweigh the costs often enough that dispersal has evolved as a universal trait for at least some individuals in all wildlife species.

We can also generalize about who disperses, and how far (Wolff 2003). Juveniles tend to disperse more often than adults. Dispersal distance tends to scale positively with body size, is greater for meat-eaters than herbivores or insectivores, and is affected by habitat preferences, social structure, and whether the organism evolved in a stable or dynamic landscape. In mammals, males tend to disperse further and at higher frequencies than females (i.e. male-biased dispersal tends to dominate), whereas in birds female-biased dispersal is more common. No trend is apparent for sex bias in dispersal of reptiles and amphibians.

Such generalizations can be useful. Carnivores tend to have relatively long dispersal with high human-caused mortality, leading to a general recommendation that reserve systems for carnivores should emphasize reduction in mortality sources among populations (Van Vuren 1998, Woodroffe & Ginsberg 1998). For species such as ground squirrels that might not have evolved strategies for widespread dispersal to isolated habitat, landscape fragmentation would likely have much more deleterious effects than for other species that are good dispersers. Likewise, the sex that disperses less might be of greatest concern if management alters connectivity (Wolff 2003).

Even though generalizations are useful, there will always be exceptions, and as the saying goes, the devil is in the details. For many (perhaps most) wildlife applications we need to know more than general expectations; rather, we need to know the likelihood of an individual to move, and how far it may go. Unfortunately, wide-ranging movements of wildlife through cryptic terrain have often defeated our ability to follow them, leading to chronic underestimates of dispersal rates and distances (Koenig et al. 1996). Most missed are long-distance dispersal events that are relatively infrequent but key to geographic range shifts and maintaining species persistence in heterogeneous and changing conditions (Nathan et al. 2003).

However, on the heels of technology things are changing fast. Next I'll give a brief overview of the main approaches to estimating connectivity with demographic and genetic methods. The demographic methods provide a direct estimate of movements over a well-defined time period. However, the intensity of work (and money) required to properly estimate movements among populations tend to restrict both the spatial and temporal scope. By contrast, most genetic measures integrate over longer time frames and can be obtained over larger sample areas, but are more fuzzy as to exactly who moved where, and when, and what their demographic contributions were. Given the benefits of combining insights from both demographic and genetic methods, I'll first shine a spotlight on each method individually and then give examples of how they may be combined.

Demographic methods: radiotelemetry and mark–recapture

Like anything that uses electronics, wildlife radiotelemetry equipment has improved enormously over the last 30 years, becoming smaller, lighter, longer-lasting, and more accurate (Fuller et al. 2005). Conventional transmitters can be implanted, glued on, or attached as radiocollars, backpacks, or ear tags. Transmitters using global-positioning satellite (GPS) technology provide locations and status around the clock, year round. Dispersal of telemetered animals is estimated using known fate models described in Chapter 4, with the potential to evaluate covariates that might affect dispersal (e.g. age, sex, or condition; Fig. 10.2). Telemetry can also show the specific pathway of movement.

After telemetry, the next most direct way to estimate connectivity among wildlife populations is to mark animals in different populations and estimate movements with subsequent captures. The complication is that we must statistically separate what we care about – movements from one population to another – from the possibility that an animal was not captured because it died or was present but not captured.

The basic statistical framework used to estimate movement with capture–recapture data is the Jolly–Seber model (Chapter 4) extended to geographic **multi-states** among which animals might move (Nichols & Coffman 1999, Kendall & Nichols 2004). In essence, apparent survival at time i (ϕ_i) is extended to include not only survival (S_i) but also the probability of moving from its home population r to another population (ψ_i). The movement probability ψ_i can be estimated directly from mark–recapture data sampled across multiple patch types or states, often using the robust design (Box 10.1). Hypotheses about directionality of movements can be tested by likelihood ratio tests

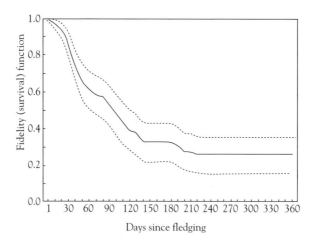

Days since fledging

Fig. 10.2 An example of estimating dispersal (1–fidelity) using Kaplan–Meier analysis of telemetry data. Data are based on 117 radiotransmittered juvenile snail kites followed for 3 years. The fidelity function can be thought of as the cumulative probability that an animal has not dispersed since the start of the study. Notice that by the end of their first year (365 days) post-fledging, 25% of juvenile kites are philopatric while 75% have dispersed from their natal wetland. The dotted lines show one SE around the estimate. Modified from Bennetts et al. (2001). By permission of Oxford University Press.

Box 10.1 Estimating movement rates from mark–recapture data and the robust design: an example with deer mice on clearcuts and forest fragments

As you saw in Box 4.6, deer mice in Oregon are known to be voracious seed predators, and a robust design was used to show that deer mice love clearcut logging: density was higher in forest fragments than on unfragmented control sites, and survival was highest in the clearcuts in the fragmented landscape. To quantify movement during the 20-day intervals, geographic states in the fragmented landscape included the clearcut, the fragment edge (all traps on the fragment within 30 m of the edge), and the fragment interior (fragment traps more than 30 m from the edge). Control trapping grids in large unfragmented sites had the same trap configuration so that there were control interior, edge, and periphery traps (50 m away from the grid, analogous to the clearcut traps) even though there was no fragmentation (Tallmon et al. 2003).

Using an information-theoretic approach (Chapter 2) and a candidate set of 15 models that constituted different hypotheses, the best-supported multi-state model indicated no variation in movement (or survival) for the three control states ($\hat{\psi}=0.24\pm0.07$ across all geographic states). This result is not surprising because on the controls there was no fragmentation, so you would not expect different movements between the traps on the grid edge, interior, or periphery. In the fragmented landscape, however, the best-supported model indicated that movement rates were greater between adjacent habitats than between distant habitats (see figure). For example, you can see that movement (ψ) from the clearcut to the fragment edge ($\hat{\psi}=0.14$) was much higher than from the clearcut to the fragment interior ($\hat{\psi}=0.02$). Emigration out of the fragment interior was the highest, with 41% ($\hat{\psi}=0.41$) of the animals in the interior moving to the edge and 17% moving to the clearcut. Thus, more than half of the mice captured in fragment interiors were recaptured in the surrounding edge or clearcut the next trapping session. In short, the high subsidized densities of mice in fragmented landscapes, coupled with high movements, have led to cascading effects on the ecosystem.

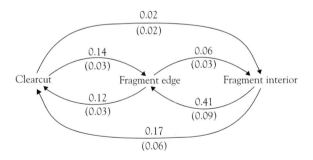

Deer mouse movement estimates ($\hat{\psi}$), with one SE in parentheses, among clearcut, fragment edge, and fragment interior habitats in the fragmented landscapes studied in southwestern Oregon. The movement rates represent the proportion of mice in a habitat type that moved per 20-day period. From Tallmon et al. (2003). Reproduced by permission of the ESA.

or by calculating information theoretic criteria (Chapter 2) for models with different movement predictions.

Genetic approaches

Given the remarkable new tools for obtaining genotypes, either from handled animals or noninvasively by collecting feces, hair, feathers, or other bits of tissue (Chapter 3), how can the genetic information be used to estimate connectivity?

Historical gene flow (equilibrium approaches)

The classic approach to estimating gene flow derives from the idea that smaller and more isolated populations lose heterozygosity within populations and accumulate more genetic differentiation among populations (Chapter 9). The among-population differentiation can be quantified by F_{ST} and its relatives[2] to measure population sub-division, ranging from near zero when populations have similar allele frequencies and toward one as isolation and genetic drift cause different populations to be fixed for different alleles. Formally, the number of migrants entering a subpopulation and breeding each generation (Nm) can be approximated by (Wright 1931, Mills & Allendorf 1996):

$$Nm \approx [1/(4\ F_{ST})] - (1/4) \tag{10.2}$$

The bigger the F_{ST}, the smaller the gene flow. These are equilibrium measures because they assume that the variance in gene frequencies represents an equilibrium between genetic drift increasing divergence and gene flow decreasing it. And that brings us to some major cautions for using these methods to quantify current gene-flow levels using calculations based on the number of migrants (Nm; Mills et al. 2003). First, the equilibrium conditions were recorded in the genetic structure dozens – or perhaps hundreds – of generations ago, so the measured F_{ST} is more likely to measure historic gene flow. Second, gene flow of all individuals in a population is assumed to be the same, without information on particular individuals. Third, important assumptions – including equal population sizes and constant, symmetrical gene flow over time – may be violated in real field cases. So, although F_{ST}-based measures do provide quantitative estimates of gene-flow levels, it is ill-advised to interpret them this way (that is, to distinguish four migrants per generation from five or six). Rather, they should be used to give qualitative or categorical assessments of historical gene flow, where, say, an Nm of less than 1.0 (corresponding to $F_{ST} > 0.2$) indicates low gene flow and an Nm of more than 10.0 (or $F_{ST} < 0.02$) represents high gene flow[3]. As an example of qualitative

---

[2]Some of the measures related to F_{ST} that you will see include θ, R_{ST}, and G_{ST}.

[3]If linked to other information, assessments of current gene flow levels are possible (see the section below on combining demographic and genetic approaches). For example, if abundance is determined to be small, with a substantial period of isolation, then a low F_{ST} value can be interpreted as being maintained by ongoing gene flow (Tallmon et al. 2002).

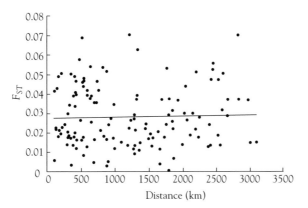

Fig. 10.3 Medium to high levels of gene flow among 17 populations of Canada lynx are indicated by very low F_{ST} values between pairs of populations ranging from nearly adjacent to more than 3000 km apart (from Schwartz et al. 2002; reproduced by permission of Nature Publishing Group). The flat line indicates no isolation by distance.

insights, Canada lynx in western North America have low pairwise F_{ST} values between pairs of populations separated by 3000 km or more, indicating medium to high gene flow across the continent (Fig. 10.3).

Recent, computationally intensive, **coalescent approaches** offer alternatives over traditional F_{ST} for many equilibrium assessments of long-term or historical gene flow (these are complicated and not intuitive, so I'll just refer you to a review by Pearse & Crandall 2004). But for estimates of current dispersal comparable to telemetry or mark–recapture estimates, you'll need to turn to nonequilibrium genetic approaches, preferably coupled with demographic information.

Assignment tests: a nonequilibrium approach to estimating current dispersal

Assignment tests do not assume genetic equilibrium and track individuals that move, thereby quantifying current gene flow more like dispersal using demographic methods (Manel et al. 2005). Here is how assignment tests work (Fig. 10.4): the genotype of each individual is compared to the genotypes of individuals from the population where they were captured and from other populations where they may have been born: an individual whose genotype is assigned with highest probability to a population other than where it was captured is deemed to be a disperser. Although offspring of dispersers will also carry the signal of parental genotypes, first-generation immigrants can be distinguished from offspring and relatives (Wilson & Rannala 2003, Paetkau et al. 2004). Thus dispersal rate is estimated by the proportion of individuals sampled whose genotype is assigned to a population other than the one from which it was captured or sampled. By focusing on genotypes of individuals, assignment methods provide higher resolution compared to equilibrium approaches, and allow for different populations to have different dispersal rates.

(a)

(b)

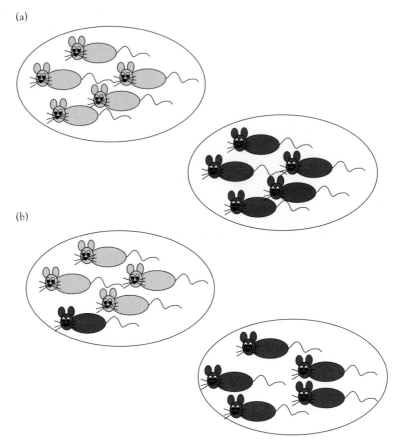

Fig. 10.4 A simplified version of how the assignment test works to identify dispersers. (a) Two populations that are strongly genetically differentiated (typically determined by allele frequencies at microsatellite loci). (b) An individual that has dispersed from the dark-gray to the light-gray population is detected as a disperser because its genotype is more similar to that of the dark-gray population, where it was most likely born.

The idea behind assignment tests has long been used in mixed-stock fisheries to estimate the population of origin of individual samples. In one of the first applications of the assignment test to estimate movement in a wildlife population, Paetkau et al. (1995) found that the assignment test was consistent with telemetry information for polar bears: most animals not assigned to where they were sampled were assigned to the nearby populations among which telemetry had shown movements. Thus the conventional wisdom that polar bears were nomadic was rejected in favor of the finding that they were relatively philopatric, with seasonal fidelity to particular areas.

In general, assignment tests improve as genetic differentiation increases (or as dispersal decreases), as the number of loci and number of individuals sampled increases, and as the variability of each locus increases (Berry et al. 2004, Paetkau et al. 2004, Piry et al. 2004)[4]. And of course, the software to carry out assignment methods is improving rapidly (e.g. Manel et al. 2002, Wilson & Rannala 2003, Piry et al. 2004).

Joint insights from equilibrium and nonequilibrium approaches

We have seen that equilibrium measures (F_{ST} and coalescent-based) assume that demography has been relatively constant, and measure long-term gene flow, whereas nonequilibrium approaches have the potential to measure current dispersal. The two approaches are complementary, telling us useful things about connectivity on different timescales. If we are interested in connectivity in a recently disturbed landscape, equilibrium gene-flow methods can look back into the period before the perturbation, providing a qualitative insight into historical gene flow based on the genetic signature arising from past differentiation. Historical connectivity can then be compared with current dispersal from assignment measures to see how it has changed. Box 10.2 shows an example of equilibrium and nonequilibrium techniques in action to assess movements of an endangered skink.

Measuring sex-biased dispersal with genetic approaches

It is also possible to use genetic markers to evaluate dispersal for different sexes. There are two main approaches (Prugnolle & de Meeus 2002). The first operates on the idea that the sex that disperses more should show greater genetic variation within – and less differentiation among – populations, as shown by a lower F_{ST} value. Similarly, a version of the assignment test corrected for different levels of genetic diversity among populations (called the **assignment index**) will be lower and more variable for the sex that disperses most, because both residents and immigrants are represented (Berry et al. 2005). For example, in greater white-toothed shrews, a monogamous species, females have lower and more variable probabilities of being assigned to the population of capture, indicating an unusual pattern (for mammals) of female-biased natal dispersal (Favre et al. 1997). Spong and Creel (2001) extended this approach for lions, using genetic similarity relative to distance among strongly philopatric females to derive estimates of actual dispersal distances for dispersing male lions.

The second type of approach to measuring sex-biased dispersal with genetic tools takes advantage of genetic markers that have different modes of inheritance in the different sexes, such as maternally inherited mtDNA or paternally inherited Y-linked markers (Chapter 3). For instance, mtDNA variation in red-backed voles was reduced in forest fragments relative to controls but nuclear DNA variation was not, implying that males dispersed among forest patches whereas females did not (Tallmon et al. 2002).

---

[4] As a rough rule of thumb, current assignment methods work best with more than nine or so highly variable loci and F_{ST} values of more than 0.05 (Berry et al. 2004).

Box 10.2 Estimating movement rates from genetic approaches: an example with grand skinks on rocks surrounded by vegetation

Grand skinks are a large, endangered territorial lizard endemic to southern New Zealand, found in patchy populations on rock outcrops surrounded by vegetation. Based on toe clips or tail tips from captured animals, genetic data were used to understand how agriculture was affecting skink movement. Equilibrium-based F_{ST} approaches indicated higher gene flow (F_{ST} around 0.05) in the two sites where populations were separated by native tussock compared to the two sites with pasture as the intervening matrix (F_{ST} around 0.1). This qualitative finding of higher gene flow in the native-tussock landscape compared to the pasture landscape was refined and extended with the assignment test, which identified the particular rocks that skinks moved from. Interestingly, every resident skink on each rock outcrop was marked, and the mark–recapture data indicated that the assignment tests were highly accurate at assigning a skink to its correct natal rock (accuracies of 100, 95, 79, and 65% at the four replicate sites); also, there was high concordance between the number of dispersers from mark–recapture and from the assignment test (where a disperser was defined as a skink assigned to a rock other than that where it was captured; Berry et al. 2004, 2005).

A grand skink on a rock outcrop. Photograph by James Reardon.

Combining demographic and genetic approaches

Using both demographic (mark–recapture, telemetry) and genetic approaches takes advantage of the direct and (relatively) easily interpretable insights into current dispersal from mark–recapture and telemetry studies, while also harnessing the strengths of genetic tools to quantify whether dispersal is accompanied by reproduction and to track dispersal in animals that are hard to capture or study over large spatial and temporal scales (Peacock & Ray 2001, Mills et al. 2003). Because equilibrium genetic methods address historic gene flow (dispersal and reproduction) while demographic-based and genetic assignment methods measure current dispersal with reproduction status unknown, combining these methods provides complementary insights.

Combining genetic and demographic information can also strengthen inference beyond what is possible for either method alone. For instance, most assignment measures for inferring current gene flow are Bayesian or partly Bayesian (Chapter 4), which means the predictions improve with the inclusion of prior information from mark–recapture records (Berry et al. 2005). Mark–recapture abundance estimates can also be coupled with assignment measures to estimate emigration rates in addition to the immigration rates that can be calculated from assignment measures alone (Wilson & Rannala 2003:1187).

Multiple populations are not all equal

For multiple populations on a landscape, the effect of human-caused perturbations depends on the response in both within-population vital rates and connectivity among populations. Understanding the different ways multiple populations interact has obvious applications, ranging from helping to decide how to prioritize conservation of particular populations to interpreting the role that a particular population (say in a national park or in a managed landscape matrix) plays in the overall persistence of the species.

This section will focus on multiple isolated populations, metapopulations, and source–sink dynamics, three potential outcomes when a continuous population becomes interspersed with an intervening matrix that decreases movement. Remember, however, that in nature not all disturbances decrease connectivity among patches, and not all changes in connectivity are problematic (Doak & Mills 1994, Mills 1996, Wiens 1996). A shrew and a raven will view a particular perturbation very differently, even if it looks like habitat islands in a hostile sea to us humans as we fly over it in an airplane.

Multiple isolated populations

The extreme situation for populations set in a hostile intervening matrix is for each to be isolated from the others. Each population experiences heightened susceptibility to extinction, without the chance of recolonization. Key to evaluating persistence of multiple isolated populations is the correlation in their dynamics caused by shared environmental controls (den Boer 1981), informally known as determining whether all of your eggs are in one basket. If isolated populations are correlated so they respond similarly to threats to persistence, then additional populations do little to reduce extinction probability of the collection of populations. Conversely, if the fates of populations are independent – perhaps because they are widely separated – then the likelihood of all populations going extinct at the same time becomes much smaller, the product of all the independent extinction probabilities. Consider how closely spaced or far apart populations would be if simultaneously affected by a hurricane, tsunami, or invasive novel predator, for instance. In more formal terms, a decoupling of environmental stochasticity among populations makes it less likely that a single bad year or series of years would cause extinction for the full set of populations (see Box 10.3). Although

Box 10.3 A simple example of how correlation in dynamics affects extinction probability for a set of multiple isolated populations

If three isolated populations had independent probabilities of extinction, as shown in panel a of the figure, then the probability of simultaneous extinction of all populations would be much less than that of any single population. By contrast, if the fates of the populations were completely correlated (shown in panel b by adjacency but still assuming isolated populations), the probability of total extinction would be the same as for one population.

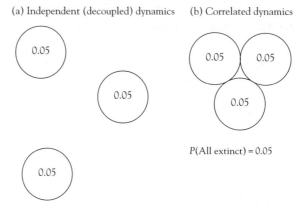

Probability of simultaneous extinction for a set of three isolated populations, each with an extinction probability of 0.05. (A) Assuming the dynamics of the 3 populations are independent, probability of extinction is 0.000125; (B) Assuming the 3 isolated populations have correlated dynamics, the probability of simultaneous extinction is 0.05, the same as for one population.

distance among populations often drives correlated dynamics, it does not always do so; for instance, pathways for deadly diseases may follow river bottoms that make distant bottomland populations more tightly linked than closer populations separated on a hillside.

Metapopulations

What if connectivity is altered, but not completely eliminated? The term **metapopulations** was coined by Richard Levins (1970:105) to refer to a "population of populations," where connectivity occurs and populations go extinct and are recolonized[5].

---

[5]Levins' original model is a form of the equilibrium logistic model we discussed in Chapter 6, except that births and deaths of individuals are replaced by recolonization and extinction of populations.

Although each population has its own dynamics, the long-term persistence and stability of the metapopulation depends on the turnover (changes in species identity) arising from population extinctions and recolonization. As an interesting aside, the metapopulation idea – which has gained so much recognition in the conservation realm for species in trouble – was originally formulated to figure out how to wipe out abundant insect pests in multiple populations across large areas!

The theory has matured with revision, expansion, and exposure to a battery of field and experimental tests (Hanski & Gaggiotti 2004). In current practice, the most important contribution of the metapopulation concept to applied wildlife population biology is the emphasis on both within-population processes as well as among-population movements. As noted by Susan Harrison (1994:117):

> It seems necessary to adopt a broader and vaguer view of metapopulations as sets of spatially distributed populations, among which dispersal and turnover are possible but do not necessarily occur. Such a definition leaves little hope for strong generalizations about the role or importance of metapopulations. A possible way forward is to ask, in each specific case, 'what is the relative importance of among-population processes, versus within-population ones, in the viability and conservation of this species?'

A firm scientific consensus now holds that movement among populations is a key component of metapopulation persistence, making the management of dispersal – including areas where the species is not currently found but which it might travel through – as important as managing extant populations.

In one of the first cases where connectivity was incorporated into metapopulation models for managing a species of concern, managers of northern spotted owls explicitly incorporated dispersal dynamics into planning documents after Lande (1988b) and a flurry of subsequent papers detailed how connectivity affected persistence in a fragmented landscape. Similarly, as urban sprawl gobbled up land in the Santa Ana mountains of southern California, threatening a cougar population in an area of 2070 km^2, Beier (1996) incorporated hard-won field data (both movements based on radiotelemetry and within-population vital rates) into a matrix model to show planners which specific combinations of patch size and connectivity would foster persistence. One very useful result of this work was that if habitat destruction was deemed inevitable, enhancing connectivity through corridor protection and highway underpass construction could help cougars to endure the habitat loss. In Australia, creation of new reserves far enough apart to make them uncorrelated with respect to fire was seen as an efficient strategy to increase the persistence of Leadbeater's possum while minimizing effects on the timber industry (Possingham et al. 2002; see also Box 14.5).

Sometimes the models that frame the empirical data for wildlife metapopulation questions are **spatially implicit**, meaning that habitat patches and local populations are assumed to be discrete and with equal connections to each other. Spatially implicit approaches simplify the math and the parameters that run the models, but at the cost of loss of realism. By contrast, **spatially explicit** or spatially realistic models give the populations locations on a grid with patch-specific connectivity, reflecting our knowledge that movements will occur more often between closer populations with a favorable intervening matrix (Taylor et al. 1993).

As implied in Harrison's "broader and vaguer view of metapopulations" (see above), the characteristics of metapopulations vary widely. Some will be dominated by turnover in species identities over time, whereas others – such as the mainland–island systems of **island biogeography** so illuminating for insights into species richness in fragmented landscapes – will have one or a few nearly immortal populations and many others that blink out often (Hanski & Simberloff 1997). Some will have a small amount of connectivity, and nearly independent fluctuations, whereas other metapopulations will have a lot of movement and coupled dynamics (Bjornstad et al. 1999). Knowing where in this spectrum a target metapopulation falls can greatly increase the efficiency and effectiveness of management efforts.

Source–sink populations

A form of metapopulation of special interest in applied wildlife population biology is captured by the metaphor of source and sink. The metaphor suggests that some populations are strong contributors (sources) to the metapopulation while others are drains on the system (sinks). Although the term sink was used to describe dispersal dynamics as early as 1975 (Lidicker 1975), Pulliam (1988) injected the term deep into ecological theory and practice. The crucial key to define sources and sinks for wildlife population biology lies in careful interpretation of population processes. You cannot use a population's relative abundance or density to define a source because abundances vary over different habitats in different seasons, and because poor habitats may have high numbers merely because subordinates are forced into them. Fundamentally, abundance alone says nothing about vital rates or population growth or how the individuals got there (Van Horne 1983). For the same reasons, you cannot assume that a large area is a source while a small area is a sink (Hanski 1996).

A reasonable alternative might be to define a sink as one with a λ value of less than 1 and a source as having a λ value of more than 1. However, within-population λ alone does not account for how much of a population's growth rate comes from immigrants, or how much a local population contributes to the metapopulation via emigration. Conceivably, a population could be a net contributor to the metapopulation even though it would have a negative growth rate without immigration. For example, most hatch-year California spotted owls disperse from natal areas, so high-quality habitats provide emigrants to other areas while depending on immigrants (Franklin et al. 2004). Alternatively, an increasing population might contribute nothing to the metapopulation and have its high growth just from immigrants. Headwater salamanders in upper reaches of first-order streams in New Hampshire have positive population growth rates only because downstream salamanders have high reproduction and preferentially move upstream (Lowe 2003).

So we need a criterion that identifies sources that are net contributors to the metapopulation and sinks that drain the metapopulation as a whole. Both within-population contributions (population growth not counting immigrants) as well as emigration to the rest of the metapopulation must be considered. Box 10.4 describes

---

Box 10.4 A criterion for distinguishing sources from sinks

Sources are distinguished from sinks in how they contribute to a metapopulation as a whole (from Runge et al. 2006). The per-capita contribution (C^r) of focal subpopulation r to the whole metapopulation is made up of survival (S^r) of both residents and emigrants from subpopulation r (with adults or juveniles subscripted with A or J), and reproductive rate (juveniles per adult; B^r):

$$C^r = S_A^r + B^r S_J^r$$

This formula captures the fact that for each individual in subpopulation r at time t there will be C^r individuals in the metapopulation at time $t+1$. Individuals in population r contribute to the metapopulation by emigrating to other populations (E^r) and through their own self-recruitment ($C^r - E^r$). The C^r metric gives us an operational criterion to define a local population as a source or a sink: if C^r is more than 1 the focal subpopulation contributes more individuals to the metapopulation than it loses to mortality, making it a source, while if C^r is less than 1 the focal population is a sink because it loses more animals to mortality than it contributes.

I noted above that usually we do not know true survival (S), because if a marked animal disappears it could have either died or left the area[1]. So for practical purposes the contribution metric can be defined in terms of apparent survival (ϕ^{rs}), the probability that an animal in subpopulation r in a year is alive and in either r or another subpopulation s the next year:

$$C^r = \phi_A^{rr} + \sum_{s \neq r} \phi_A^{rs} + B^r \left(\phi_J^{rr} + \sum_{s \neq r} \phi_J^{rs} \right)$$

The C^r metric can be extended to multiple stages or ages.

In the case of a single population with no connectivity, the growth rate of local population r (λ_{local}^r), represented by its change in abundance each time step, will equal its own contribution, $\lambda_{local}^r = C^r$. For multiple populations, the λ_{local}^r of each population depends on its own net contribution to itself ($C^r - E^r$), plus contributions to the population from immigration (I^r):

$$\lambda_{Local}^r = \frac{N_{t+1}^r}{N_t^r} = C^r - E^r + I^r$$

Meanwhile, the growth of the entire metapopulation is the average of the contributions from all populations, weighted by the relative abundance of each population.

[1] In fact, many modeling exercises based on vital rates from field studies are plagued by the fact that apparent survival incorporates all losses from the population, including emigration, but immigration into the population is not likewise accounted for (Nichols et al. 2000b, Franklin 2004).

such a framework, where demographic information is used to develop a practical criterion for distinguishing source and sink populations.

An example of source–sink dynamics can be found in Iberian lynx, a critically endangered species restricted to the Iberian Peninsula of southwestern Europe. One lynx metapopulation consisting of nine populations is self-sustaining only because the

---

Box 10.5 An example of an ecological trap for indigo buntings

Many forest bird species require disturbed areas. Historically, these areas would have occurred along or within natural forest openings or fire-maintained successional habitats, where forage was good and predators minimal. Edges created by humans via logging, or clearing for agricultural or urbanization purposes, mimic natural disturbances and attract many nesting birds. However, anthropogenic edges often harbor high numbers of predators and brood parasites (such as brown-headed cowbirds in the USA). The preference for edges that are filled with enemies sets the stage for an ecological trap.

Indigo buntings in South Carolina show strong preferences for edges, probably because the highly territorial males prefer the multitude of elevated perches for observation and territorial singing and females prefer to place their nests along edges (for reasons not really known). In a controlled field experiment comparing patches of the same size but different amounts of edge, indigo buntings not only preferred edges, but also suffered predation that reduced fledging success by 50% (Weldon & Haddad 2005). The trap is sprung because historically determined preference for edges evolved in a context without the high number of subsidized and introduced predators that accompany human-created edges (Chapter 11). Forest fragments in such a landscape would have less impact as ecological traps if their shape were as simple as possible, without convex corners that contain a lot of edge.

---

populations within a national park have high survival and send emigrants to populations outside the park; persistence of the metapopulation would be best served by restoring habitat at the within-park sources (as opposed to the sinks), and decreasing mortality due to illegal harvest in the sinks (Gaona et al. 1998).

Restoration of a population deemed to be a sink will depend on knowing what causes intrinsic vital rates to be low and why immigrants keep going there. Sometimes a sink habitat is preferentially chosen over better habitats because formerly reliable cues are mismatched with current fitness consequences in a landscape modified by humans (Box 10.5). Such **ecological traps** lead to a particularly insidious form of sink because the drain created when "good animals love bad habitats" (Battin 2004) can overwhelm even strong sources and cause the whole metapopulation to decline toward extinction. Because ecological traps are in essence an evolutionary lag between the quality of a patch and the cues that lead to its preference, leading to a dispersing animal making a mistake in where they settle, traps probably occur more often in recently modified environments[6]. Likewise, species that evolve or learn slowly, and that are

---

[6]Ecological traps can be thought of as a specific case of more general **evolutionary traps**, where any behavioral or life-history decision might be inappropriate because it is based on environmental cues that are maladaptive in a changed environment. To stress the generality of the evolutionary trap concept, Schlaepfer et al. (2002) note that our human craving of fatty foods is a relict of an evolutionary past where fat was limited in supply and quite necessary in the small amounts that could be obtained; now we consume massive amounts of fatty foods preferentially over healthier alternatives, despite the fact that the choice takes us down the road toward obesity, diabetes, and other problems.

severely affected by the trap, will suffer the most from ecological traps (Schlaepfer et al. 2002, Battin 2004).

As with any sink, the good news is that mitigating the mechanism causing the trap to be a sink will increase not only local, but also regional population growth. The negative effects of ecological traps can be ameliorated by either increasing the quality of the trap, so that it no longer serves as a sink, or by decreasing the trap's attractiveness. If a grassland bird of concern preferentially nests in fields with abundant fences or poles as perches, but mowing of the fields destroys nests, solutions would include adjusting mowing schedules to increase quality of the area or eliminating the perches or scaring off the birds to reduce attraction to the trap (Battin 2004).

Options for restoring connectivity

Having established that connectivity is essential to understanding wildlife distributed as multiple populations, the next question is what to do if that connectivity becomes severed and reconnections are desired. Although many options are possible, I will briefly cover two main areas: facilitating movement through corridors or a modified intervening matrix, and physically moving animals via translocations.

Corridors and managing the intervening matrix

When someone asks how to reconnect isolated fragments, the image that probably pops into most of our minds is that of a corridor, perhaps a line of trees winding through a clearcut between two intact forests, or a pleasant green strip through the heart of an urban area. Indeed, the corridor concept has popularized the importance of connectivity, not only for biologists but also land planners and the general public (Beier & Noss 1998). For that reason, corridors are useful for denoting large, regional connections to facilitate animal movements across the landscape (Dobson et al. 1999).

But sometimes **corridor** gets used as a simple mental crutch to substitute for hard thinking about the best way to impose connectivity. A linear patch of habitat will, in fact, act as a conduit for some species, funneling individuals from one population to another (Beier 1996). A corridor added between populations may also add habitat per se, which in itself can have obvious benefits; indeed, corridor-related increases in habitat may be more politically palatable to the general public than a more mundane call for increased reserve size. And a corridor might act like a giant drift fence, scooping up wanderers as they stumble across the modified matrix, a lifeline that directs lost souls back to the appropriate habitat. All of these benefits have been found for certain wildlife species under certain conditions (Rosenberg et al. 1997, Beier & Noss 1998, Haddad et al. 2003). In a powerful demonstration of corridors as conduits, Josh Tewksbury and colleagues (2002) controlled for patch size and drift-fence effects via experimental manipulations, and showed that movements through corridors were increased not only for two butterfly species, but also for animal-dispersed pollen and seeds, an important reminder of cascading effects from a beneficial management action.

So yes, corridors can work. But by now it should be an ecological first principle for you to realize that corridors would be a bad investment for certain species, corridor configurations, or intervening matrix types. The increased edge created by long linear corridors, and especially the corners created where the corridor joins the main patches, could create ecological traps (Weldon & Haddad 2005; see Box 10.4). If certain introduced or strongly interacting species prefer the corridors, they could change species composition through their interactions; for example, rodents eat more seeds of certain species in corridors compared to unconnected patches (Orrock & Damschen 2005). Likewise, disease transmission in or along a corridor can create havoc. And where the corridor goes makes a difference. A corridor in a hostile matrix may be more effective (Rosenberg et al. 1997) or less effective (Baum et al. 2004) at enhancing dispersal than one through a mellow intervening matrix. Also, locating corridors solely in creeks and gullies, a common practice because those areas are already set-aside or protected, should be resisted as a one-size-fits-all strategy (Claridge & Lindenmayer 1994); either a network of corridors should be implemented across topography or, even better, the best location for the species of concern should be decided upon (as where likely recolonization routes of black-tailed prairie dogs in Colorado were determined using genetic methods; Roach et al. 2001).

Thus the productive question about corridors must go beyond asking whether corridors work?, to be more specific. Is the intended role of a corridor to induce emigration from a fragment, or simply to direct movements? There is not yet much evidence that corridors actually do the first, but strong evidence that they do the second (Haddad et al. 2003). If augmented dispersal is the objective, then the question to ask is how much dispersal is desirable for a particular species in a particular place, and how best to achieve it. In some cases a linear corridor across the landscape, or a wildlife passage under a highway or railroad, may be exactly the best solution, improving dispersal and gene flow, despite potential costs. In other cases, a corridor is an expensive trap that takes away funding from the acquisition of larger reserves that does more harm than good for the species of concern. Sometimes a corridor will function better when we know more about the animals involved; Michael Reed (2002) notes that if we knew as much about wild animal behaviors as we do domestic ones (e.g. domestic sheep do not like walking in their own shadow), we could design better corridors.

An alternative, or perhaps complementary, way to increase connectivity among fragments is to manage the **intervening matrix**. Hardly a page in this chapter goes by without noting the importance of the quality of matrix in affecting connectivity. If the entire intervening matrix can be made less hostile and more permeable to movements of species of interest, then the connectivity payoffs will be much greater than any single corridor could be. Plus, there is so much disturbed matrix already out there! As Jerry Franklin (1993:205) noted:

> *Human activities can either produce very hostile conditions in the matrix – deep seas full of sharks, barren of food, lethal temperatures, etc. Or activities can be designed to enhance dispersion and in-place survival of organisms.*

Again, the movement behavior of individual species emerges as a key control on connectivity. We know that animals sometimes experience psychological isolation, failing to cross gaps, ecotones, or other features they are perfectly capable of traversing (e.g. Desrochers & Hannon 1997). Similarly, species whose dispersal depends on the presence of conspecifics (the phenomenon of **conspecific attraction**) may require more intensive management than just managing the matrix: nest boxes, song playback, or even white paint to simulate droppings may suffice (Reed et al. 2002).

Physically moving animals: translocations

If connectivity needs to be restored, but improved movement through corridors or a modified intervening matrix is not possible, a second choice might be physical translocation from one population to another. Translocations can occur as an augmentation (or supplementation) into an existing population, as a reintroduction into an area where the species existed previously but is now absent, or an introduction to an area not previously occupied by the species (in the next chapter I will beseech you to avoid introductions).

Establishing connectivity through translocations involves many of the ideas discussed so far in this chapter, with a few differences. One difference is more sociological and political than biological: a translocation puts the biologist into a goldfish bowl, with a lot of people watching. If you do not pay attention to public sentiment, the translocation is likely to fail, especially for potentially controversial and high-profile species such as carnivores and ungulates.

The primary biological differences between animals moving themselves and being translocated by humans is that with translocations one often has the luxury – and responsibility – of deciding where the migrants come from, where they go, the number to be released, their sex and age characteristics, and how they are released (Wolf et al. 1996). If multiple wild populations can be chosen from, one must first ensure that removal of animals will not negatively affect growth of the donor population. Next, genetic tools can help inform the decision of which donor population to use (see Chapters 3 and 9). If inbreeding depression is a concern in a supplemented population then individuals from a population with different genotypes might be most beneficial. In contrast, for populations highly adapted to their local environmental conditions, outbreeding depression is a possibility because adaptive phenotypic traits (perhaps in behaviors, coat colors, molt timing, etc.) shaped by both genetic differences and the local environment may not be well suited to the new area (Ruggiero et al. 2000).

If the donating population is captive, as is often the case for translocations of endangered species, another suite of issues comes into play. First, one must avoid selecting for traits that serve the animals well in captivity (e.g. docility or willingness to eat foods different from those in the wild) but undercut their vital rates in the wild (Frankham et al. 2002, Allendorf & Luikart 2006). Second, intensive on-site management may be necessary. Predators might be excluded or removed to help naïve introduced animals survive, or released animals might be behaviorally conditioned. There is strong evidence that animals can be taught how to fear and avoid predators, as when captive-raised Siberian polecats learned to hide from badgers after several training sessions where stuffed badgers were presented at the same time that experimenters shot at the polecat with elastic bands (Miller et al. 1990), or hare-wallabies learned to fear models of foxes or cats operated as a puppet that leapt at the hare-wallabies and squirted them with water (McLean et al. 1996). Such behavioral conditioning can be used to trigger appropriate responses to predators, mating rituals, obtaining foods, and finding shelter, all of which will increase survival of captive animals released to the wild (Griffin et al. 2000, Reed et al. 2002).

The number of individuals chosen for translocation, and their characteristics, will depend on why the translocation is being done. Population-projection models (Chapter 7) and population viability analysis (Chapter 14) can help determine the optimal number of individuals. Optimal characteristics of the translocated dispersers, and the timing of their release, will depend on logistics (who can you get?) but can also be informed by behavioral and demographic analysis. Sex, age, territory status, and other characteristics will influence whether the translocated disperers stay put or try to go home or somewhere else, and natural history and behavioral knowledge can help you predict those events (Van Vuren 1998). Stephens and Sutherland (1999) tell the fascinating story of how simultaneous release of male and female bush-tailed phascogales led to males failing to find females when they dispersed; however, if females were released first so they could establish territories before males were released, the males dispersed less and found mates. Once behavioral subtleties are considered, animals with high reproductive value could be considered as the seeds with the highest potential to jump-start population growth (Chapter 7).

I will end this section on translocations with two examples that underscore several of the issues that must be dealt with when using translocation as a tool for imposing connectivity. Although woodland caribou were historically distributed over much of western Canada and the northern USA, by the early 1980s only one small herd of about 25 animals persisted in the USA, and in 1984 the species was listed as endangered under the US Endangered Species Act (Compton et al. 1995, Zager et al. 1995). The recovery plan for woodland caribou called for addressing the perceived causes of decline (habitat degradation, poaching, and vehicle collisions), as well as translocation to support the remaining population of 25 animals in the Selkirk Mountains of northern Idaho, northeastern Washington, and southern British Columbia, Canada. Between 1987 and 1990 60 caribou were translocated to the south of the existing resident herd to increase persistence by establishing a second herd with environmental stochasticity that was decoupled from the first. Three died during capture and transport, and survival over the next few years was below 80%, so that by 1995 total metapopulation size was about 55, with only 13 of the 60 translocated caribou still alive[7]. What happened? Many of the caribou moved away, especially when there were no resident caribou in the area (Compton et al. 1995). Also, some of the translocated

---

[7]In the mid-1990s a second translocation of 43 caribou was conducted; success was again lukewarm. The total number of Selkirk caribou in 2005 – including the original 25 and the 103 animals translocated between 1987 and 1998 – was approximately 35 (P. Zager & W. Wakkinen, personal communication).

animals were from a different ecotype (northern, as opposed to mountain), and experienced higher mortality in part because in winter they fruitlessly pawed through deep snow in search of lichen to eat on the ground instead of consuming the plentiful arboreal lichens favored by the resident ecotype (Warren et al. 1996). But, more fundamentally, the lack of success in the Selkirk caribou translocations may have been a failure to address the primary driver of decline, which was that intensive logging and large wildfires had created extensive white-tailed deer habitat and increased deer numbers, subsidizing larger mountain lion populations. Fixing this problem is a long-term, politically volatile challenge on large spatial scales, and it may have been that translocations were seen as an easier step than tackling habitat restoration (Zager et al. 1995). Like any form of connectivity, translocations will only successfully increase population size in the long term if the original cause of decline is ameliorated; otherwise you only create a demographic sink that sucks down animals from the donor populations.

The recovery of the American peregrine falcon is one of the great successes of the US Endangered Species Act. As a species whose reproduction was badly affected by the pesticide DDT (see Chapter 11), falcons were listed as endangered in 1970. In 1999, 25 years after the banning of DDT in the USA, American peregrine falcons were delisted. Although the primary driver of recovery was improved reproduction following the banning of DDT, another major contributor was the success of thousands of captive-bred falcons translocated into the wild. Because the assessment of species recovery was based on count data, an interesting question arose: was the recovery solely due to improved vital rates within populations, or did the increased human-mediated connectivity via translocations help? Matthew Kauffman et al. (2003) tackled this question for peregrine falcons in California by building matrix population models with habitat and time-specific vital rates estimated from field data. Although counts through the 1980s showed a steady increase in the breeding population in rural areas, intrinsic vital rates in that area indicated that λ was less than 1 during that time; thus, the apparent increase in falcons in rural areas was likely due to the augmentation per se. More recently, peregrine falcons in both rural and urban areas have achieved vital rates such that λ was greater than 1. In short, the range-wide recovery of peregrine falcons was due not only to the reversal of factors limiting fecundity and survival, but also to augmentation in rural areas during the 1980s.

Summary

Dispersal and gene flow into and out of populations – broadly called connectivity – fundamentally shape wildlife distribution and abundance across the landscape. Connectivity determines taxonomic distinctiveness, colonization of new sites in a changing environment, and persistence of both local populations and metapopulations of linked populations. Although we can make generalizations about which sex, age, or life histories will be most likely to disperse, for most management applications specific dispersal rates must be quantified, a daunting task given the mobility and cryptic habits of many species. Luckily, a dazzling array of new technology and techniques are available to help, ranging from radiotelemetry and multi-state mark–recapture techniques to molecular tools from which historic gene flow and even recent dispersal can be inferred from tiny bits of nonlethally collected samples.

With measures of connectivity in hand, we can understand the role it plays for a particular wildlife species, and predict the consequences of changes in a human-altered landscape. Human perturbation can increase connectivity for certain species, leading to transfer of diseases, colonization of new areas, or changes in genetic structure. Or it can be neutral, creating a perforated or patchy landscape traversed by dispersers. At the other extreme, multiple populations could be isolated, with each separate population subject to extinction without recolonization; the fate of the set will depend on both population-specific extinction probabilities and on how correlated their dynamics are due to shared environments (the eggs-in-a-basket phenomenon).

If the populations are distributed as a metapopulation, with some relatively small level of connectivity, persistence becomes a question of weighing within-population vital rates (births and deaths) against among-population rates (immigration and emigration). Source populations which contribute to metapopulation persistence can be distinguished from sinks that drain the metapopulation. The most insidious sinks are ecological traps, with cues that attract animals but then lead to low reproduction or survival.

If connectivity is important, and if connectivity is being reduced to the peril of particular species in a human-dominated landscape, how should it be restored? Corridors are one option, as is modifying the intervening matrix that originally broke up the connectivity. As a last resort, physical translocations might be necessary, applying all that you know about demography, genetics, and behavior of movement to figure out the optimal donor population, number of individuals to translocate (and their sex, age, reproductive value, territorial status, etc.), and specifics of release. If your back is against the wall and you have to do a translocation, know that your actions will be scrutinized by the public, but take solace in the wealth of ecological insights that can underlie your decisions.

Further reading

Clobert, J., Danchin, E., Dhondt, A.A., and Nichols, J.D. (2001) *Dispersal*. Oxford University Press, Oxford. An edited volume with detailed treatment of dispersal-related topics.

III

Applying knowledge of
population processes to problems
of declining, small, or
harvestable populations

11

Human perturbations: deterministic factors leading to population decline

> *Growing up in cutover Vermont, [George Perkins] Marsh had observed what deforestation did to streams, fish, birds, animals, the land; and serving for many years in the Mediterranean basins and the Fertile Crescent, he had studied man-made deserts, the barren mountains of Greece, the ruins of once-great civilizations. He saw calamity coming to America and wrote his book [in 1864] 'to point out the dangers of imprudence and the necessity of caution in all operations which, on a large scale, interfere with the spontaneous arrangements of the organic or the inorganic world.'*
>
> **Wallace Stegner (1992:123), *A capsule history of conservation***

> *Sometimes, walking in the country, one comes upon an abandoned flower garden overtaken by wild flowers. Is it still a garden? The natural and artificial orders intermingle, and ready definition is lost.*
>
> **Fred Chappell (1987:97), *I am One of You Forever: a Novel***

Introduction

Why did a butterfly species go from perhaps 100,000 individuals to being extinct in England in 30 years? How did vultures that were one of the most common raptors in the Indian subcontinent decline by 95% in 20 years? In a blink of 200 years, how could New Zealand lose half of its bird species? How could turtle populations become all male? What caused Allegheny woodrats in the USA and red squirrels in Britain to decline precipitously, while white-tailed deer and deer mouse populations increased to the point that they halt the recruitment of perennial flowers?

In three words: deterministic human perturbations. **Deterministic** is used in the sense of Chapter 5 to refer to factors that change population in relatively predictable ways, as opposed to stochastic or random factors. Despite the fact that much of wildlife population biology is nonintuitive, illuminated by relatively arcane concepts including mathematical models and genetic sampling, there really is no substitute for understanding the deterministic factors known to influence vital rates.

The big five human-caused deterministic perturbations are habitat loss and fragmentation, invasive species (including disease), pollution, over-exploitation, and a relatively recent but looming fifth factor: global climate change. Obviously, all of these interact with each other and with the populations they affect, and I will address those synergisms at the end of the chapter.

General effects of deterministic stressors on populations

As you will see throughout this chapter, four responses can occur when wildlife populations are confronted by human-caused perturbations or stressors. The biologically trivial response is no response because the stressor is within the physiological or behavioral norm for that species. The second is that individuals making up a population could move, thereby shifting geographic range. The third response is an adaptation to the change without moving, either through plasticity – changing physiology, morphology, or behavior without a genetic change in those traits – or through evolving new mechanisms to adapt. Finally, a deterministic stressor can modify vital rates, potentially causing populations to decline toward extinction. Because some species will do well and some poorly in the modified conditions created by deterministic stressors, species composition will change.

Of most concern is the potential for deterministic stressors to cause extinction. For animal species worldwide, the most prominent causes of recent extinctions – several of which operate concurrently – include invasive species (being a cause in 54% of cases), habitat destruction and fragmentation (48%), and hunting or gathering (45%; Clavero & García-Berthou 2005). A similar review for the USA pointed toward habitat destruction and degradation as the most pervasive stressor, with pollution, exotic species, and disease next in importance (Wilcove et al. 1998). Climate change is coming on strong, with predictions that by the year 2100 it will be second only to land-use changes as a driver of extinctions in terrestrial ecosystems (Sala et al. 2000). Little solace can be gleaned from speculation that the deterministic stressors will cause speciation that will offset extinctions, because speciation is not likely to occur as fast as extinction (Box 11.1).

As you will see, some of the extinctions are direct consequences of the perturbations. In addition, linked interactions among species can cause changes to pirouette into unexpected abundance changes in species at multiple trophic levels, leading to **indirect**, **cascading**, or **knock-on** consequences. Indirect effects can arise from the bottom trophic levels and spiral up through herbivores and predators, or top-down from predators initiating trophic cascades (see Chapter 8). Most often, cascading indirect changes come from a combination of both bottom-up and top-down effects. Now let's consider the five major deterministic factors, and how they affect wildlife populations.

Habitat loss and fragmentation

Look outside. It should come as no surprise that habitat loss and fragmentation constitutes the leading human-caused deterministic factor affecting wildlife populations, driven by activities from agriculture and forestry, commercial and residential

---

Box 11.1 Could extinctions due to deterministic factors be replaced by speciation?

Because we know that a major driver of speciation is the geographic isolation of small **founder populations**, it is not unreasonable to ask: could human-caused deterministic stressors actually increase speciation rates and thereby compensate for extinctions by promoting new biodiversity? It turns out that very restrictive conditions must hold before a founding event is likely to trigger speciation, with one of the most important conditions being rapid population growth after founding. This condition is of course exactly the opposite of what happens in fragmented populations, when genetic diversity is decreased and population growth is stalled.

As a tangible example, Templeton et al. (2001) describe the dynamics of the eastern collared lizard in the Missouri Ozarks. Like other desert-adapted species, collared lizards were cut off from their southwestern ancestral range about 4000 years ago as the climate changed; the lizards persisted in open glades surrounded by open savanna forests and grasslands. Following logging and fire suppression (beginning in earnest about 1950), the glades both shrank – as they were invaded by fire-sensitive junipers and other woody plants – and were surrounded by a very different matrix as dense oak-hickory forests took over. Small population size and greatly reduced gene flow led to high population subdivision ($F_{ST}=0.40$), and extinctions with little recolonizations on the remaining glades. The conditions necessary for speciation were contravened by decreased genetic diversity, by allele frequencies changing randomly due to drift (as opposed to selection; see Chapter 9), and by stagnant population growth due to changes on and around the fragments. "An extinction ratchet, not speciation, is the primary impact of human-induced fragmentation" (Templeton et al. 2001:5431).

I like this story because in addition to making robust points about the consequences of habitat fragmentation it has an encouraging ending. A commitment by the State of Missouri to restore fundamental evolutionary and ecological processes led to reintroduction of fire, both on the glades and in the surrounding forests. As a result, collared lizard population sizes, movement among glades, and recolonization of extinct glades have all dramatically increased. Populations now have lower short-term extinction probabilities, and the capacity to maintain high levels of genetic diversity for local adaptation.

---

development, water development and fire regimes, and infrastructure development including roads. We'll examine habitat loss and fragmentation separately, then together.

Habitat loss can reduce populations

The first and foremost impact of habitat loss is the direct reduction in population size leading to decline for certain species. If habitat loss causes large continuous populations to become small, then all of the specters affecting small population persistence could exact their toll (Chapters 9 and 12). As a case study, Brook et al. (2003) estimated that in the last two centuries, deforestation in Singapore leading to habitat loss exceeding 95% has caused extinction of at least a third of all species of butterflies, fish, birds, and mammals.

Interestingly, two different mechanisms can lead to a transient increase in density on remaining fragments following habitat loss. First, animals may be displaced from the modified habitat so that they crowd into the fragment, leading to a short-term increase in numbers on the fragment (Lovejoy et al. 1986, Hagan et al. 1996); this **displacement effect** is sometimes referred to with the metaphor: crowding on the ark (Fig. 11.1). A second mechanism leading to temporary increase in density following habitat loss occurs if the surrounding matrix is hostile enough to prevent dispersal so that density on the fragment climbs due to a **fence effect** (Ostfeld 1994). In both cases, displacement and fence effects are likely temporary (Debinski & Holt 2000).

Habitat fragmentation adds to the problems of habitat loss

In addition to habitat loss decreasing population size, habitat fragmentation can cause the populations that remain to be broken apart. This process of habitat fragmentation has two primary components: altered connectivity among populations, and changes in vital rates for populations that remain.

Altered connectivity

As we have seen, connectivity could be affected when fragmentation increases distance and changes the matrix between fragments. Some classic examples of reduced connectivity following fragmentation involve red squirrels (Bakker & Van Vuren 2004), grand skinks (Berry et al. 2004, 2005), and rainforest possums (Laurance 1990). Even birds, which you might think could just fly across the intervening matrix, may move less following habitat fragmentation [e.g. brown treecreepers and blue-fairy wrens in Australia (Brooker & Brooker 2002) and several passerine birds in North America (Desrochers & Hannon 1997)].

Altered conditions on remaining fragments: edge effects

One of the primary ways that habitat fragmentation can change the vital rates of populations that persist is through creation of ecological edges where the habitat fragment and the intervening matrix meet. The prescient Aldo Leopold identified the importance of edges long ago, noting that for many harvested species the juxtaposition of different types of food and cover leads to wildlife riches.

> Every grouse hunter knows this when he selects the edge of a woods, with its grape tangles, haw-bushes, and little grassy bays, as the likely place to look for birds. The quail hunter follows the common edge between the bushy draw and the weedy corn, the snipe hunter the edge between the marsh and the pasture, the deer hunter the edge between the oaks of the south slope and the pine thicket of the north slope, the rabbit hunter the grassy edge of the thicket.
>
> **(Leopold 1933:131)**

For species experiencing **positive edge effects**, Leopold (1933:132) concluded that densities will be increased relative to the degree of interspersion of different required

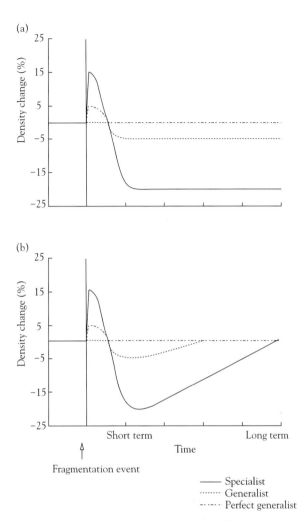

Fig. 11.1 Some potential changes in density for populations on habitat fragments (modified from Hagan et al. 1996:fig. 2; reproduced by permission of Blackwell Publishing Ltd.). In the short term, densities could increase due to displacement effects or fence effects. Over longer periods, densities may decline due to density dependence and perhaps negative edge effects. The extent of change will depend on the specifics of how species respond to fragmentation (a range of possibility is captured here by specialist and generalist categories). For contrast, the response for a species that is unaffected by fragmentation is shown as a flat line (perfect generalist); of course, other species will benefit from the fragmentation. The differences arising from the type of modified matrix are shown in the two panels, where habitat loss and fragmentation are either (a) permanent; or (b) dissipating over time, as in regenerating forests.

habitat types. The phenomenon seemed obvious to Leopold, but he added: "I am not sure that the scientific ecologists know this law as well as woodsmen do."

Leopold (1933) also realized that not all species are positively benefited by edges, and subsequent research has identified several types of **negative edge effect** (Laurance et al. 2002, Batáry & Báldi 2004). Edges can create strong changes in humidity, wind patterns, light intensity, and spectral quality, as well as in soil temperatures and surface albedo. In turn, these abiotic changes can change vegetation structure and plant species composition (Laurance 2004), leading to fundamental shifts in habitat for certain animal species. Negative edge effects can also occur more directly, as when protected area edges bring large carnivores into direct conflict with people on the reserve borders (Woodroffe & Ginsberg 1998).

As abiotic conditions and vegetation changes, the relative abundance of predators and parasites may also change, causing additional edge effects. In a classic case, Gates and Gysel (1978) found more nests near the field-forest edge for 21 species of open-nesting passerine birds, but these nests were less successful because of high predation and nest parasitism from brown-headed cowbirds. Thus edges may act as ecological traps (Chapter 10), with cues that prompt nesting even though fitness is low.

Overall, edge effects can happen but are not ubiquitous, and will vary from positive to negative to none at all across time and space and for different species. The characteristics of the surrounding matrix, and the larger landscape context the fragments are found in, will have huge effects (Harper et al. 2005). As you might expect, larger clearings will result in higher wind velocities, with more potential for increased structural damage, plus more extreme dessication and temperature fluctuations (Laurance 2004). A fragment imbedded in an agricultural or urban matrix may be more likely to experience nest predation at edges than fragments in a logged landscape, and edge effects in heavily fragmented landscape contexts may be amplified compared to more contiguous landscapes (Marzluff & Restani 1999, Chalfoun et al. 2002, Driscoll & Donovan 2004).

Whether positive or negative, edge effects mean that all habitat fragments – and therefore all wildlife populations on those fragments – are not created equal. Smaller, more linear fragments will have proportionately more edge than will larger or more circular fragments (Fig. 11.2). Therefore, a $2\,km^2$ circular fragment has 21 times as much interior habitat as a circular fragment that is one-tenth as large [from Fig. 11.2, $(2\,km^2 \times 0.76$ interior$)/(0.2\,km^2 \times 0.36$ interior$)=21$], and more interior habitat than a rectangle of the same size (how much depends on the skinniness of the rectangle). Furthermore, small or irregularly shaped fragments may experience amplified edge effects because any spot within the fragment is exposed to multiple nearby edges instead of a single edge (Malcolm 1994). These consequences of size and shape of fragments are not of mere theoretical interest: in human-dominated tropical landscapes, most fragments are smaller than $1\,km^2$ and far from round, and so have a high proportion of edges (Laurance 2004).

Habitat loss and fragmentation operate concurrently

Although the distinction between habitat loss and habitat fragmentation is useful to clarify ecological responses for wildlife populations (Fahrig 2003), for practical purposes they tend to occur simultaneously. Unlike shattering a glass plate where the pieces separate but the total amount of plate stays the same, habitat fragmentation nearly always occurs due to loss of intervening habitat.

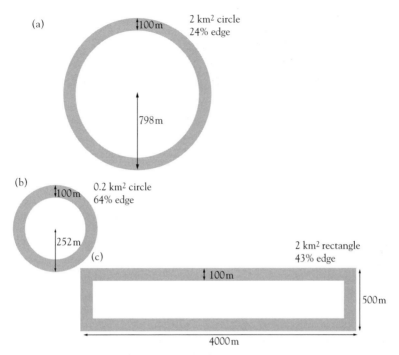

Fig. 11.2 How the proportion of interior compared with edge habitat in a fragment is affected by fragment size and shape. Assume a 100 m edge effect, shaded in each of the fragments. (a) In a 2 km² circular fragment, the radius would be 798 m and the fragment would be 76% interior and 24% edge; (b) a 0.2 km² circular fragment would have 36% interior and 63% edge; (c) a 2 km² rectangular fragment that is 4000 m long by 500 m wide would have 57% interior and 43% edge.

Roads may be the form of habitat modification that best fits the glass-plate analogy, in that connectivity can be affected with less direct habitat loss or reduction in population sizes. Reduced connectivity among populations on either side of roads has been shown for flightless ground beetles in a Swiss forest (Keller & Largiader 2003), for bank voles in Europe (Gerlach & Musolf 2000), and for Amazonian forest birds (Laurance et al. 2004). Even with roads, however, there will be some loss of habitat and certainly edge effects. If you assume that roads affect ecological systems for several hundred meters either side – via edge effects and access by humans and invasive species – then over 20% of land in the USA is affected by roads (Forman 2000, Ritters & Wickham 2003).

In summary, for some wildlife species habitat loss and fragmentation can produce transient increases in numbers (due to displacement and fence effects) followed by decreases in population size and connectivity among patches. Edge effects can affect conditions for populations that remain on fragments. Of course, the modified matrix will not usually be barren, but rather occupied by species that profit from the disturbance. Thus, habitat loss for certain species will benefit others, leading to shifts in community structure. For example, in Oregon, clearcut logging leads to a hot, dry

microclimate with a shallower soil organic layer, causing drastic decreases in both truf-
fles (below-ground fruiting bodies of mycorrhizal fungi) and a primary truffle-eater,
the western red-backed vole, in the modified matrix (Mills 1995, 1996, Tallmon & Mills
2004). Meanwhile, a sympatric native species, the deer mouse, benefits strongly from
the burst of annual plant growth in the logged areas and greatly increases its survival
and density there (see Boxes 4.6 and 10.1). Overall, population consequences due to
habitat loss and fragmentation depend very much on the specifics: what the original
habitat is converted to, and the degree, speed, and pattern of loss. With data and population biology knowledge in hand, the species most likely to suffer can be identified,
and potential negative effects can be addressed (Fig. 11.3).

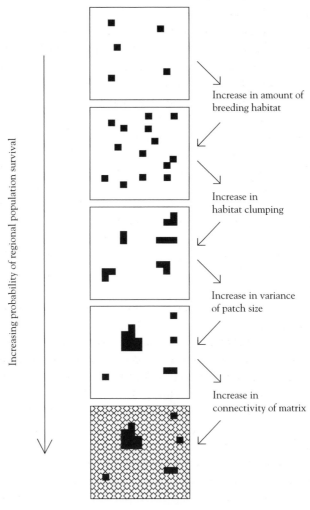

Increase in amount of
breeding habitat

Increase in
habitat clumping

Increase in variance
of patch size

Increase in
connectivity of matrix

Increasing probability of regional population survival

Fig. 11.3 Ways to increase population size, vital rates, and connectivity in order to increase
metapopulation persistence for particular target species in the face of habitat loss and fragmentation. From Harrison and Fahrig (1995). With kind permission of Springer Science and
Business Media.

Introduced and invasive species

Introduced species are nothing new. Species composition in every ecological community has always been dynamic, shifting in response to environmental changes and chance events. And throughout our history, humans have not only responded to but even facilitated those shifts to make our lives safer and more comfortable. But the number of introduced species that become invasive – spreading in numbers and range and causing ecological or economic harm (Lodge & Shrader-Frechette 2003) – has exploded in recent decades. While we will mostly focus on ecological consequences, the economic costs of invasive species are not trivial: in the USA alone, the cost of damage and losses from invasive species has been estimated at $137 billion per year (Pimentel et al. 2000).

One reason for the proliferation of invasive species is that global transportation is moving species around like never before (Mooney & Cleland 2001). Even though only a small fraction of transported introduced species become established, and of these perhaps 1% become invasive or considered as pests, the sheer numbers have soared. Here are some examples. In the continental USA, pairs of states now share on average 15 more species than they did before European settlement (e.g. Arizona and Montana previously had no fish in common; now they share 33 species). In New Zealand introduced vascular plants outnumber natives, and 34 introduced mammal species have invaded as well (Craig et al. 2000). Nearly 100% of the imperiled plants and birds in Hawaii are threatened by exotic birds and by the introduction in 1826 of the mosquito that carries avian malaria (Wilcove et al. 1998). In an interesting irony, invasive species can actually increase local biodiversity, especially on depauperate island ecosystems (e.g. New Zealand now has more bird species than it ever has had in the past, despite massive extinctions of native species), even as global biodiversity becomes decreased due to biotic homogenization and loss of endemic species (Brook 2004).

In addition to increased globalization and transportation, more invasive species are emerging as a major force by overcoming the time lag between introduction of a species and when it becomes an invasive pest. The environment may change favorably, perhaps due to some other deterministic factor, as when collared doves simmered along for centuries after colonizing parts of Asia and Middle East but suddenly (within 50 years) colonized temperate Europe and northwest Africa via a positive response to urbanization and longer breeding seasons afforded by climate change (Crooks & Soulé 1999). A lag between introduction and invasiveness may also occur simply because it takes time to increase and spread to damaging levels, and even more time may be needed if there are Allee effects to overcome (see Chapter 6).

Another reason for a delay in introduced species becoming invasive is that it can take a while for adaptive changes to tip community interactions towards the invader (Crooks & Soulé 1999, Ashley et al. 2003). For the introduced species, responses can occur through phenotypic plasticity and genetic (evolutionary) changes. A classic case of adaptive differentiation occurred in *Anolis* lizards in the Bahamas, whose optimal hindlimb length is closely tied to perch diameter; within 15 years of introduction to islands with very different vegetation, the lizard's hindlimbs became shorter, exactly as expected given changes in average perch diameter (Losos et al. 2001)[1].

Of course, native species also have the potential to adapt to the invader. Consider a pair of interactions between toads and snakes. On the Mediterranean island of Mallorca, the native midwife toad has become critically endangered due to the viperine snake introduced about 2000 years ago. However, it appears that the toads have been evolving inducible defenses (longer tails with deeper tail muscles) against the snake (Moore et al. 2004). The battle between toad and snake is reversed in Australia, where cane toads have become spectacular invaders since 1935, affecting up to 49 species of snakes with a lethal toxin that kills most individuals that try to eat them. For two native snakes, strong selection by cane toad poison has led to increased body sizes and decreased relative head sizes, decreasing the likelihood of ingesting a toad large enough to kill it (Phillips & Shine 2004). Regardless of whether such changes represent evolution in action or merely phenotypic plasticity being turned on, they demonstrate that the invasiveness of an introduced species, and the effects it has on its community, depend in part on dynamic adaptive responses among the players.

Although some wildlife species will clearly adapt to introduced species, extinction can occur if predation or competition from the exotic species is overwhelming (Case & Bolger 1991, Blackburn et al. 2004). Predators such as cats, rats, and brown tree snakes (Box 11.2) famously wreak havoc on native prey both directly - often exploiting ecological naiveté of native wildlife - and indirectly, through hyperpredation (Chapter 8). Introduced species may also displace natives through competitive dominance, as when superior feeding efficiency by the invasive North American gray squirrel helped to replace native red squirrels in Britain (Mack et al. 2000). Interestingly, an additional mechanism for why invasive gray squirrels displaced red squirrels may be apparent competition, as a parapox virus benign in the invasive gray squirrel can jump into the native red squirrel, where it is fatal (Prenter et al. 2004).

Invasive species can also affect natives by hybridizing them out of existence. Chapter 3 mentioned a couple of examples: (i) the greatest threat to recovery of red wolves in the USA is the potential for hybridization with recently arrived coyotes; (ii) the introduction of mallard ducks to New Zealand and Hawaii has led to hybridization with and subsequent decline of endemic duck species. In addition to reducing distinctiveness of a species, hybridization with invasives can create demographic dead-ends because cross-breeding reduces the number of new offspring added to the native population (Mack et al. 2000, Schwartz et al. 2004). One such instance involves native European mink, which are faring poorly both due to habitat destruction and because when they hybridize with introduced North American mink the hybrid embryos are invariably aborted, wasting reproductive effort.

---

[1]Although such changes could be driven by either evolution or phenotypic plasticity in hindlimb length, hindlimb length in *Anolis* is known to be highly heritable (Calsbeek et al., unpublished data), indicating that in this case evolution is occurring.

Box 11.2 Brown tree snakes as an example of an invasive predator with deadly efficiency

As with other Pacific islands, Guam began to endure extinctions during early human colo-
nizations around 1500 BC, with habitat destruction, hunting pressures, and pesticide use
amplified from around 1940. However, it was inadvertently introduced brown tree snakes that
led to one of the most remarkable extinction spasms ever recorded. Brown tree snakes on
Guam arrived sometime after World War II, probably in tanks, jeeps, and airplanes salvaged
by the US Navy, from islands near New Guinea. The buildup in snake abundance was relatively slow, but by about 1970 they were distributed throughout the island with densities of
up to 100 snakes/ha. Brown tree snakes are voracious generalist predators that can climb caves
and trees and operate at night when birds are roosting. Many of the species on Guam were
ecologically naïve. Furthermore, a suite of other species introduced after World War II could
sustain high snake densities even as native prey declined (a good – if grim – example of hyperpredation; see Chapter 8; Fritts & Rodda 1998, Wiles et al. 2003).

   The result was a massive loss of native species. By the 1990s, less than 40 years after being
detected on Guam, the brown tree snake had caused the extinction or near-extinction of virtually every one of the 18 native bird species, about four of the native reptiles, and at least
one bat species. There are still plenty of alternative introduced species to support the brown
tree snake even with a depauperate native fauna; the food web for the island is now entirely
different than it was a half century ago. It is an ecological mess made by humans, and now
humans are trying to fix it: in the fiscal year 2005 the US interagency budget included a total
of $4,247,000 for brown tree snake control (Simberloff et al. 2005).

   And finally, even if you don't like what it does, you've got to admire the brown tree snake
(Quammen 1996:333–4):

> It's not an evil animal, after all. It's just an amoral and earnestly stupid crea-
> ture in the wrong place. What it has done here in Guam is precisely what
> Homo sapiens has done all over the planet: succeeded extravagantly at the
> expense of other species. Encountered in New Guinea, encountered in Queens-
> land, encountered in Sulawesi or Guadalcanal, B. irregularis is a handsome
> and sleek native reptile, constrained by the natural boundaries of its station
> in life.

The special case of human-subsidized native species

As a variant on the theme of introduced species, **human-subsidized native species**
reach superabundant densities because they benefit from human perturbations. In the
USA, two such species are the white-tailed deer and the deer mouse. Nearly hunted to
extinction in the late 19th century, white-tailed deer have rebounded strongly to record
high numbers. Currently, deer are transforming vegetation and contributing to plant
community changes including the potential extinction of American ginseng, the
premier medicinal plant harvested from the wild in the USA (Rooney et al. 2004,
McGraw & Furedi 2005). Similarly, deer mice – a voracious seed predator capable of

---

Box 11.3 Release from parasites as one explanation for which species become invasive pests

Going back at least to Elton (1958), ecologists have felt that certain species become invasive – increasing in numbers to where they can negatively impact other species – because they are released from their natural enemies (pathogens, predators, competitors) and do not encounter such effective foes in their new communities. Parasites appear to play a large role in driving this phenomenon. For 473 plant species that are native to Europe but have invaded the USA, the species were infected on average by 77% fewer fungal and viral species in the USA than in Europe; also, more noxious weeds were more free from parasites. Similarly, for 26 invasive animal species ranging from common periwinkle to the black rat, host species in their native range had an average of 16 parasite species compared with seven where they were successfully introduced. For a specific example, the ubiquitous European starling was founded in the USA by only 100 transplants released in New York City in 1890; the small founding size, coupled with the lack of appropriate intermediate hosts, resulted in current starlings having only nine parasite species compared with 44 species in their native range. Thus invasive species may escape control from parasites and pathogens where they are introduced. This may hold potential promise for controlling exotics by introducing parasites from their native home, although any such introduction always holds considerable risk of unexpected interactions (Mitchell & Power 2003, Torchin et al. 2003).

---

halting recruitment of some perennial flowers (Tallmon et al. 2003) – are heavily subsidized by clearcut logging in some areas (see Box 4.6) and in general love just about every perturbation humans make; they even thrive on introduced biological control agents (gall flies; *Urophora* spp.) brought in to control spotted knapweed (Ortega et al. 2004). These human-subsidized native species become deterministic stressors ecologically analagous to invasive species.

The special case of parasites and disease

The processes of disease outbreaks and their effects on wildlife populations also parallel those of invasive or subsidized species. Furthermore, parasites and diseases can determine whether an introduced species will become invasive (Box 11.3).

Many of the worst diseases for wildlife are brought in with domestic and other animals introduced by humans (Daszak et al. 2000). In a classic case, the highly pathogenic rinderpest virus was introduced via livestock to Africa in 1889. In 10 years it spread 5000 km, extirpating over 90% of Kenya's buffalo population and causing secondary effects on predator populations. Likewise, endangered African wild dogs are under siege from canine distemper and rabies (see Fig. 12.5). Finally, wildlife can suffer from diseases transmitted directly from humans, as when wild mountain gorillas habituated to tourists contracted measles (Daszak et al. 2000).

The susceptibility of wildlife populations to disease is exacerbated by the same ecological naiveté that allows other invasives to flourish: Wilcove et al. (1998) found that more than 70% of the imperiled bird species in Hawaii were affected by disease, compared with fewer than 10% of imperiled birds in the continental USA. If an introduced species is asymptomatic for a parasite or disease but a related native species is negatively affected by it, then parasite-mediated apparent competition can occur (Prenter et al. 2004). For example, in the UK, introduced pheasants are unaffected by a caecal nematode, but the transmittal of the nematode from the pheasant reservoir to native grey partridge has contributed to the partridge's decline (Tompkins et al. 2000).

Diseases can also affect wildlife populations when a native host is subsidized and acts as a reservoir to maintain high levels of a parasite that in turn affects other native species. This seems to be the case where raccoons sustain high densities around humans in New York and New Jersey, inflating numbers of raccoon roundworm, which in turn cause the decline of the Allegheny woodrat (LoGiudice 2003). Similarly, feeding, captive breeding, or fencing native species in small reserves can increase transmission of pathogens and parasites (Ezenwa 2004).

Rules of thumb for dealing with invasive species

Because introduced invasive species are decreasing both species diversity (by causing extinctions) and genetic diversity (through hybridization), they are leading to biotic homogenization across the globe (Olden et al. 2004). Here are some rules of thumb that apply to managing invasive species.

Rule 1: keep them out. Although seemingly obvious, rule 1 is a reminder that consideration of bringing in introduced species – for biological control, pleasure, or any other reason – should be tempered with forethought. About half of all problems with invasive species come from deliberate introductions (Simberloff et al. 2005), which is a sobering point.

Rule 2: if they do get in, try to quickly find and eradicate (or at least limit) them. The typical lag phase in establishment of an invasive species means that control is much easier if done early rather than after the explosive growth phase occurs. Success requires sufficient legal and logistical willpower to eradicate the population completely to avoid wasting money or – even worse – selecting for evolved resistance by the pest to future actions. Mack et al. (2000) describe the classic case of imported fire ants in the southern USA as "the Vietnam of entomology" and "a 200 million dollar disaster" because eradication efforts were too tentative and lacking in biological grounding to be successful. Similarly, the monk parakeet, a known agricultural pest, was introduced as a pet in the USA in the 1960s; although it could have been eradicated early by shooting, its charisma led to public sentiment against killing it and by the year 2000 it occupied 15 states and numbered at least 100,000 in Florida alone (Simberloff 2003). On the other hand, when the giant African snail was brought to Miami and discovered to have infested 42 city blocks some 30 years later, the state of Florida mounted a successful control campaign that included poison, publicity, and hand picking. Simberloff (2003:88) articulates the point: "Of course, the methods deployed in such a rapid response are likely to resemble a blunderbuss attack rather than a surgical strike. But because of their population growth and dispersal abilities, introduced species are one

---

Box 11.4 Control of introduced rats and recovery of native wildlife

Rats [Pacific rat, Norway rat, and ship rat] have been introduced to about 90% of the world's islands, making them among the planet's most successful mammal invaders. Furthermore, rats have caused extinctions of many native vertebrate species (more than 50 species globally), and suppression of many more. However, they can be and have been successfully controlled, and if control happens fast and is complete enough, native species can recover (Towns & Broome 2003, Towns et al. 2006).

New Zealand provides a useful case study. When early poisoning efforts proved to not only eradicate rats but also stem the loss of biodiversity, more formal tests of poison types, dosages, and applications began in earnest. By the mid-1980s, the call to conserve native wildlife from invasive rats was so strong that poisoning went large-scale, including aerial spread of baits by helicopter. Rats were removed from more than 90 islands, recreating rat-free habitat on almost 19,000 ha. Dozens of native animal species either recovered on the islands or were successfully reintroduced; cascading effects of vegetation regeneration also occurred as the ecological communities were restored. In large part, the reasons for success include careful attention to minimizing and documenting effects on nontarget species, and educating the public in the ecological consequences that would occur both with and without rat eradication.

---

target of resource management at which it is often better to shoot first and ask questions later."

Rule 3: if they become established, don't give up. Eradication or sustained reduction of numbers may still be possible as new political and biological opportunities arise. Even long-established and hard-hitting species like rats have been decreased hundreds of years after their introduction, with subsequent impressive recovery of native fauna (Box 11.4). If eradication is not possible you may need to settle for longer-term reduction of population size below an economic or ecological threshold. Realize that if multiple introduced species have arrived you will need to account for interactions among species; hyperpredation and trophic cascades, for instance, mean that often suites of species must be removed simultaneously (recall from Chapter 8 that removing introduced pigs without removing introduced eagles could actually cause more harm to the endemic Channel Island fox). In some cases, long-established introduced species may naturalize such that they play critical roles in the ecosystem, and to remove the invader completely would do more harm than good. Edith's checkerspot butterfly has begun to evolve feeding preferences for introduced species subsidized by cattle ranching and logging; if trends from 1983 to 1990 continue, the genetic changes may make the butterfly unable or unwilling to feed on their native plants (Singer et al. 1993). In such cases, gradual reduction of the introduced plant species would be both more successful and appropriate for the butterfly than trying to rapidly eradicate the invasive foods (Myers et al. 2000).

Rules 2 and 3 bring up the issue of how control should be implemented. Eradication can be a tricky business because the control agents are themselves deterministic drivers of population dynamics; pesticides are poisons and biological control agents are introduced species. Many pesticides and biocontrol agents do not actually control the species of interest, they may end up negatively affecting non-target species, they may artificially subsidize certain native or nontarget introduced species, and they may spread to places where they are not wanted. An example of a control agent gone horribly wrong is the cane toad mentioned above; introduced to Australia in 1935 to control cane beetles, cane toads are now one of the most sensational invasive species in the world, as they are toxic at all life stages, reach densities of up to 2138 individuals/ha and are predicted to extend their range to 2 million km^2 by 2030 (B.L. Phillips et al. 2003).

As an alternative to poisons or introduced biocontrol agents, some invasive species can be reduced by hunting and trapping (Mack et al. 2000). In any case, when weighing the disadvantages of each method of control of an invasive species, remember that becoming paralyzed with indecision and doing nothing is tantamount to making a decision to endure the ecological consequences of the invasive stressor.

Pollution

Pollutants can be synthetic chemicals or natural substances (e.g. nutrients) that reach unusually high concentrations. The most high-profile effects of pollutants on wildlife have involved pesticides; 50,000 are registered for use in the USA alone (Stiling 2002). Pesticides travel around the world, are long-lasting in animal bodies and the environment, affect multiple species directly, and have cascading effects across trophic levels. Although the classic case study comes from organochlorine pesticides causing raptors and other birds to decline (Box 11.5), other pesticides can also cause problems. Two illustrations include common organophosphorus pesticides altering the ability of white-throated sparrows to migrate (Vyas et al. 1995), and the glyphosate herbicide Roundup ravaging frogs (Relyea 2005).

Importantly, effects of pesticides on wildlife may depend as much on how the poison is applied as it does on the amount. At the peak of the 2001 foot and mouth outbreak in England, up to half a million livestock were killed per week in an attempt to contain the disease. Fear of scavengers spreading the disease from culled to live animals led to widespread rat control, using up to 20 times the amount of rodenticide bait in several days as normally used in a year. Surprisingly, a major rat predator, barn owls, actually had lower exposure to the rodenticide in rat-control areas during the outbreak, probably because the rodenticides were professionally and carefully applied during the outbreak compared with the more casual applications outside of outbreaks (Shore et al. 2006).

Of course, many chemical pollutants other than pesticides may affect wildlife populations. For example, livestock around the world are treated with pharmaceuticals to increase weight gain and reduce disease outbreaks, and consumption of the livestock by scavengers can have devastating effects. For example, oriental white-backed vultures, once one of the most common raptors on the Indian subcontinent, experienced a mysterious decline by more than 95% starting in the 1990s, and are now considered critically endangered. The mystery of their decline was solved by the discovery that in

---

Box 11.5 Organochlorine pesticides and predatory birds

In 1948, a Swiss chemist named Paul Müller won the Nobel Prize for Physiology and Medicine for discovering that a remarkable chemical named DDT [1,1,1-trichloro-2,2-bis-(p-chlorophenyl)ethane] killed insects, including mosquitoes, while being considered perfectly safe for humans. Meanwhile, other forms of organochlorine compounds (e.g. cyclodienes such as aldrin, dieldrin, and heptachlor) began to be used in agriculture after 1955. On the heels of Rachel Carson's (1962) landmark book detailing the ecological damage from DDT, most organochlorine pesticides were banned throughout much (but not all) of the world by the end of the 1970s.

And what were the ecological effects? The toxic cyclodienes, typically applied to protect seeds against insect attack, killed both seed-eating birds and their predators outright. DDT, and its main metabolite DDE [1,1-dichloro-2,2-bis(p-dichlorodiphenyl)ethylene], kills birds only at very high exposures; its primary demographic effect was to decrease reproduction by reducing the availability of calcium carbonate during eggshell formation, leading to thin-shelled eggs that broke easily during incubation or dehydrated the embryo. Because DDT is fat-soluble, it tended to accumulate in birds of upper trophic levels, including predatory birds that ate birds and fish (mammals have a more effective detoxification system for organochlorines, so neither mammals nor birds that ate primarily mammals were strongly affected). Through effects on vital rates, the cyclodienes and DDT caused bird species around the world to decline. The good news is that since these pesticides were banned in the USA their presence in the environment has dropped and species including peregrine falcons, stock doves, sparrowhawks, bald eagles, and pelicans have experienced remarkable rebounds in population size (Newton 1998).

Pakistan and elsewhere the anti-inflammatory drug diclofenac is widely used for all types of hoofed livestock; when vultures scavenge the dead livestock, diclofenac causes fatal renal failure (Oaks et al. 2004).

In addition to direct poisoning, pollutants can act indirectly on wildlife populations by affecting components of food or cover (Newton 1998). For instance, the insect prey base of grey partridge was decimated in agricultural fields both by insecticides and by removal of broad-leaved weed species via herbicides (Potts & Aebisher 1994, Boatman et al. 2004). The decrease in insect populations reduced partridge chick survival, leading to widespread decline (disease, harvest, and introduced predators also contributed), and to management recommendations to leave an unsprayed strip of crop around fields to enable noncrop plants and insects to grow.

High levels of nutrients (subsidized by human activities) can also be deterministic pollutant stressors. For example, siltation and nutrient inputs from agricultural runoff are often included as a form of water pollution, and indeed are a leading threat to aquatic organisms in North America (Richter et al. 1997). Another example – enrichment of CO_2 leading to global warming – will get its own section shortly.

Overharvest

Population biology provides deep insight into the harvest of wildlife for subsistence, sport, and commercial purposes, and numerous success stories exist where harvests by humans have been and are sustainable (Chapters 1 and 14). However, harvest of wildlife can also be a strong deterministic factor capable of causing extinctions.

Some of the most drastic cases of overharvest in terrestrial systems come from eras before enforced hunting regulations, tracing back at least 50,000 years (Brook & Bowman 2002, Milner-Gulland et al. 2003a). As an example for one bird group, the moas, matrix model projections coupled with ^{14}C dating of remains indicated that harvesting by Polynesian settlers (and concurrent habitat destruction) drove 11 species of moa to extinction in less than 100 years (Holdaway & Jacomb 2000).

Today, unregulated hunting continues to be a major driver of wildlife population declines in many tropical forests of the world, as wild animals are harvested for subsistence in local rural communities, for protein exported to urban consumers, and for commercial products including oil, and leather, skins, and feathers for fashion (Redford 1992, Fa & Peres 2001, Milner-Gulland et al. 2003a). The raw estimates of take of bush meat, or wild meat in the tropics, are staggering: weekly sales of 1500 forest rats in one Sulawesi market, 25 tons of turtles exported weekly from Sumatra, Indonesia, annual takes of 23,500 tons of wild meat in Sarawak, approximately 100,000 tons in the Brazilian Amazon, and more than 1 million tons in Central Africa. We've seen that numbers killed do not necessarily imply population decline, but harvests of wild meat have been shown to be unsustainable in many areas, as in Central Africa where harvest rates average six times the maximum sustainable rate and in Vietnam where harvest was the main driver (in conjunction with development and forest loss) of the extinction of 12 large vertebrate species in the past 40 years. The effects of wild-meat harvest can reverberate well beyond the species taken, because the large-bodied species typically targeted first (e.g. tapirs and primates) are often strong interactors whose loss can affect other species. Solving the **wild-meat crisis** is complicated because it interacts with other factors, such as the need to feed colossal increases in human populations, road building, forest loss and fragmentation, improved weaponry, global commerce, and complex socioeconomic and cultural norms. For example in Ghana, fish are the primary source of animal protein and wild terrestrial mammals are secondary; as fish harvest has declined due to commercial export and local consumption the harvest pressure has transferred to terrestrial wild sources, exacerbating the decline in biomass and species richness of large mammals (Brashares et al. 2004).

In addition to the obvious direct effects of harvest, nontarget species can decline due to harvest of target species. For example, several albatross species are being killed as by-catch during long-line fishing, especially in the southern blue fin tuna fishery where 120 million hooks were set in 1980, leading to marked population declines (Weimerskirch et al. 1997). This example also serves to remind us that overharvest issues also exist in temperate regions, particularly in fisheries.

Global climate change

Climate has varied throughout the history of life on Earth. As atmospheric gases and Earth surface temperatures have changed, so too has sea level risen and fallen, glaciers advanced and retreated, and species dispersed, evolved, or gone extinct. However, by burning prodigious amounts of fossil fuels, which in turn release greenhouse gases (primarily CO_2), humans have changed conditions quickly and drastically. Because CO_2 and other gases absorb infrared radiation and reflect it back to the ground, the Earth is experiencing unprecedented rates of increase in mean global temperature (Crowley 2000, Beedlow et al. 2004).

How much has the temperature increased, and how much more than expected compared with natural fluctuations? The answers are strong and consistent across a variety of sources, including bubbles trapped in Antarctic ice sheets dating back 400,000 years[2], worldwide records on changes in glaciers since the 1600s (Oerlemans 2005), and global temperature-monitoring instruments in operation since about 1850. The temperature increases of the past two decades are beyond background levels in both rate and magnitude. The 1990s were the warmest decade in the past 150 years, and so far the first decade of 2000 follows the same trend. Averaging across space and time, global surface temperatures have increased by 0.6°C, with nighttime temperatures increasing more than daytime temperatures, daily minima more than maxima, winter warming more than summer, and land warming more than the sea (Easterling et al. 2000). Snow cover has decreased and sea level has risen (IPCC 2001). Precipitation changes are more complex, but in general land precipitation has increased in the mid-to-high latitudes but decreased in the tropics and subtropics. Furthermore, streamflow patterns are changing around the world (Milly et al. 2005).

Of course, all of these changes are enormously complicated, with feedbacks among them and unequal patterns across the globe. So although the climate will continue to change, exact predictions are impossible. The best science indicates that over the next 100 years the planet will warm on average by 2–5°C, along with an increase in the number of extreme precipitation events.

Is this enough of a change to really affect wildlife populations? Well, consider that during the last ice age (74,000–14,000 years ago) temperatures were only about 5–10°C colder than they are now; the temperature change predicted over the next century approaches that amount. Another way to ask what changes might occur under climate change is to review the changes that have already occurred. Based on the ecological first principle that animals, and the plants and other organisms upon which they depend, are directly and indirectly affected by temperature and precipitation, it should not be surprising that wildlife is responding to global climate change. In particular, a growing literature documents the effects of climate change on phenology (timing of

---

[2]Ice cores provide a frozen sample of air and water that goes as far back in time as you can drill down (Petit et al. 1999). CO_2 can be analyzed from bubbles trapped in the ice. Temperatures can be determined by analyzing the ratio of two isotopes of water (^{16}O and ^{18}O); the warmer the temperature when the ice is deposited, the higher the proportional abundance of the heavy isotope.

life-cycle events), geographic range, and vital rates of many wildlife species (reviews by Root et al. 2003, Parmesan & Gailbraith 2004, IPCC 2001). We will consider each of these in turn.

Climate change has initiated phenological shifts in the timing of migration, reproduction, and dormancy patterns for hundreds of species. For example, increasing spring temperatures led to earlier nesting for 28 migrating bird species on the east coast of the USA (Butler 2003), earlier egg laying for Mexican jays (Brown et al. 1999) and tree swallows (Dunn & Winkler 1999), earlier breeding by red squirrels in Canada (Reale et al. 2003), and earlier breeding calls by frogs in New York (Gibbs & Breisch 2001). Overall, across 677 species about 62% showed trends towards spring advancement in breeding, flowering, budburst, or seasonal migration, while 27% of the species showed no trends and only 9% showed the opposite trend (Parmesan & Yohe 2003). Similarly, Terry Root and colleagues (2003) found that across species and studies, spring phenology is 5.1 days earlier per decade, with larger shifts at higher latitudes where warming is exacerbated.

What are the consequences for wildlife that shift phenology and those that do not? Both abundances and community composition will change as some species deal with or adapt to the new conditions better than others (Berteaux et al. 2004). Endangered red-cockaded woodpeckers in North Carolina are laying eggs earlier, so that their hatchlings are in synchrony with temperature-driven changes in food availability (Schiegg et al. 2002). However, inexperienced females and birds that have become inbred (as is likely in these small, fragmented populations) do not track the changes with earlier breeding, indicating that isolated and endangered species may be less able to make the shifts necessary to adapt to climate change.

Climate change has also shifted geographic ranges. A review of 99 species in North America and Europe indicates that birds, butterflies and alpine herbs are shifting their range limits by an average of 6.1 km northward or several meters upward in altitude per decade (Parmesan & Yohe 2003). The Edith's checkerspot butterfly has undergone massive local extinctions in the southern part of their western North American range and at low elevations, resulting in a northward range shift of 89 km and an upward elevation shift of 125 m (Parmesan 1996, Parmesan & Gailbraith 2004; see also Crozier 2003 for northern expansion of the sachem skipper butterfly in the western USA). The exact consequences of range shifts will vary, but certainly community composition will change.

Furthermore, range expansions by exotic species could wreak havoc on community dynamics. For example, as red foxes expanded northward into northern Canada with warming temperatures, competitively subordinate arctic foxes contracted their range (Hersteinsson & Macdonald 1992). Although cascading effects from this range expansion have not been documented, red foxes expanding into other areas are often considered pests because they extirpate prey (including endangered species) and carry diseases including rabies. Similarly, fire ants have spread throughout the southeastern USA, damaging crops and other plants, displacing native ants and other invertebrates, causing nest failure and mortality in birds (including bobwhite quail, a popular game species) and mammals, and disrupting mutualistic interactions (Holway et al. 2002). Because fire ants are limited by temperature, climate change will likely spread this pest to many parts of the USA (Korzukhin et al. 2001). Diseases would fall into this category as well, as both humans and wildlife will be affected as a suite of nasty tropical diseases and their hosts move into currently more temperate latitudes.

What if wildlife species are unable to adapt or shift their range in the face of global climate change? Vital rates may be affected. Of intense concern for global temperature increases are reptiles and amphibians with temperature-dependent sex determination (Chapter 4), which could become all female (in the case of many turtles) or all male (in the case of tuatara in New Zealand) as temperatures increase by just a few degrees. For another case where adaptations to global warming are unlikely, consider North American wood warblers (Strode 2003) and European blue tits (Thomas et al. 2001): their migration time is fixed by photoperiod, and so cannot shift with global temperature change, but their invertebrate prey are emerging earlier, leading to lower vital rates in the birds as the young hatch after the peak availability of prey. Similarly, for amphibians whose production of eggs and movement to breeding ponds is intimately tied to temperature and moisture, mismatches between breeding phenology, pond drying, and arrival at the pond can lead to changes in community composition and nutrient flow in ponds (Beebee 1995, Wilbur 1997). In general, mountain regions may be especially vulnerable because they support endemic species that depend on strong environmental gradients that can disappear or shift abruptly with climate change: for 65 regionally endemic vertebrate species in the mountainous west tropics of Australia, core environments are expected to be lost with subsequent extinctions (Williams et al. 2003). It is unknown whether polar bears (whose geographic range is fixed because they are already in the extreme north) will be able to adapt to predicted changes in distribution and abundance of sea ice (Laurance & Cochrane 2001, Derocher et al. 2004). Box 11.6 gives a more detailed case study of the potential consequences of not making phenological shifts.

Synergistic effects among deterministic stressors

Although I've described each of the five main deterministic stressors with separate headings, as if they are distinct, many of the examples included interactions among different stressors. Indeed, all of these factors interact simultaneously and synergistically (Didham et al. 2005), and a couple of final examples will stress that synergisms among deterministic factors are probably the norm. First, the nest predators of greatest consequence for forest birds following habitat loss and fragmentation are typically introduced or subsidized species (as with American crows, cats, dogs, raccoons, rats, and nonnative squirrels in western USA forest fragments; Marzluff & Restani 1999). Second, outbreaks of disease that harm wildlife arise from interaction with other stressors (Fig. 11.4). For instance, the pathogenic fungi – most famously the chytrid group and *Saprolegnia ferax* – that cause mass amphibian mortality in the Americas, Australia, Europe, and Africa may be exacerbated by factors including climate change and pollution from heavy metals (Daszak et al. 2003, Pounds et al. 2006).

A final example of synergisms occurs in tropical forests. Greenhouse gas emissions, and subsequent global warming, has been increased as direct destruction of tropical

---

> **Box 11.6** *A challenge to adapt to global warming: waterfowl in the USA*
>
> Waterfowl in the prairie pothole region (PPR) of North America present a case study in how changes in climate can overwhelm phenological shifts (Poiani & Johnson 1991, Johnson et al. 2005). The majority of the continent's ducks are produced in the PPR, and breeding activities here determine 90% of the variation in growth rate for mid-continent mallard populations (Hoekman et al. 2002; see also Chapter 7 in this volume). Thus, waterfowl production in the PPR supports the 13 million waterfowl harvested annually in the USA by 1.5 million sport hunters, with a total annual economic output of $1.6 billion (Williams et al. 2002). Temperature and precipitation – and subsequent wetland abundance and water levels – directly influence waterfowl reproductive effort, clutch sizes, renesting propensity, and brood survival. Climate-change scenarios predict that the future PPR will have fewer wetlands for breeding waterfowl in what historically have been the most productive portions of the PPR (central and western portions including the Dakotas and southern Saskatchewan). Waterfowl are known to rapidly re-colonize drought-stricken landscapes when water returns, but changes in land-use practices further limit options that birds have to adapt to a changing climate. Simulations by Johnson et al. (2005) suggest that the most favorable climate for waterfowl production will shift to the eastern PPR (Minnesota and Iowa) where unfortunately nearly all wetlands have been drained and grassland nesting habitat converted to row-crop agriculture. The prediction by Sorenson et al. (1998) that waterfowl populations in the PPR could be cut in half by 2050 as a result of climate change would strike an economic blow to states that depend on revenues from sport hunting to support local economies.

forests releases stored carbon in the trees, and by the fact that at the forest edges big trees are dying (and decomposing, releasing their stored carbon) and being replaced by smaller trees and lianas that store less carbon (Laurance et al. 2002). Furthermore, habitat fragmentation and loss has synergistic effects by facilitating invasive species, by increasing hunting pressure from logging roads, and by increasing destructive fire effects as the land dries and collects more fuels adjacent to intentional fires in cleared cattle pastures (Laurance 2004). In dealing with cases where multiple factors operate concurrently, remember to use the tools you have to evaluate the relative effects of different management actions (as in Chapter 7).

Summary

We've covered some heavy and depressing ground in this chapter on human perturbations as deterministic stressors on wildlife populations. Don't let it lead you to despair, and don't freeze up or walk away because of worry or uncertainty. Remember that the first step in fixing a problem is to identify it. Many of our most spectacular wildlife success stories come from recognizing and reversing deterministic stressors.

For any human-caused deterministic factor or combination of factors, some wildlife populations will benefit while others will be harmed. For those negatively affected,

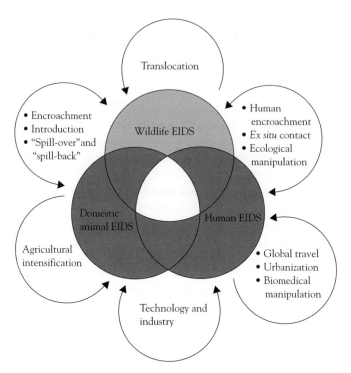

Fig. 11.4 Wildlife emerging infectious diseases (EIDs) as synergistic stressors on wildlife populations that interact with human movements, introductions, urbanization, and domestic animals. From Daszak et al. (2000:fig. 1). Copyright (2000) AAAS.

populations can respond by adapting in place, by shifting geographic range, or in the worst case by declining toward extinction. The response of wildlife populations can be driven directly by the factor(s), or indirectly through cascading effects across trophic levels.

Habitat loss and fragmentation is probably the worst threat worldwide to wildlife populations. The consequences of habitat loss and population reduction are fairly obvious; more subtle are the subsequent fragmentation effects as continuous populations are broken apart. Connectivity can be altered, as can the quality of the patches that remain, particularly through positive and negative edge effects. The overall effects of habitat loss and fragmentation depend on how much and how fast it occurs, and on the nature of the intervening matrix separating the remaining fragments.

Introduced species that become invasive are another global killer. Native wildlife species are often susceptible to invasive species because they have not evolved defenses to them and because the invasives are released from factors that limited them in their native range. Although subsequent evolution by both the native and the invasive can be spectacular, all too often the endpoint of an invasion is extinction of native species. Likewise, human-subsidized native species can also reach such high densities that they become ecologically similar to invasives, and parasites and diseases can both influence the outcome of invasions and act as marauding invaders themselves.

Pollutants and overharvest also act as strong deterministic drivers of wildlife population dynamics in certain times and places. Finally, global climate change is causing fundamental shifts in life-history traits and geographic ranges of wildlife species, and promises to be a formidable foe to wildlife if the international community does not recognize and act on it.

Deterministic factors will rarely act in isolation; most often, several will operate concurrently. But knowing about them gives us the potential to act. All of the population-biology tools in this book can be used to help wildlife populations confronted by these deterministic factors, even in the face of uncertainty. Dan Simberloff (2003) – speaking about invasives but making points that apply to any of these deterministic stressors – charges that far too often decisions to deal with problems get held hostage to a call for more population-biology science before action can be taken. Many success stories do include population-biology knowledge, but often the key to success lies simply in decisive action early on. All of the other concepts in this book – ranging from genetic information to estimated vital rates to population models – form critical links in management decision-making, and can help prioritize efforts. But when the heat is on, step number one is to use basic natural history and field knowledge to identify and reverse deterministic stressors.

Further reading

Caughley, G. and Gunn, A. (1996) *Conservation Biology in Theory and Practice*. Blackwell Science, Cambridge, MA. The book has a strong emphasis on deterministic factors of decline, with excellent case studies.

IPCC (2001) *Climate Change 2001: the Scientific Basis*. Contribution of Working Group I to the third assessment report of the Intergovernmental Panel on Climate Change. Cambridge University Press, Cambridge. The comprehensive scientific assessment of global climate change, and its interactions with other deterministic factors.

12

Predicting the dynamics of small and declining populations

There's an old adage, translated from the ancient Coptic, that contains all the wisdom of the ages – 'Life is life and fun is fun, but it's all so quiet when the goldfish die'.

Beryl Markham (1983:218), *West with the Night*

Introduction

How well is a species doing? How likely is it that particular stressors will cause a wildlife population to decline or go extinct? What is the best way to reverse the trend? These are the big-picture questions commonly asked of applied population ecologists. Having built the foundation for understanding population biology and discussed factors causing large populations to decline due to human perturbations, we can now consider ecological tools for predicting risks to small and declining populations[1]. Although interesting ecological questions can be generated from species that "naturally" persist at low abundance (e.g. Gaston 1994, Brown 1995), here the focus will be on applied situations where populations that were historically abundant are currently small and/or declining.

What is a small population (Mills et al. 2005)? Smallness is a meaningful concept only in relation to other species, to historical population sizes, or even to arbitrary management standards. For an exploited species such as the canvasback, conservation plans may call for corrective management to be implemented when population sizes decrease to the tens of thousands. In contrast, conservation efforts for threatened species may be delayed until the population falls below 100 or is putatively extinct (e.g. the Hawaiian honeycreeper with the two letter name: 'ōʻū).

---

[1] I will avoid the terms **rare** or **rarity**. These are often considered synonymous with low abundance, but can also sometimes refer to high habitat specificity, ecological specialization, and limited geographical distribution (Rabinowitz et al. 1986, Gaston 1994). When using rare or rarity the context should be defined precisely in terms of abundance, range size, and habitat use (Mace & Kershaw 1997).

In this chapter I will first review ecological characteristics that might predispose populations or species to extinction due to humans. The rest of the chapter will describe viability assessment for small populations, including quantitative approaches of population viability analysis (PVA).

Ecological characteristics predicting risk

Are there broad ecological characteristics that can help us classify population dynamics – and therefore risk – for particular populations or species? Obviously, smaller populations will be more vulnerable to extinction (all else being equal) than larger populations, an idea rooted in the classics of applied ecology (e.g. Leopold 1933, MacArthur & Wilson 1967). For example, Mace and Kershaw (1997) found that population size was the best predictor of extinction risk in a global survey of birds.

In some (but not all) cases, having a restricted range or being endemic can serve as a proxy for small population size and therefore be a predictor of vulnerability to extinction for a species (Channell & Lomolino 2000, Purvis et al. 2000). There are no simple thresholds of population size that guarantee persistence (Box 12.1), and later in the chapter we will explore the best ways to assess risk for populations of different sizes.

Another predictor of extinction is the ratio of the variance in population growth rate (σ) to the mean growth rate (r), where variance represents the environmental stochasticity of Chapter 5 (Fagan et al. 2001). Species with low variability relative to growth rate (low σ/r) are most resistant to extinction, while a high σ/r implies that local extinction is likely without refuges or dispersal among populations; carrying-capacity-dependent species have intermediate σ/r ratios, so that persistence is increased in larger populations.

A third predictor of extinction is body size, where larger-bodied animals tend to be more vulnerable to extinction both historically (as in the Pleistocene era) and currently (Brook & Bowman 2005, Cardillo et al. 2005). As body size increases for animal species, population growth rate and density tend to decrease while home-range size increases; furthermore, larger animals are more vulnerable to harvest and other human-caused threats.

Of course, all such rules of thumb for vulnerability are tempered by the reality that simple predictions may be overwhelmed by the specific situation (Tracy & George 1992, Belovsky et al. 1994). For example, vulnerability of primates and carnivores was underestimated by a model based on species characteristics in cases where the species had lost habitat, been commercially overexploited, or suffered from problems created by exotic species (Purvis et al. 2000). Similarly, extinction vulnerability for 145 Australian marsupial species depended more on geographical-range overlap with sheep than it did on species characteristics such as body size, reproductive rate, or habitat specialization (Fisher et al. 2003). Vulnerability may also derive in part from particular behavioral attributes such as Allee effects or naiveté toward predators (e.g. the passenger pigeon; Reed 1999).

Box 12.1 The 50–500 rule

As an historical footnote while considering characteristics that may predict extinction vulnerability, it is worth reviewing the famous **50–500 rule**. This rule of thumb emerged from the application of conservation genetics to wild species (Franklin 1980, Soulé 1980, Frankel & Soulé 1981), and over the next decade was swept up into management plans (e.g. the Puerto Rican parrot recovery plan) and a number of biological opinions (Mills et al. 2005). In essence, the rule provides a minimum genetic effective size for short- and long-term protection.

An effective size of 50 was proposed as a minimum to protect against short-term loss of fitness due to inbreeding, based on empirical observations of the decrease in fitness-related traits with incremental inbreeding in a variety of animal species. Several caveats implicit in the original rule were lost as it became applied in management (Soulé 1987, Soulé & Mills 1992). For example, the 50 is the **genetic effective size** (N), which is often one-fifth to one-third that of the total population size (Chapter 9); thus an N_e value of 50 translates to 150–250 or so actual animals. Second, the rule was proposed as a short-term guideline for captive breeding and similar holding operations, not to the long-term survival of wild populations which would have many other factors affecting their persistence. Third, this was a rule based purely on genetic factors, not incorporating the other factors that would again increase the minimum necessary size for persistence. Based on these considerations, it is untenable to argue that an actual population size of 50 is sufficient as a rule to support any wildlife population into the future.

A value of 500 was proposed as the minimum size (N_e) necessary to ensure long-term maintenance of genetic variation, thereby preserving evolutionary options for future adaptation. In more formal terms, the number was the estimated minimum genetic effective size where the loss of additive genetic variation of a quantitative character due to genetic drift would be balanced by new variation due to mutations. This number has received serious scrutiny by population geneticists, with arguments to increase it to as large as 5000 or more (Frankham et al. 2002). As Allendorf and Ryman (2002) note, this debate will likely continue, but there is little doubt that the *actual population size* (as opposed to the genetic effective population size) necessary to maintain evolutionary potential for the long term should be thousands of individuals and not hundreds.

The extinction vortex

In managing small and declining populations, the overriding factors to consider are what caused the population to become small, and how to reverse the decline. Whether the cause of decline was habitat loss or fragmentation, overharvest, exotic species, climate change or some combination of these or other causes, reversing the human-human-caused deterministic perturbations that led to the decline is of paramount importance to successful recovery (Chapter 11).

Unfortunately, when a population becomes small, it becomes particularly susceptible to a host of stochastic threats that interact with and exacerbate problems caused by deterministic factors (Shaffer 1987). So, even if the deterministic problems were reversed so that a small population achieved a positive average population growth, the population could still stumble toward extinction. There are three main types of stochasticity that affect persistence: demographic, environmental, and genetic (see Chapters 5 and 9). As a reminder, demographic stochasticity causes variation because mean vital rates are probabilistic, so as numbers of individuals become few there can be large deviations from mean survival, fecundity, and sex-ratio expectations. Environmental stochasticity refers to random changes in the mean vital rates for the population, often driven by weather. Genetic stochasticity arises from genetic drift; alleles that are harmful accumulate as homozygotes and thereby reduce demographic rates.

Both deterministic and stochastic factors interact in small populations to drive the **extinction vortex** (Fig. 12.1). The extinction vortex makes it very clear that when

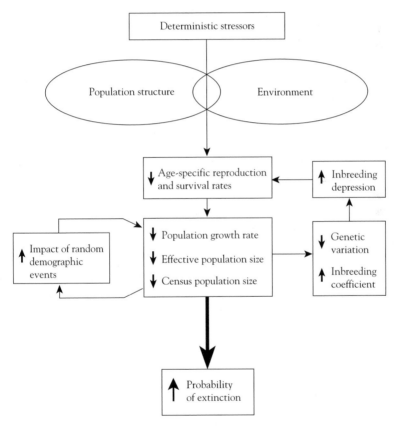

Fig. 12.1 A simplified representation of the extinction vortex. The effects of deterministic stressors are filtered by the population's environment (habitat as well as variable extrinsic factors such as weather, competition, predators, and food abundance) and by its structure (including age structure, sex ratio, behavioral interactions, distribution, physiological status, and intrinsic birth and death rates). Each turn of the feedback cycle increases extinction probability (Gilpin & Soulé 1986). The extinction vortex model predicts that some small populations are more likely to become smaller and eventually go extinct with each generation due to the interaction of genetic and nongenetic factors. Modified from Soulé and Mills (1998). Copyright (1998) AAAS.

evaluating persistence we should not emphasize one factor (e.g. the cause of decline, genetic stochasticity, or environmental stochasticity) and disregard the others. Rather, management actions are best judged against the relative importance of the different factors, and how they interact in any particular case (Lande 1988a, Mills & Smouse 1994).

Because the extinction vortex is by definition an interaction among a number of ecological factors (including virtually every process discussed in this book), it is difficult to see it working in the field. For that reason, much of the best support for the extinction vortex has come from manipulative laboratory studies (Box 12.2).

There are, however, some instructive examples from the field, such as the case of the Illinois population of the greater prairie-chicken discussed in Chapter 9. Despite reasonable success in reversing the deterministic factors that had affected it, the population continued to decline, at least in part due to genetic and perhaps demographic stochasticity. A management action of translocations has decreased inbreeding depression and helped the population increase. Of course, continued vigilance in addressing the deterministic causes of decline remains necessary (Maehr & Lacy 2002).

Predicting risks in small populations

To confront the extinction vortex requires a formal framework to assess viability. The intellectual roots for assessing viability in wildlife biology go back at least to the 1930s

---

Box 12.2 Insights into the extinction vortex from a model system

In a clever test of the extinction vortex, fruit flies were exposed to changing and stressful environments, not unlike those experienced by endangered species (Bijlsma et al. 2000). The replicated worlds were vials 22 mm in diameter holding up to about 120 flies. High temperature and different levels of ethanol produced treatments with stressful conditions. For each treatment of environmental condition and inbreeding level 50 populations were followed over eight generations (imagine the difficulty of doing this experiment with your favorite wild mammal or bird species).

As the results show (see figure), control populations (top panel) had relatively low extinction probabilities over time, although inbred populations (with inbreeding coefficients, F, greater than 0) had higher extinction rates. Under stressful conditions extinction rates increased (middle and lower panels), with inbreeding exacerbating the extinction risk. The researchers repeated the experiment 2 years later and found the same result: extinction rate was elevated by environmental stress conditions, and highly inbred populations were much more likely to go extinct when under environmental stress than the less inbred populations. Inbreeding and environmental stress act synergistically, making a convincing general case that the extinction vortex is real and that genetic effects should not be considered independently from environmental and demographic effects.

Box 12.2 Continued

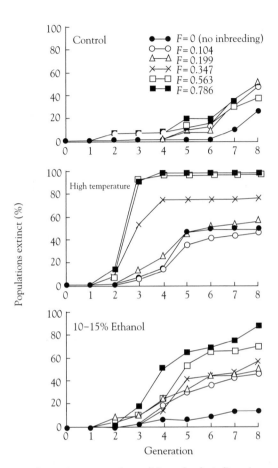

Extinctions over time under various sets of conditions for fruit flies that are not inbred ($F=0$) or are inbred to various levels. Modified from Bijlsma et al. (2000). Reproduced by permission of Blackwell Publishing Ltd.

(Reed et al. 1998), when Aldo Leopold (1933:47) noted the importance of recognizing "the minimum number of individuals which may successfully exist as a detached population." More recently, viability assessment has centered on quantitative models of **population viability analysis (PVA)**, linking science to specific public-policy applications (Shaffer et al. 2002).

Assessment of viability should place intuition, theory, and field data into an operational framework to allow insights into factors that caused decline and that may cause further decline in the future. At its best, viability assessments make both hard data and best guesses explicit, so that the input and output can be honestly debated. Because it is a framework for incorporating multiple, interacting processes, viability assessment can reveal non-intuitive and non-obvious outcomes that can assist management in surprising and important ways. As the quantitative arm of general viability assessment, PVA methods can go further to quantify these risks explicitly.

Population viability analysis: quantitative methods of assessing viability

PVA defined

PVA can be defined as: the application of data and models to estimate probabilities that a population will persist for specified times into the future (and to give insights into factors that constitute the biggest threats). In a real sense, PVA incorporates virtually every concept of applied population biology that we have discussed so far in this book.

Notice that this definition says nothing about a **minimum viable population** (MVP), a term that dominated much of conservation biology in the 1980s (see Gilpin & Soulé 1986, Soulé & Mills 1992). This is because PVA approaches embrace the idea behind MVP, while making MVP itself obsolete. MVP is problematic for both philosophical and scientific reasons. Philosophically, it seems questionable to presume to manage for the minimum number of individuals that could persist on this planet. Scientifically, the problem is that we simply cannot correctly determine a single minimum number of individuals that will be viable for the long term, because of the inherent uncertainty in nature and management, including ecological processes, management scenarios, and measurement of vital rates and trends in wild populations. Finally, the number of individuals required to carry out ecological functions – including nutrient cycling or limitation of prey numbers – may be much bigger than the minimum needed for that species to persist (Soulé et al. 2003). Therefore, instead of a futile focus on a single number (MVP), a much more constructive and reliable philosophy evaluates a range of effects for a range of possibilities.

Three components of PVA

As defined above, PVA has three central concepts: persistence, time, and probabilities.

Persistence

Persistence is commonly considered to be **not extinct**, implying that a population remains above zero individuals (or one mating pair). Although extinction per se is indeed an important threshold, there are often other thresholds that are useful to track for biological or management-based reasons. These **quasi-extinction thresholds** might include, for example, biological thresholds below which Allee effects occur or where strongly interacting species become unable to carry out critical ecosystem functions. The quasi-extinction threshold may also include management thresholds such as the triggering number to bring a wild population into captivity, or the abundance below which a threatened species would receive special management (Ginzburg et al.

1990, Burgman et al. 1993). For the rest of this section, therefore, extinction is meant in the broad sense, including both true extinction as well as management or quasi-extinction thresholds.

Time

Time is a second important component of the PVA definition. As with many other predictions (e.g. weather or stock markets), the assumptions used in PVA will be less reliable further into the future, reducing predictive accuracy. Therefore, PVA in endangered species recovery plans should incorporate short-term projections evaluated against a long-term goal (Scott et al. 1995, Goodman 2002). The long-term viability assessment includes management goals relatively free of political and legal considerations (i.e. they should be biologically based). The short-term projections show trade-offs for a range of options explicitly incorporating political, legal, social, and economic constraints; monitoring and the iterative use of short-term PVAs evaluate how well the goal of long-term persistence is being met. Thus public input (and political tradeoffs) can be incorporated in choosing short-term management strategies, but ultimate success is judged against the yardstick of the long-term, biologically based goal. An analogy is useful for thinking about the relative merits of short-term, more reliable projections compared with long-term projections: "We can see only as far as our headlights reach, but we need to be concerned about what lies beyond their reach" (Allendorf & Ryman 2002:77).

Probabilities

A key underlying component of the definition of PVA is that it involves predicting probabilities. Obviously, a higher probability of persistence over a given time frame will require a larger initial population size, all other things being equal. In quantitative PVAs, probabilities are often visually displayed with **quasi-extinction curves** (see Burgman et al. 1993, Groom & Pascual 1998, Akçakaya 2000). There are many ways these can be portrayed, but a common approach uses **cumulative distribution functions**, representing the cumulative probability of reaching a quasi-extinction threshold over a range of time periods (Fig. 12.2a), or of declining to different population sizes during a fixed time period (see Fig. 12.4, below, for an example)[2]. A related metric expresses the risk of decline by a given amount relative to the initial population (Fig. 12.2b).

---

[2]The median quasi-extinction time or population size can be read off the graph. For example, in Fig. 12.2a the median quasi-extinction time is 25 years, the time that it takes to reach a 50% quasi-extinction cumulative distribution function. When it accompanies a cumulative distribution function the median is a useful summary metric; avoid reporting only the mean extinction time because extinction-time distributions have a long tail, consisting of the small probability of lasting a very long time, which causes the mean to be much higher than the median and overestimates the probability of safety for the most likely population (see Burgman et al. 1993, Akçakaya 2000).

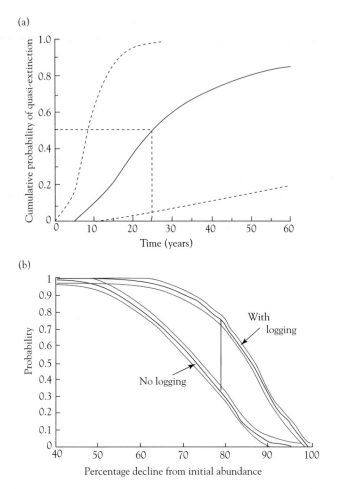

Fig. 12.2 Likelihoods of quasi-extinctions may be portrayed in many different ways. (a) An example of a plot of cumulative distribution function showing the cumulative probability of 12 California condors declining to one bird (the quasi-extinction threshold) as time passes (from Dennis et al. 1991; reproduced by permission of the ESA). The dotted lines represent the 95% confidence interval. As the dashed line shows, the median time to extinction is 25 years, the time that it takes to reach a 50% quasi-extinction probability. (b) A different way of portraying risk: the probability of declining by some percentage of the original population size. The graph shows the risk of decline of a northern spotted owl metapopulation, simulated with a demographically explicit model (modified from Akçakaya & Raphael 1998; reproduced by kind permission of Springer Science and Business Media). The top curve gives risk under a timber-harvest scenario, and the bottom curve assumes no habitat loss. Each point on the curve indicates the probability that the overall abundance will decrease by some percentage from the initial abundance during a 100-year interval. In this example the maximum difference between the curves is at a 78% decline (marked by a vertical line); the probability of this level of decline from the initial abundance is approximately 77% with logging and 33% without logging.

Consideration of all of these PVA components – viability thresholds, persistence times, and probabilities of persistence – argue for examining alternatives, with a range of inputs and outputs, instead of a single analysis with X data input for Y probability of persistence over Z years. Also, it should be clear that PVA has a strong biological basis but the selection of goals requires a social component. Obviously, issues such as for how long we want to evaluate persistence and how secure that persistence should be require social, cultural, economic, and political considerations (Shaffer 1987, Ludwig & Walters 2002).

Finally, remember for any PVA that the scale of the analysis should be linked to the population being analyzed and the management perturbations being considered (Ruggiero et al. 1994, Gärdenfors et al. 2001). These different scales may not match, for example if the analysis is for only one National Forest while the perturbations are occurring regionwide and the species persists across an even larger scale (say, continent-wide for a large carnivore species).

How to conduct a PVA

There are two main methods to perform a formal, quantitative PVA for a single population. One type uses time series, or counts over time. The other uses demographic rates, such as reproduction, survival, age structure, and density dependence. Both approaches can be extended to multiple populations, and presence/absence data can be used for a third type of multiple-population PVA.

Time-series PVA

A series of abundance estimates can be used to estimate probabilities of a population reaching quasi-extinction. The mathematical approaches become complicated (for readable details see Morris & Doak 2002), but the bedrock underlying all of these approaches is relatively simple. The method builds on the average trend ($\hat{r}$, often denoted in PVA as $\hat{\mu}$) and its variance (σ^2), estimated from a series of abundance estimates (see Oli et al. 2001 for an example for endangered Gulf Coast beach mice). Assuming that the future will have similar growth and bounce (variation) as the past, one can calculate the future probability of a population bouncing its way down to some threshold quasi-extinction or management threshold[3]. The math captures the non-intuitive but important fact that stochasticity can cause even populations with positive growth rates to decline to extinction (Chapter 5), implying that managing variability in population growth can be as important as managing the mean growth rate (Burgman et al. 1993:73).

To see how time-series PVA works, consider estimates of abundance for gray whales off the central California coast from 1967 through 2001 (Fig. 12.3a). The population

---

[3]How long a time series is needed to estimate extinction risk for a single population? At the very least, 10–15 time steps (e.g. years) are needed to be able to characterize population growth and correlation structure, although considerably more may be necessary to properly capture variance in growth rates or determine density-dependence structure (Holmes 2001, Brook & Bradshaw 2006).

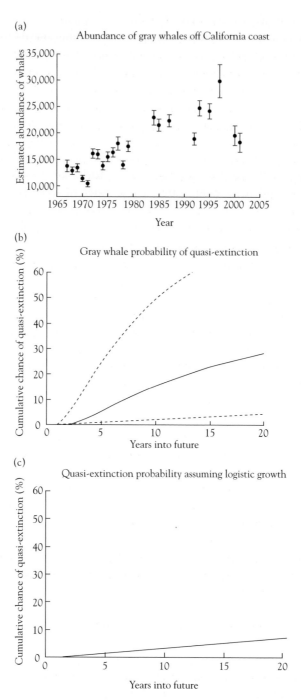

Fig. 12.3 Time-series PVA based on gray whales off the coast of California. (a) Abundance estimates over time, with SE bars representing sample variance (data from Rugh et al. 2005; see also Wade 2002a). (b) The cumulative distribution function (and its 90% confidence interval) of the density-independent quasi-extinction probability of decline from the 2001 abundance of 18,178 to a quasi-extinction threshold of 10,000 or fewer whales. (c) Quasi-extinction probability, as in (b) except that a logistic growth model of negative density dependence is assumed.

was depleted by commercial whaling in the late 1800s. Legal protection began in 1946, and by 1994 the population had recovered enough to be removed from the list of endangered and threatened wildlife (Gerber et al. 1999). The abundance estimates used in the time series were collected while whales were migrating south during the winter. Multiple simultaneous observers independently viewed whales from a bluff above the water, facilitating an estimate of probability of detection, abundance, and sample variance (Reilly 1992, Rugh et al. 2005). Looking at the time series (Fig. 12.3a), it appears that the population had been increasing over time. More formally, trend analysis using the regression method demonstrated in Chapter 5 gives an $\hat{r}$ and 90% confidence interval of 0.0082 ± 0.043. Translating $\hat{r}$ to λ provides an estimated λ value of 1.0082 with a 90% confidence interval of (0.97, 1.052). Thus the most likely trajectory of the gray whales has been an increase of just under 1% per year, with up to a 3% decrease or a 5% increase being consistent with the data.

If the time series is assumed to be density-independent, the so-called **diffusion approximation model** of Dennis et al. (1991) can estimate quasi-extinction probability. For the gray whale example, I set the quasi-extinction threshold at 10,000, a little smaller than the lowest number of whales during the 40 years of the time series; obviously, if you were interested in the probability of reaching a much lower quasi-extinction threshold, of say 1000 or 500 whales over 20 years, the likelihood would be much lower[4]. In this case, the best estimate of the probability that the population will decline from its abundance of roughly 18,000 whales (in 2001–2) to 10,000 is about 15% in 10 years and 28% in 20 years (Fig. 12.3b).

Elaborations of this basic density-independent PVA of time-series data can be extended to account for many real-world complexities, including correlations or changes in average trajectory over time, outliers, and density dependence (Morris & Doak 2002). Density dependence is often important to model because of its impacts on real populations. In general, positive density dependence (e.g. Allee effects) will tend to increase extinctions in time-series PVA, while negative density dependence has more complicated effects: at low numbers it tends to reduce extinction probability because population growth increases, but the regulatory effect of negative density dependence will also cap population growth, which could increase extinction probability.

Despite its effect on model outcome, using field data to detect either the form of density dependence or the parameters that determine its shape is exceptionally difficult, leading to three options, as follows.

- Estimate density-dependent structure and parameters from the data using a model-fitting framework (e.g. Akaike's information criterion, AIC); this is undoubtedly the best approach, but it requires high quality and quantity of data.
- Use a density-independent model in the hope that it performs well enough to say something useful about real populations whether or not they are experiencing density dependence.

---

[4]Of course, my choice of 10,000 for the quasi-extinction analysis of this population is for demonstration purposes only.

- Use an array of models with and without density dependence to bracket what might actually be occurring and see if management alternatives are robust to the model form used (Pascual et al. 1997).

A recent analysis by Sabo et al. (2004) argues for a combination of the second and third approaches using time-series data. They found that simple diffusion analysis (density-independent) models can characterize risk in cases where:

- populations subject to density dependence are at abundance levels where density effects are not strong (e.g. at abundances well below K in cases of negative density dependence);
- effects of density are similar in the past – when data are used to estimate the parameters of population growth rate and its variance – and the future for which predictions are being made;
- the form of density dependence is a ceiling (Chapter 6), as might occur under competition for space such as nesting sites;
- the goal is to detect large declines as opposed to small ones; or
- the population of interest is declining, or only slowly recovering.

These conditions cover many instances where we would be interested in performing a PVA in the first place. In such cases, a density-independent time-series PVA model is most likely to either correctly predict future dynamics, or err toward the side of caution by over-estimating the probability of reaching a quasi-extinction threshold (Sabo et al. 2004). If, however, you are interested in estimating chances of moderate declines, or if the population shows signs of recovery, or if biological intuition or sound data indicates the operation of more complex forms of density dependence with feedback across all population sizes (such as logistic; see Chapter 6), then the best approach is to use a density-independent model as well as different forms and shapes of density dependence. Figure 12.3c shows that the probabilities of quasi-extinction for the gray whales decrease with logistic-type density dependence.

Demographically explicit PVA

This class of PVA models uses estimates of demographic vital rates, including age- (or stage-) specific survival and reproduction rates, their variances and covariances, and other biological information such as age structure of the population, density dependence, and costs of inbreeding depression. Despite the data-hungry needs of this method, it has the strong benefit of being able to transcend simple prediction of viability and point towards actions that will most effectively reverse a declining population (Beissinger & Westphal 1998, Reed et al. 2002).

A striking example may be found in matrix-projection-model analysis of highly endangered northern right whales (Fujiwara & Caswell 2001). A large part of the right whale decline to about 300 worldwide can be explained by the steep decline in survival of adult females due to collisions with ships, entanglement with fishing gear, and fluctuations in food (perhaps exacerbated by climate change). The small size of the population means that every individual death has a large effect on survival. Ironically, therein also lies the hope: increasing current survival rates by preventing the deaths of just two or three individual female whales each year could actually reverse the decline of right whales and put them on the road to recovery. When populations are very small, individuals matter (some more than others) and demographically explicit analysis can help show the most efficient path to recovery.

The inclusion of multiple interacting factors in a demographically explicit PVA almost always requires computer simulations, often following the framework of matrix models (Chapter 7). Demographic and environmental stochasticity can be applied to each time step of each replicate (Chapter 5). Modeling demographic stochasticity mimics the phenomenon where small populations can bounce to extinction even if mean vital rates are constant and expected population growth is positive. If environmental stochasticity occurs on a scale outside the typical fluctuations, such as catastrophes, these should be included in the PVA by specifying the magnitude and average timing for their occurrence (Mangel & Tier 1994).

Stochastic fluctuations of vital rates are often not independent over time or among each other. The correlations or covariance that can occur both among vital rates (e.g. a good year for survival implies a good year for reproduction) and through time (e.g. a bad year is likely to be followed by another bad year) can affect both variance in population growth and probability of extinction (Ferson & Burgman 1995, Groom & Pascual 1998). As an example of the effects of correlation (Fig. 12.4), consider data from the endangered Australian possum (see also Box 12.4, below). Positive correlations among vital rates or over time tend to increase extinction probability because bad years are bad for all rates and are likely to be followed by bad rates the next year. Figure 12.4 shows that if all vital rates fluctuated independently, there will almost certainly be more than 40 possums after 5 years; if, however, there were perfect positive correlation in rates over time and among vital rates the population could be smaller

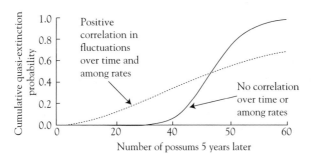

Fig. 12.4 How correlations among vital rates and over time can affect extinction probability. Based on data from the endangered Australian possum. Beginning with 30 reintroduced possums, the population is projected 5 years; the quasi-extinction probability gives the risk that the population will fall to a given number of possums or fewer after 5 years. The solid line assumes independence (no correlation) both among rates and over time; the dotted line assumes perfect correlation in rates over time and among vital rates. With positive correlation there is a higher likelihood of declining to fewer than 30 possums 5 years later. Modified from Ferson and Burgman (1995). Reproduced with permission from Elsevier.

Box 12.3 Incorporating inbreeding costs into demographically explicit PVA models

As we saw in Chapter 9, genetic drift in small populations can decrease heterozygosity, leading to inbreeding depression. The cost of inbreeding depression on fitness is incorporated in PVA models via **lethal equivalents** expressed either per gamete (per haploid genome) or per individual (per diploid genome; twice the number per gamete). Lethal equivalents include the effects of independently acting lethal alleles as well as the cumulative effects of partially deleterious alleles that would kill the individual if made homozygous. To bring this closer to home, humans have been found to carry enough deleterious alleles – lethal equivalents per individual – to kill each of us between two and five times over (Keller 1998). These deleterious alleles tend not to be expressed in individuals in large populations because natural selection holds them at low frequencies and they are usually recessive, so nondeleterious alleles mask their effects. But inbreeding can express the deleterious recessive alleles, leading to reduced survival or reproduction.

Lethal equivalents are estimated by determining (usually through regression) the relationship between the inbreeding coefficient and fitness. The difficulty of measuring both inbreeding level and fitness means that in many PVAs a range of values from other species are used to bracket possible effects. For example, for 40 different nondomestic mammal species in zoos the lethal equivalents per diploid individual for juvenile survival ranged up to 30.3, with a median of 3.1 (Ralls et al. 1988). Over the full range of lifetime reproductive success, field estimates of lethal equivalents may be closer to 12 (O'Grady et al. 2006).

When decrementing vital rates with loss of genetic variation in a PVA model, another consideration includes whether or not the cost of inbreeding is constant. The shape of the curve relating inbreeding to fitness is a complex topic that includes whether fitness interactions among genetic loci are synergistic and whether there is a threshold level of inbreeding above which the costs get worse (see Frankham 1995), as well as the extent to which inbreeding depression is purged over time (see Chapter 9). Different PVA programs account for synergistic effects and purging in different ways (e.g. Mills & Smouse 1994, Lacy 2000).

---

than 10 after 5 years. In the face of uncertainty about correlations, one should model worst-case scenarios with perfect positive correlations among rates and over time, and best-case scenarios based on fluctuations that are independent or negatively correlated with each other and over time[5]. The real biology will almost certainly be somewhere in between.

In addition to stochasticity, demographically explicit PVAs can include consequences of inbreeding due to genetic drift in a small population (Chapter 9). Genetic stochasticity is incorporated by specifying how demographic rates decline with increased inbreeding (Box 12.3). The range of inbreeding expected for any species, as well as uncertainty in the shape of the curve relating inbreeding to fitness, means that incorporating genetic stochasticity is the same as with any other uncertain parameter

---

[5] The generalization that positive temporal autocorrelation increases extinction probability does not hold for all forms of density dependence (see Morris & Doak 2002).

in PVA (including, for example, dispersal rates, density dependence, breeding structure, correlations among rates and over time, and so on): you should include a range of plausible possibilities, including worst-case and best-case scenarios.

Demographically explicit models also accommodate density dependence. As with time-series PVA approaches, density dependence is one of the hardest functional relationships to specify for field populations, yet it can drastically affect the PVA predictions. Density dependence could save populations from extinction by increasing population growth rate at very low numbers, or it could increase long-term extinction probability via Allee effects or because a cap on population size limits the potential to escape from low numbers. Again, the recommendation is to include at least one set of runs without density dependence, to provide a baseline understanding of extinction risks (Ginzburg et al. 1990, Mills et al. 1996).

The framework for building demographically explicit PVAs varies widely. In some cases, commercial or shareware PVA programs suffice. Two of the most popular are the matrix-based RAMAS (Akçakaya et al. 2004) and the individual-based VORTEX (Lacy 2000). In other cases, particular aspects of proposed management options or of the animal's life history or behavior require development of species-specific PVA programs [e.g. for African wild dogs (Vucetich & Creel 1999), cheetahs (Kelly & Durant 2000), sage grouse (Johnson & Braun 1999), and red-cockaded woodpeckers (Daniels et al. 2000, Walters et al. 2002)]. Figure 12.5 gives an example.

As mentioned above, demographically explicit PVAs allow the user to take the next step of playing out multiple what-if-type scenarios in sensitivity analyses (Chapter 7), evaluating how different management actions affect population recovery, potentially including human demographic, economic, and social systems (Lacy & Miller 2002). That is one of their biggest strengths. The obvious tradeoff is that demographically explicit approaches are data-hungry. Clearly, there comes a point where ignorance of input values makes it an exercise in misleading futility to try to parameterize a complex model with poor or non-existent data; in such cases the time-series approach or one of the other approaches described below may be more appropriate. However, in the zone where the match between model needs and data availability is reasonable, one can embrace uncertainty by acknowledging it explicitly and considering scenarios across a range of plausible values.

PVA with multiple populations

Any PVA approach for a single population (including the two described above) can be scaled up to consider multiple populations across the landscape. With sufficient data, multiple-population PVA models can be spatially explicit, incorporating exact locations of populations or individuals or other features (Reed et al. 2002). Levels of connectivity – and its positive and negative effects – as well as correlation in dynamics among the populations (Chapter 10) can be incorporated. The extent of correlation, or coupling, in environmental stochasticity is often related to the distance among populations, because similar climatic events or other perturbations (invasions by exotics, deaths by predators or disease, etc.) are more likely to occur simultaneously in populations that are close together. Correlations among population dynamics are

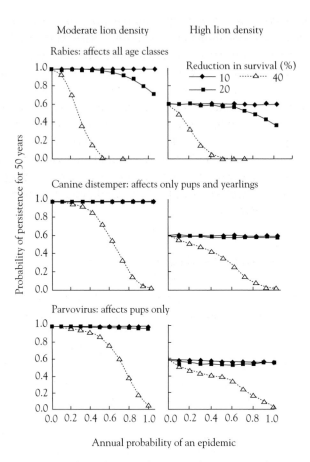

Fig. 12.5 An example output for a demographically explicit PVA. In this case, one question of interest was how both predators (lions) and diseases of differing virulence would affect African wild dog populations. The panels (modified from Vucetich & Creel 1999) show probability of persistence for 50 years for a population of 98 wild dogs, in cases where disease reduces wild dog survival by 10, 20, or 40%. The left-hand panels give results when wild dogs are in the presence of moderate lion density (100 adult lions/1000 km²) and the right-hand panels represent high lion density (131 adult lions/1000 km²). The top panels show results with rabies, which affects all age classes equally, the middle panels show the outcome of canine distemper which primarily affects pups and yearlings (and adults only to a lesser extent), and the bottom panels are for canine parvovirus, affecting only pups. Both lions and disease affect wild dog persistence.

also facilitated by movement of individuals; for example, Canada lynx populations across western North America are connected by gene flow (Chapter 10), which may facilitate the relatively synchronous dynamics of lynx populations at the continent-wide scale.

In addition to application of time-series or demographically explicit models across more than one population, a different type of multi-population PVA uses occupancy data (presence/absence) for a species on multiple patches. These are broadly called **patch-occupancy models**, including incidence function models and logistic regression modeling of colonization and extinction events (Sjögren-Gulve & Hanski 2000, Hanski 2002). The key data are whether a patch is occupied and the size of distinct patches. Connectivity data include rate of dispersal and distance between patches (metrics of connectivity other than distance can be used if available). Patch-occupancy models infer persistence from the minimum amount and distribution of occupied habitat that will support the species. I will not elaborate on their complexities (Moilanen 2000, Hanski 2002, Morris & Doak 2002) because application to most species and applied situations may be fairly limited: patch-occupancy approaches ignore local population dynamics and are data-intensive, with more than 20 patches needing to be sampled for at least two presence/absence surveys.

Other approaches to assessing viability

The worst-case scenario for a biologist is to conduct an assessment of viability when time is short and data are scarce to non-existent. And yet, it is not unusual for biologists to be asked to conduct a PVA with neither the time nor the data to conduct quantitative PVA using the time-series analysis, demographic rates, or multiple population models as described above. For example, in 1993, President Clinton appointed a Forest Ecosystem Management Assessment Team (FEMAT) to evaluate the effects of large-scale timber-harvest options on wildlife species in western Washington, Oregon, and northern California (Meslow et al. 1994, Thomas 1994). More than 1000 plant and animal species were to be included in the assessment, including many species that were (and are) little known. The team had 3 months to complete the job.

In the case of the 1000 species assessed as part of FEMAT, the best that could be done was a subjective expert-panel-type approach to assess viability. This method had evolved from earlier use in analyses in the Pacific northwest of the USA (e.g. Thomas et al. 1993), and continued to evolve after FEMAT (Marcot et al. 1997). **Expert opinion** or other subjective approaches to assess viability are problematic because humans are inherently bad at guessing risks, even when they are informed guesses. We are led astray by how visible or controllable the risk appears, and by the consequences of the risks (Burgman et al. 1993). Thus, we overestimate many low-level risks (e.g. death by tornado or anthrax) and underestimate high-level risks (e.g. death by heart disease or automobile accident). Also, subjective decision-making is idiosyncratic to the experiences of the expert making the decision: the term severe risk will mean different things to different people. It is difficult to make transparent or testable the logic, mechanisms, predictor variables, sources of uncertainty, or other processes that go into the outcome of a subjective judgment (McCarthy et al. 2004). In short, expert-opinion assessments of viability remain an uncomfortable and insufficient last resort.

Therefore, to close the discussion of viability assessment, we will consider two less data-intensive methods that are not part of PVA per se, but that can be used to assess viability when detailed population data are not available.

Rules of thumb

Rule-of-thumb approaches assign qualitative ranks of risk using specified, operational criteria, such as those developed by the Nature Conservancy (Master et al. 2000, Samson 2002) and the World Conservation Union (the International Union for the Conservation of Nature and Natural Resources; IUCN) Red List Categorization system (Mace 1995, IUCN 2001a). As an example I will focus on the IUCN system, which forms the basis for **Red Lists** assessing the conservation status of more than 18,000 plant and animal species worldwide. The IUCN approach assigns species to one of nine categories (Fig. 12.6a). To be placed in one of the three categories at risk of extinction

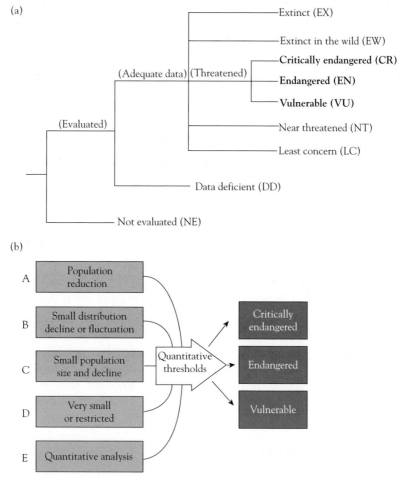

Fig. 12.6 The IUCN population-assessment procedures. (a) Evaluated species are classified into one of nine categories. The three categories at risk of extinction are shown in bold. Modified from IUCN (2001b). (b) Five rule-of-thumb criteria are used to place a species in one of the three categories of extinction threat (the highlighted categories in panel a). At least one of the quantitative thresholds for the criteria must be met (from Gärdenfors et al. 2001; see also IUCN 2001b). Reproduced with permission of Blackwell Publishing Ltd.

(critically endangered, endangered, or vulnerable), at least one operational rule-of-thumb criterion must apply: (i) deep declines in population size, (ii) reduction or fluctuation in geographic range or number of populations, (iii) small population size coupled with decline or fluctuations, (iv) very small or restricted population, or (v) quantitative analysis (Fig. 12.6b). The fifth criterion for assigning species – quantitative analysis – includes a direct quantitative estimate of extinction probability within specified time frames using a PVA, although data limitations mean that this criterion is rarely implemented (Mace & Lande 1991, Gärdenfors 2000). Table 12.1 lists the rule-of-thumb thresholds for several species categorized as critically endangered (one of the extinction-threat categories).

Sophisticated methods for making uncertainty explicit in the risk-assessment procedure have been proposed for IUCN categorization (Todd & Burgman 1998, Akçakaya et al. 2000, Taylor et al. 2002). The IUCN uses the **precautionary principle** as one way to deal with uncertainties: the credible estimate that gives the highest risk of extinction is used, so that uncertainties favor more cautious management approaches.

Another key philosophy behind the IUCN approach underscores an important general point about management of small populations: a distinction is made between assessing the severity of threat and setting conservation priorities (Mace 1994, 1995, Gärdenfors et al. 2001). Categories of threat established by the rules of thumb are just one piece of information used to set conservation priorities. Other important criteria might include the likelihood of success in restoring the species, the number of other threatened species occupying the same habitat, taxonomic uniqueness, availability of funds, and the legal and political framework for conserving a particular species.

There are obvious limitations to any rule-of-thumb approach, yet they may be the best available method at times. Mace and Hudson (1999:244) report that

> . . . although the IUCN system may be efficient at picking up different species facing diverse threats, it is not designed to be an accurate tool for measuring extinction risk, for projecting population status, or for designing population management plans. Its role is to highlight species exhibiting one of several symptoms of pending extinction and to classify species according to the relative severity of the apparent threat. The Red List is a useful conservation tool only when listing leads to measures to assess the causes of threat and to develop, where necessary, appropriate management responses and species recovery plans. In short, the IUCN Red List criteria are designed to be robust and precautionary across a wide range of circumstances, to operate when data are scarce, and to pinpoint species in need of attention.

Approaches based on habitat and other information

The presence of habitat alone cannot constitute an assessment of population viability. As Kent Redford (1992) noted for wildlife in tropical forests: "The presence of soaring, buttressed tropical trees, however, does not guarantee the presence of resident fauna . . . although satellites passing overhead may reassuringly register them as forest, they are empty of much of the faunal richness valued by humans. An empty forest is a doomed forest." Habitat is necessary, but not sufficient, to guarantee population

Table 12.1 Examples of assigning species into one of the IUCN risk categories. To demonstrate the system, the table shows the operational criteria for listing into one of the IUCN at risk categories, critically endangered (IUCN 2001b). Notice that the criteria (A–E) are the rule-of-thumb thresholds based on population-ecology principles shown in Fig. 12.6(b). As examples, the characteristics that led to the assigning of this category are shown for three species: bactrian camel, Iberian lynx, and bawbaw frog (see the IUCN website, www.iucnredlist.org). In some cases a species triggers more criteria than necessary for listing. Note that the same approach would be taken for the other categories of extinction threat, but the thresholds differ. A species is critically endangered when the best available evidence indicates that it meets any of the following criteria (A–E), and it is therefore considered to be facing an extremely high risk of extinction in the wild.

IUCN criteria	Species
A. Reduction in population size based on any of the following	
1. An observed, estimated, inferred, or suspected population size reduction of ≥90% over the last 10 years or three generations, whichever is longer, where the causes of the reduction are clearly reversible and understood and ceased, based on any of the following:	
(a) direct observation;	
(b) an index of abundance appropriate to the taxon;	
(c) a decline in area of occupancy, extent of occurrence, and/or quality of habitat;	
(d) actual or potential levels of exploitation,	
(e) the effects of introduced taxa, hybridization, pathogens, pollutants, competitors, or parasites.	
2. An observed, estimated, inferred, or suspected population size reduction of ≥80% over the last 10 years or three generations, whichever is the longer, where the reduction or its causes may not have ceased or may not be understood or may not be reversible, based on (and specifying) any of (a)–(e) under A1.	Frog
3. A population size reduction of ≥80%, projected, or suspected to be met within the next 10 years or three generations, whichever is the longer (up to a maximum of 100 years), based on (and specifying) any of (b)–(e) under A1.	Camel
4. An observed, estimated, inferred, projected, or suspected population size reduction of ≥80% over any 10-year or three-generation period, whichever is longer (up to a maximum of 100 years in the future), where the time period must include both the past and the future, and where the reduction or its causes may not have ceased or may not be understood or may not be reversible, based on (and specifying) any of (a)–(e) under A1.	Camel
B. Geographic range in the form of either B1 (extent of occurrence) or B2 (area of occupancy) or both	
1. Extent of occurrence estimated to be less than 100 km², and estimates indicating at least two of (a)–(c).	
a. Severely fragmented or known to exist at only a single location.	Frog

Table 12.1 Continued

b. Continuing decline, observed, inferred, or projected, in any of the following:	
(i) extent of occurrence;	Frog
(ii) area of occupancy;	Frog
(iii) area, extent, and/or quality of habitat;	Frog
(iv) number of locations or subpopulations;	
(v) number of mature individuals.	
c. Extreme fluctuations in any of the following:	
(i) extent of occurrence;	
(ii) area of occupancy;	
(iii) number of locations or subpopulations;	
(iv) number of mature individuals.	

2. Area of occupancy estimated to be less than 10 km², and estimates indicating at least two of (a)–(c).

a. Severely fragmented or known to exist at only a single location.	Frog
b. Continuing decline, observed, inferred, or projected, in any of the following:	
(i) extent of occurrence;	Frog
(ii) area of occupancy;	Frog
(iii) area, extent, and/or quality of habitat;	Frog
(iv) number of locations or subpopulations;	Frog
(v) number of mature individuals.	Frog
c. Extreme fluctuations in any of the following:	
(i) extent of occurrence;	
(ii) area of occupancy;	
(iii) number of locations or subpopulations;	
(iv) number of mature individuals.	

C. Population size estimated to number fewer than 250 mature individuals and either

1. An estimated continuing decline of at least 25% within 3 years or one generation, whichever is longer (up to a maximum of 100 years in the future) or

2. A continuing decline, observed, projected, or inferred, in numbers of mature individuals and at least one of the following (a or b):

(a) population structure in the form of one of the following:	
(i) no subpopulation estimated to contain more than 50 mature individuals or	Lynx
(ii) at least 90% of mature individuals in one subpopulation.	
(b) extreme fluctuations in number of mature individuals.	

D. Population size estimated to number fewer than 50 mature individuals

E. Quantitative analysis showing the probability of extinction in the wild is at least 50% within 10 years or three generations, whichever is the longer (up to a maximum of 100 years)

persistence or to predict what will happen to the population in the future (Belovsky et al. 1994).

Nevertheless, for certain species at certain times much more information exists for habitat relations than for demographic vital rates, and the habitat information alone can contribute useful information to assessments of viability (Boyce 1992). Recently, researchers in federal land management agencies in the USA have combined habitat associations with other information for region-wide or species-wide assessments in a **Bayesian belief network** (Lee 2000, Marcot et al. 2001, Raphael et al. 2001). The inputs to the Bayesian belief network include associations with habitat and other variables, as well as expert opinion and ancillary models (including true PVA models). When expert opinion is included, it is incorporated in a way that can be scrutinized easily. Using a Bayesian statistical framework (Chapter 2), input variable values are combined with conditional probability tables to estimate the probability of a response relevant to population status. Risks associated with alternative courses of actions can be explored. As an example, viability of 28 species has recently been assessed as part of land planning for the 58-million-ha Interior Columbia River basin (see Wisdom et al. 2002 for assessment of greater sage-grouse and Rowland et al. 2003 for wolverine).

Marcot et al. (2001:29–30) describe both the utility and limitations of these approaches, noting that when scant scientific data are available but decision-making is nevertheless moving forward:

> . . . The experts must provide their best professional evaluation or step aside and let activities proceed without their input . . . Our [Bayesian belief network] models of wildlife population response, however, do not substitute for empirically based, quantitative, stochastic analyses of population demography, genetics, and persistence such as those used in population viability analysis . . . [the Bayesian belief network approach is] most useful when empirical data on population trends, demography, and genetics are unavailable.

Some closing thoughts about assessing viability

The primary benefit of assessing viability is that it forces us to be explicit about the threats to a population. It puts assumptions on the table so that people can honestly debate and disagree. Sometimes the important factors for management focus turn out to be non-intuitive, emerging only when multiple factors are analyzed in the synthetic framework of a viability assessment. For the same reasons, PVAs also help identify surprising gaps in knowledge to target with research.

Some recent references with excellent practical lists of dos and don'ts for PVA include Akçakaya and Sjögren-Gulve (2000), Burgman and Possingham (2000), and Reed et al. (2002). Distilling from these works and others, and from personal experience, I will close with three overlapping take-home messages.

First, remain acutely aware of the quality of the data available, and the match between the data and the model. One of the hottest topics in PVA and related disciplines is the effect of data quality and sampling error on model performance (e.g.

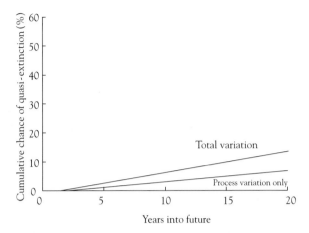

Fig. 12.7 An example of sample variance inflating quasi-extinction probabilities for the whale analysis in Fig. 12.3. Here I plot extinction probability under logistic growth if we had simply included total variation (process variance + sample variance) in the analysis of quasi-extinction under logistic growth. Notice that it is higher than the quasi-extinction probability under logistic growth using process variance only (which is the line from Fig. 12.3c).

Ludwig 1999, Fieberg & Ellner 2000, Sæther & Engen 2002). As we have seen, variation in vital rates, abundances, and carrying capacities are a function of both process variance (true temporal and spatial variation), and variance that arises from the fact that we must sample nature (sampling variance or observation error). Sampling error will make nature seem more variable than it really is, which will tend to overestimate the predicted probability of extinction. Figure 12.7 shows how the cumulative distribution function for quasi-extinction probabilities for whales under density dependence would be inflated if I had used the total variance in growth rate generated by the abundance estimates instead of discounting total variance by the estimated average sample variance. To be sure, wildlife biologists should heed suggestions on how parameters of PVA models can be estimated such that sampling error can be quantified (White 2000, White et al. 2002). And if your attempt to parameterize a complex model hemorrhages with missing data, do not try to force a model that requires those data.

Second, remember that viability assessments are more useful as a comparative tool for ranking management options than for making precise predictions of extinction. Although the scientific process underpinning PVA can provide a sound basis for predicting actual population trajectories, such precise predictions will typically be limited by the quality and quantity of data (Brook et al. 2000, 2002). Testing the absolute accuracy of PVA models is a complex yet important topic (McCarthy et al. 2001, Belovsky et al. 2002, Ellner et al. 2002). In general, the quality of data will rarely be ideal, especially for the more realistic (detailed) models, and we will almost always be ignorant about the specific future changes (natural or human-caused) that should be included in the predictive model. Using PVA to evaluate the relative merits of different management options allows it to be incorporated into the decision-making that guides management action and policy (Maguire 1991, Noon & McKelvey 1996). Box 12.4 provides a case study of how a PVA was used to improve decision-making for Leadbeater's possum, an endangered marsupial at the center of one of the most contentious forestry debates in Australia.

Third, embrace uncertainty by considering a range of possibilities for every step for which there is doubt about a process, a functional relationship, or a measured parameter. The worst PVAs are those that take one set of input data and provide one point estimate of extinction probabilities, while the best are those that consider a range of biological and management-based inputs, and a range of predictions (projection

---

Box 12.4 An example of how PVA can be an input to decision analysis

The primary threat to Leadbeater's possums is their requirement for nest sites in trees over 150 years old. Early this century, fires burned more than 60% of the forest within the range of the species, and clearcut logging has more recently occurred over 75% of its known distribution. The species now occupies an area 60 km by 50 km in the central highlands of the state of Victoria in southeastern Australia. Current management is to avoid cutting in certain areas (including old growth patches) while clearcutting continues in other areas; areas that burn in the future may be salvage logged.

The viability of this species was assessed subject to current and potential future management options. For each option, the number gives the probability of extinction over the next 150 years (shown in parentheses in the figure) and over a typical 150-year period in the future when the forest has reached an equilibrium with the management actions (think of this as the period 500–650 years from now, assuming constant conditions).

Under current management, possums would be expected to persist for the next 150 years (only 38% chance of extinction), but not into the future (100% chance of extinction). If existing old-growth forest were not allowed to be salvage logged (see figure, second box), possum viability increases because it prevents the removal of trees that are damaged but alive after a fire. The next option also prevents salvage logging of other areas that can grow into old growth (see figure, third box), further increasing possum viability.

Two popular suggestions for further aiding this species are to increase the rotation time and to make more reserves, so these possibilities were considered next. Although increasing rotation time does reduce extinction probabilities, it requires an almost complete halt of logging for the next 150 years, hardly a politically realistic possibility. By contrast, setting aside reserves improves viability even more but reduces logging very little; for example, setting aside just six 50-ha reserves (5% of the forest block) reduces extinction probability to 18% over the long term but reduces logging by only 5% at most. With the identification of additional permanent reserves as a viable approach, a number of scenarios were considered, trading off size and number of reserves. The authors assessed the sensitivity of conclusions by modeling a range of possibilities for processes about which they were uncertain. The recommendations emerging from this work are currently being implemented on the ground (Possingham et al. 2002).

Box 12.4 Continued

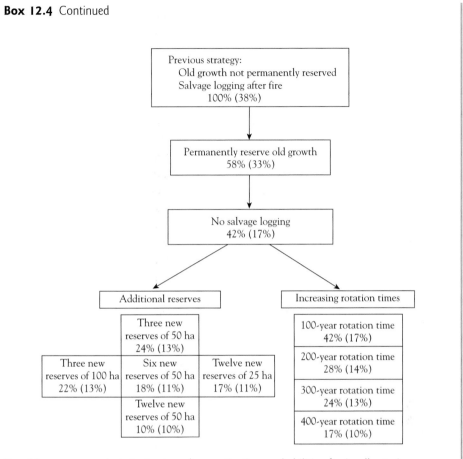

Possible management strategies to reduce extinction probabilities for Leadbeater's possums. For each management option, the percentage chance of extinction is given for the long-term of 500–600 years, and over the next 150 years (in parentheses). From Possingham et al. (2002). Reproduced with permission of the University of Chicago Press.

period, probabilities of persistence, varying scenarios, etc.). Ideally, in addition to the range of input values and output metrics, a PVA should be performed with multiple methods (Gärdenfors 2000, Kindvall 2000). Sensitivity analysis in the broad sense, evaluating what inputs most affect the output, is an essential part of viability assessment (Reed et al. 1998, Mills & Lindberg 2002). Also, Bayesian approaches to viability assessment directly incorporate parameter uncertainty (Taylor et al. 1996, 2002, Goodman 2002, Wade 2002b).

Burgman and Possingham (2000:104) recall the epithet that "all models are wrong but some are useful" (from Box 1979:202) to emphasize their point that:

> . . . the only correct model is an entire reconstruction of the actual system –
> whereupon it ceases to be a model. The utility of a PVA is determined by several

things, including the care taken to include all ecological intuition faithfully, the care taken to represent all views (hypotheses) as structural alternatives, the detail and transparency of statements about assumptions, and the role of the model within the decision-making framework. One of the most important steps in establishing the credibility of a PVA is to communicate the uncertainties embedded in the model and its assumptions.

Summary

The most vulnerable wildlife populations are those that have high susceptibility to human-caused stressors, have gone from being large to being small, and have high variability relative to growth rate. In such cases, the most important actions are to reverse the decline and increase numbers. However, a population that has become small may also be sucked into the extinction vortex, whereby demographic, environmental, and genetic randomness exacerbate the potential for extinction even if the causes of decline are reversed. PVA and related viability assessment procedures provide a framework to capture the intuition, theory, and field data comprising an assessment of the extinction vortex. Because it incorporates multiple, interacting factors, viability assessment can reveal non-intuitive outcomes of management; it also makes assumptions transparent to debate.

A quantitative PVA may be conducted with time series and with demographically explicit models, and other approaches can assess viability outside the quantitative PVA framework. Time-series analysis estimates quasi-extinction probability (decline to numerical thresholds of importance) based on counts of abundance over time. By contrast, demographically explicit PVAs account for a full range of ecological data using estimated rates of survival, reproduction, density dependence, inbreeding costs, correlations among vital rates, and other information to assess likelihood of quasi-extinction. Although considerably more data-hungry than the other methods, demographically explicit scenarios can efficiently use what-if scenarios or sensitivity analysis to target specific management changes of greatest benefit.

In more data-sparse situations, other classes of methods may be used to assess viability. For example, the World Conservation Union (producer of Red Lists for 18,000 species worldwide) assigns qualitative ranks of risk using specified, operational rule-of-thumb thresholds based on trend, abundance, fluctuations, and degree of connectivity. Bayesian belief networks offer another approach by providing a framework to consolidate and make transparent field data, expert opinion, and PVA models for management decision-making. Although these methods do not fall under the formal rubric of quantitative PVA, they accommodate uncertainty, make input explicit, and display risks associated with alternative courses of action.

In all cases, viability can be assessed both for single populations and for multiple populations distributed across the landscape. Movement among populations and the degree to which dynamics of multiple population are correlated will help determine the persistence of a suite of populations. Presence or absence of the species in multiple patches provides information that can offer another approach to assessing viability (patch-occupancy models).

Small-population management includes peeking into the crystal ball to fathom how particular scenarios will affect the likelihood of extinction. In so doing, we should (i) remain acutely aware of data quality, (ii) use PVA as a comparative tool, and (iii) embrace uncertainty, using all of the data and population-biology concepts and theory at our disposal to make good predictions while acknowledging and making transparent the assumptions underlying the assessments. Through viability assessment, population biology can and should be a vital part of decision-making for small and declining populations.

Further reading

Akçakaya, H.R., Burgman, M.A., and Ginzburg, L.R. (1999) *Applied Population Ecology: Principles and Computer Exercises using RAMAS EcoLab*. Sinauer Associates, Sunderland, MA. Contains helpful hints and hands-on exercises for many aspects of PVA.

Beissinger, S.R. and McCullough, D.R. (2002) *Population Viability Analysis*. University of Chicago Press, Chicago. An edited volume containing timely syntheses and cutting edge analyses of concepts across the social and biological spectrum.

13

Bridging applied population and ecosystem ecology with focal species concepts

> *Father worm sat back, stretching himself out to his full, glorious three and a half inches. 'Take us worms, for example. We till, aerate, and enrich the earth's soil, making it suitable for plants. No worms, no plants; and no plants, no so-called higher animals running around with their oh-so-precious backbones!' He was really getting into it now. 'Heck, we're invertebrates, my boy! As a whole, we're the movers and shakers on this planet! Spineless superheroes, that's what we are!' And since Father worm didn't have a fist to bring down on the table, he just yelled, 'BANG!'*
>
> **Gary Larson (1998:53),** *There's a Hair in my Dirt! A Worm's Story*

Introduction

Where should new reserves be located, and what should be their design? How well is a reserve, park, or ecosystem maintaining its ecological integrity? How should a degraded area be restored? What should be the priority ranking for conservation efforts when money is limited but needs across the world are so great?

These are all vexing questions being tackled by a lot of very smart people. What this book can offer is a tiny slice of the type of information needed to answer these large-scale, synthetic questions. In addition to global sociopolitical insights, and the ecosystem process and habitat composition databases generated by Geographic Information Systems (GIS) databases and synthesized by supercomputer analyses, one component of these large-scale management decisions will incorporate population biology for particular wildlife species. There are no shortcuts to avoid population biology if we want to restore or monitor ecosystems, or implement conservation actions (Simberloff 1998, Noon 2003). Ecosystems are hard to define, and retaining land or habitat per se alone is necessary but by itself not necessarily sufficient. As just one example, the shattering 50% decline in the last decade of gorillas and chimpanzees in western equatorial Africa is poorly predicted by the amount of intact forest habitat because the devastating mortality comes from hunting by humans (facilitated by logging roads) and Ebola hemorrhagic fever (Walsh et al. 2003).

So, information on wildlife population responses and dynamics are an essential complement to the human dimensions, habitat, and ecosystem structure and function information that go into conservation planning and monitoring (Noss 1999, Linden-mayer et al. 2002). But with 1.5 million described species, how does one incorporate population biology in a way that is biologically, logistically, financially, and politically possible? One way is to make inferences about the larger system (community or ecosystem) based on a subset of the species in the system. This in itself is not such a new idea: even thousands of years ago humans used seasonal migratory movements of certain wildlife species or flowering by particular plants to inform them of environmental conditions (Niemi & McDonald 2004).

In applied ecology, **focal** or **surrogate species** have been proposed as a practical bridge between single- and multiple-species approaches to wildlife conservation and management. Ideally, a suite of species will allow inference to the state of the ecological system of interest and provide information about ecosystem integrity (Johnson et al. 1999, Noon 2003). Debate roils on the questions of exactly what focal species are, and what role they should play in restoration, monitoring, or management issues of ecosystems (e.g. Lambeck 1997, Andelman & Fagan 2000, Lindenmayer et al. 2002, Roberge & Angelstam 2004). My point in this brief chapter is to describe the main concepts discussed in the context of focal species to span single- and multi-species objectives. Four terms, in particular, mean distinctively different things but collectively can be used to move from single-species conservation planning to multiple-species or ecosystem assessment: flagship, umbrella, indicator, and keystone species.

Flagship species

The flagship species concept does not even pretend to relate to a species' interactions or its response to human perturbations. Rather, it is purely a strategic concept for raising public awareness and financial support for broad conservation action. Flag-ships will often be animals that are huge, ferocious, cuddly, cute, or of direct benefit to humans; they are the charismatic animals most likely to make people smile, feel goose-bumps, and write a check for conservation (Fig. 13.1). Pandas and primates are classic flagship species, whose promotion has increased financial contributions to conservation and heightened global public awareness for their own conservation as well as forests in general. An interesting exception to the generalization about charismatic animals as flagships is the utilitarian but not-so-charismatic wild maize. Out of 10,000 species, including charismatic orchids, ocelots, and jaguars, the maize was chosen as the strategic flagship to galvanize the formation of the Sierra de la Manantlan Biosphere Reserve in Mexico (Leader-Williams & Dublin 2000).

Umbrella species

An umbrella species (or population) can be broadly defined as one "whose conservation confers protection to a large number of naturally co-occurring species" (Roberge & Angelstam 2004:77). Typically, umbrellas have large area requirements or specialized habitat needs, and the idea is that conservation of sufficient habitat

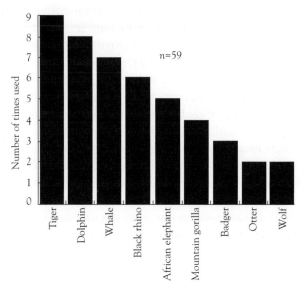

Fig. 13.1 The number of times that advertisements featured particular animals in *BBC Wildlife Magazine* from January to November 1997. Modified from Leader-Williams and Dublin (2000).

for the umbrella will embrace the needs of many others. Large animals and carnivores typically have the largest home ranges, making them common umbrella candidates.

Black rhinos have been proposed as umbrellas for conservation in Africa because their large bodies and home ranges lead to extensive habitat requirements. Berger (1997) focused on areas used by Africa's only unfenced population of rhinos with more than 100 animals, and asked whether other organisms of a similar trophic level might also be conserved on these reserves. The space used by 28 sampled black rhinos would not be sufficient to support five of six sympatric herbivores; in part this was because the other species shifted their habitat use more than rhinos during drought periods. Extending the analysis to include the region likely to be used by additional rhinos in the population increased the number of other species whose movements overlapped rhino space use, but clearly a strategy of protecting only areas currently used by rhinos and ignoring adjacent lands would be questionable for maintenance of other species. Using rhinos as an umbrella species is further undercut by the fact that they have been reduced to precariously small numbers, so their extinction may not actually indicate very much about the trends for other species.

Caro (2003) acknowledged the pitfalls of the umbrella species concept but argued that it remains a useful practical conservation tool. Noting that East Africa has one of the most comprehensive networks of protected areas in the world, and that umbrella-species protection (mostly big game species of economic value) was the implicit driver in how the reserves were set up, Caro asked how the reserves have fared over a half century of umbrella species management. With some disturbing exceptions for poached animals, both the umbrella species (e.g. buffalo) and many (but not all) other mammal species in the reserves are faring reasonably well, or at least as well as they would be doing with management under any alternative strategy.

The bottom line is that umbrella species have a more biologically based underpinning than flagship species, and can serve a practical role as one component of focal species selection. However, enveloping the space used by one or more umbrella species can never embrace all the factors underlying population growth of all (or even most) of the other species using that space and responding differently to human actions, habitat types, environmental stochasticity, and so on (Lindenmayer et al. 2002, Roberge & Angelstam 2004). The umbrella species concept has been extended by Fleishman et al. (2000, 2001), who identify umbrella species if they (i) tend to occur with many other species, (ii) have an intermediate degree of ubiquity across the landscape, and (iii) have a high sensitivity to human disturbance. This definition incorporates more biology than simply a large home range, and may therefore provide more protection.

Indicator species

The indicator species concept has been used in many different ways (see Landres et al. 1988, Caro and O'Doherty 1999, Niemi & McDonald 2004). One way is as a **signpost**, where the presence of one species is likely to indicate others. This application makes some sense for plant ecologists, who often sort out plant communities using indicator species whose distribution closely correlates with biophysical characteristics such as moisture availability, soil fertility, or length of growing seasons. In Douglas-fir forests in the northern Rocky Mountains of the USA, for example, even traces of dwarf huckleberry define a habitat with relatively high moisture that tends to contain a predictable suite of other species and characterizes potential tree productivity following disturbance.

The signpost application begins to fall apart when predicting species distribution across very different taxonomic groups, as in trying to use birds to predict insect distribution (Prendergast et al. 1993). And it stretches toward the absurd if the dynamics of the indicator in the face of human stressors is perceived to be the same as all other species in the community. As wildlife ecologists, we immediately squirm if someone proposes that one (or even several) species will represent the responses of all others to some perturbation. Wouldn't it make you nervous if someone told you that appropriate management and monitoring of 414 species of forest vertebrates would be achieved by managing for elk and three species of hawk? We know that species are not alike in how and what they eat, the habitat they use, their generation times, territory use, response to predators and competitors, and so on. So a blithe, casual, non-critical assertion that any one species will indicate the status of others will almost always fail (Niemi et al. 1997).

Instead of choosing indicators based on species that you hope (against all odds) will respond, behave, reproduce, and die like other species, a more promising definition of indicators includes species that are **most sensitive** to a perturbation or stressor of concern. These, then, are analogous to the historic use of canaries in coal mines, where miners worked with an eye toward a caged canary, knowing that if the bird passed out or died then the air quality for humans would be deteriorating.

Thus, a reasonable use of indicator species identifies a perturbation or stressor of concern, then chooses species whose behaviors and life histories predispose them to be especially sensitive to the perturbation. From this list, one would eliminate those that are logistically impossible to monitor given available funds, and those whose likely responses to other changes would confound interpretation of a response by the indicator (Hilty & Merenlender 2000). Examples of this form of indicator as focal species include the regular use of diversity or abundance of aquatic invertebrates or fish by environmental toxicologists to evaluate effects of pollutants, temperature, sediment loading, and other stressors (Rosenberg & Resh 1993). Also, plethodontid salamanders are classic indicators of stressors in forest ecosystems (Box 13.1). In both examples, we are not assuming that the indicator's dynamics are the same as or represent other species. Rather, the indicator species is being used in the same way canaries were used in mines: their decline or loss is a bellwether, a foreshadowing, of responses of other species if the perturbation is not minimized.

Box 13.1 Plethodontid salamanders as an indicator species

Plethodontid salamanders (family Plethodontidae) tend to be entirely terrestrial (they have no aquatic larval stage), and hatchlings resemble miniature adults. Although they apparently originated in the streams of Appalachia in the eastern USA, they have colonized moist forest areas throughout the USA and into the New World tropics. Plethodontid salamanders may be excellent choices as indicators for some human perturbations, including acidification and timber harvesting, because their physiology predisposes them to being highly sensitive to environmental conditions. Because they are entirely terrestrial, and must maintain moist skin to respire, they can quickly be affected by forest conditions that change pH, or that dry or compact the upper soil layers (Welsh & Droege 2001).

Plethodontid salamanders are often extremely abundant, with a biomass exceeding that of all other small vertebrates combined in some eastern forests. Indeed, at Mountain Lake Biological Station in southwest Virginia, densities of one species approach three per m^2, perhaps the highest density in the world (H. Wilbur, personal communication). The large populations, as well as the relatively stable forest environments inhabited (bypassing the aquatic stage which increases variability for many amphibians) means that variability associated with plethodontid time-series counts is relatively low and statistical power to detect changes over time is relatively high. Of course, the range of species in this family, and the ecological context they are found in, means that plethodontids will not always be ideal indicators of stressors such as forest fragmentation or acidification. Nevertheless, plethodontid salamanders may often be appropriate focal species serving as sensitive indicators of change.

Keystone species and strong interactors

In a set of experiments that forever changed the way ecologists viewed how species interact with each other – and interjected an important tool to the task of choosing focal species – Bob Paine (1969) noted that certain intertidal carnivores could drive species richness in a community. When carnivorous starfish were experimentally removed, the dominant mussel species increased in abundance and competitively excluded other species, leading to the local loss of seven out of 15 species. Thus, Paine (1969:92) concluded that:

> *. . . the species composition and physical appearance [of the system] were greatly modified by the activities of a single native species high in the food web. These individual populations are the keystone of the community's structure, and the integrity of the community and its unaltered persistence through time . . . are determined by their activities and abundances.*

The term **keystone** was borrowed from an old architectural word referring to the wedge-shaped stone at the highest point of an arch that locks the other stones in place and keeps the arch from collapsing. Thus the metaphor sent a powerful message to both basic and applied ecologists: some species have profoundly strong interactions in their communities, and losing these keystone species could lead to the collapse of community structure and a hemorrhaging of species losses. Keystones may not be particularly beautiful (flagships), or have wide space use (umbrellas), or be especially sensitive to perturbations (indicators), but if they are lost they could take with them many other species.

Sea otters are another classic case of a keystone predator. Abundant until the Pacific maritime fur trade decimated their numbers, by the early 1970s sea otter populations had returned to high abundances in some areas but were absent in others, allowing for a natural experiment into their ecological role. Otters limit density of sea urchins that in turn eat kelp, which forms the basis of a different community than is present in their absence (Estes & Duggins 1995). Thus, local extinctions of otters release the numbers of a primary consumer (urchins) that demolishes a plant (kelp) that harbors other organisms. Community composition changes.

Despite the utility of the term to capture and popularize the very real phenomenon that some species play disproportionate roles in maintaining community integrity, by the 1980s keystone species had become an overused, vague synonym for a species deemed somehow important (Mills et al. 1993, Hurlbert 1997). The vague use of the term, coupled with growing momentum to include the keystone species concept in global policy applications (Power & Mills 1995), led to development of an operational definition that could be objectively applied: a keystone species is "one whose impact on its community or ecosystem is large, and disproportionately large relative to its abundance" (Power et al. 1996:609; see Fig. 13.2). Community or ecosystem impact may be on species richness, or on other properties such as biomass of other species, primary productivity, or nutrient or soil retention. Mathematically, an index of the

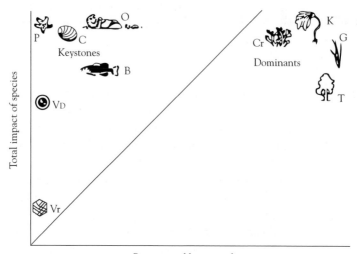

Fig. 13.2 A keystone species is one whose total impact is large (upper half of graph) and large relative to its proportional biomass in the community (upper left of graph). Notice that the second criterion separates keystones from dominant species, whose high total impact arises from their high proportional biomass. The diagonal line represents impact if it were exactly proportional to its biomass in the community. The diagram shows species that have small impacts and low proportional biomass (V_r, a rhinovirus that infects wildebeests; V_D, distemper virus), dominant species with large impacts due to high proportional biomass (T, trees; K, giant kelp; G, prairie grass; Cr, reef-building corals), and keystone species whose impacts are large and disproportionately large relative to biomass (P, *Pisaster* starfish; O, sea otter; C, the predatory whelk *Concholepas*; B, freshwater bass). All species in the upper half of the graph are strong interactors. From Power et al. (1996). Reproduced by permission of the American Institute of Biological Sciences.

extent to which a species might be considered a keystone species would be based on a measure of an ecosystem trait (i.e. species richness or primary productivity) before a species was removed (t_N) and after (t_D), scaled by the proportional abundance or biomass of the species before it was deleted (p_i; Mills et al. 1993, Power et al. 1996):

$$\text{Keystone index} = \left[\frac{(t_N - t_D)}{t_N}\right] * \left(\frac{1}{p_i}\right) \tag{13.1}$$

A species whose effects are in direct proportion to its abundance would have an index of an absolute value of 1, while for a keystone the absolute value of the index would be much greater than 1.

Kotliar (2000) considered the two main criteria of this definition of keystone species – that a species have effects on its community or ecosystem that are both large and large relative to abundance – and concluded that a third criterion was also necessary: that keystone species perform roles not performed by other species or processes. For example, prairie dogs serve as prey for a number of predators (including endangered

black-footed ferrets), dig burrows that are used as nest sites and shelter for both invertebrates and vertebrates, and alter nutrient cycling, plant species composition, and
plant structure through their grazing; some of these roles are not redundant, meaning
they are performed only minimally by other species.

Although the original starfish and classic examples such as sea otters exhibited their
keystone effects as predators inducing top-down trophic cascades (Chapter 8), predation is not the only way a keystone species could exhibit disproportionately strong
interactions in their communities. **Keystone mutualists**, or **mobile links**, include pol-
linators (including bees, moths, and hummingbirds) and seed dispersers (various
birds, bats, and small mammals) supporting plant species that in turn support otherwise separate food webs. **Keystone modifiers** (or **ecological engineers**), such as the
North American beaver, alter hydrology, biogeochemistry, species composition, and
productivity on a wide scale. The term keystone species has even been borrowed for
other applications, including **culturally defined keystone species** for plants or animals
essential to the survival of a human culture (Cristancho & Vining 2004), and **keystone
ecosystems** that have greater importance than would be predicted based on their area,
such as salt marshes providing essential resources to adjacent estuaries (deMaynadier
& Hunter 1994).

Keystone species are not the only **strong interactors** having large impacts on other
species and processes in a system (Fig. 13.2). **Dominant species** may also be strong
interactors, with pervasive effects arising from their high biomass (deMaynadier &
Hunter 1994). Consider *Spartina* cordgrass in a salt marsh, corals in a reef, or oak trees
in a forest; certainly the dominants are crucial to the system's persistence, but in a qualitatively different way than keystones. Introduced (and human-subsidized) species
often become strongly interacting dominant species when they become abundant;
recall the discussion in Chapter 11 about the importance of stopping these species
before they increase to the point where they become hard to control and wreak ecological havoc. Thus, all keystone species are automatically strong interactors, but strong
interactors are not necessarily keystones because they may be dominant species.

A focus on interaction strengths emphasizes that not all species are equal: species
playing keystone and dominant roles interact both directly and indirectly with a
variety of species, so perturbations to them can propagate throughout the community
and affect species far removed, both taxonomically and ecologically, from the strong
interactor. At a basic level, this means that simple food webs, with equal lines
connecting a number of species, are too simple; some species should have few interactions while some have many, some have bold lines connecting them while some
have none. The existence of strong interactors tells us that certain species are particularly crucial to maintaining the organization and diversity of their ecological
communities.

Interaction strength for a particular species is context-dependent, affected by population size, community composition, abiotic conditions, and successional stage
(Fauth 1999, Kotliar 2000, Fleishman et al. 2001). Such context dependence is typical
for pretty much any ecological process or phenomenon: embrace uncertainty! But a
practical way to identify strong interactors (keystones or dominants) at a given place
and time is to ask whether the loss or decrease in abundance of a species would likely

lead to reduced local species diversity or vital rates for other species, or to substantially changed productivity, nutrient dynamics, or habitat structure or composition (see eqn 13.1 for keystone species). Ultimately, interaction strength of species of concern should be one factor incorporated into policy decision-making (Redford 1992, Soulé et al. 2005); one instance mentioned in Chapter 12 was that ecologically effective sizes for strong interactors may well be much larger than the population size needed to avoid extinction (Soulé et al. 2003)[1].

Summary

Ecological effects of humans on ecosystems are ultimately judged against how they affect plants and wildlife populations (and humans). And yet it is impossible to track the viability or population dynamics of all species in any system. Therefore, the main point of this brief chapter is to point out that – like vital rates, age classes, inbreeding costs, movement rates, and other wildlife population metrics – all wildlife species are not equal in the role they play as targets of conservation and management. Some species touch people's heartstrings to facilitate raising of money and awareness (flagships), some have wide ranges that embrace the needs of other species (umbrellas), some are especially sensitive to perturbations (indicators), and some have particularly strong interactions (keystones and dominants). These concepts can be used to select focal species that help bridge single-species methods with multi-species management objectives.

Certainly other conservation targets exist to complement flagship, umbrella, indicator, and keystone species as focal species for management at the community or ecosystem level (Andelman & Fagan 2000, Groves et al. 2002). For example, monitoring might focus on species that are most threatened, most narrowly endemic, or that have the most data available (thereby increasing statistical power to detect effects).

Some species may qualify for several focal species categories. For example, flying foxes (Old World fruit bats, Pteropodidae) in the Philippines may be considered flagships because they are important to indigenous cultures and are a charismatic draw for the public, umbrellas because they use wide swaths of relatively undisturbed forest and riparian areas, keystones because they fulfill critical pollination and seed dispersal roles, and – in some cases – are threatened species (Mildenstein et al. 2005).

In other cases, a species in one focal species category will not necessarily fulfill the criteria for the others. For example, northern spotted owls are a threatened species, an attractive big-eyed flagship found in a beautiful place (old-growth forests), and possi-

---

[1]Although the US Endangered Species Act has no reference to species interactions, the US Marine Mammal Act 1972 does, stating that "population stocks should not be permitted to diminish beyond the point at which they cease to be significant functioning elements of the ecosystem of which they are a part and, consistent with this major objective, they should not be permitted to diminish below their optimum sustainable populations."

bly an umbrella species whose protected habitats will also help at least some other species. However, there is no reason to think that they serve any keystone or dominant function. Because different species with conflicting responses to perturbations or management will likely fulfill each of these categories, bridging single-species and ecosystem approaches should include a suite of focal species.

Further reading

Committee of Scientists (1999) *Sustaining the People's Land: Recommendations for Stewardship of the National Forests and Grasslands into the Next Century.* www.fs.fed.us/emc/nfma/includes/cosreport/Committee%20of%20Scientists%20Report.htm. This contains the detailed report that is abridged and summarized in the article by Johnson et al. (1999) cited in this chapter; it represents the most important thinking to date on the application of the focal species concept.

Simberloff, D. (1988) The contribution of population and community biology to conservation science. *Annual Review of Ecology and Systematics* **19**, 473–511. An articulate and provocative summary of focal species concepts and their utility.

14

Population biology of
harvested populations

Aldo Leopold was a hunter who I am sure abjured freeze-dried vegetables and extrusion burgers. His conscience was clean because his hunting was part of a larger husbandry in which the life of the country was enhanced by his own work. He knew that game populations are not bothered by hunting until they are already too precarious and that precarious game populations should not be hunted. Grizzlies should not be hunted, for instance. The enemy of game is clean farming and sinful chemicals, as well as the useless alteration of watersheds by promoter cretins and the insidious dizards of land development.

Thomas McGuane (1997:168), *The heart of the game*

To go fishing with your father: that is an ancient and elemental proposition, and if it is not as overwhelming as sex or death or the secret lives of animals, still there are legendary shadows about it entrancing to a boy twelve years old.

Fred Chappell (1987:150), *I am One of You Forever: a Novel*

Introduction

Deer, ducks, lions, quail, foxes, kangaroos, turtles: all harvested wildlife species. Wildlife harvest has been practiced and regulated – to varying extents – for nearly the full history of human civilization (Chapter 1). And hunting will continue, because it can provide commercial profit, food for subsistence, control of invasive species, and a profound philosophical connection between people and the hunted animal. Harvests can be sustainable, yet can also contribute to species' decline and extinction.

In the past, most wildlife harvest strategies could be characterized as a process of trial and error (Caughley & Sinclair 1994), with little connection between population-biology principles, data, and harvest regulations. Such casual methodologies may be acceptable or even appropriate where hunting pressure is not intense, so that errors in harvest strategy have little impact on the population. However, for harvests with a strong commercial interest (including both classic fisheries examples as well as some game animals, especially those harvested for trophies), there will often be pressure to determine the maximal sustainable harvest for economic benefit of the hunters and their communities. In fact, even for many relatively non-intense sport harvests, intuition or trial and error alone may be dangerous when habitat fragmentation or other stressors make overharvest more likely.

In short, there is a growing need to understand what population biology can tell us about harvest management. The entire battery of knowledge and techniques discussed in this book can be brought to bear on harvest strategies for subsistence, sport, or profit, or to decrease population size of exotic or pest populations. In this chapter I will first explore how to determine whether hunting is likely to affect either the demographic (abundance and growth rate) or evolutionary trajectory for a harvested population. From this foundation, I will next describe models to determine sustainable harvest levels, starting first without age or stage structure, then proceeding to more complex approaches using demographic information. Because waterfowl management stands out as a shining example where population biology has intersected with and influenced harvest management, I will specifically address the adaptive harvest-management approaches applied in the waterfowl world. Finally, I will end with a mention of the special case of harvest of overabundant or pest wildlife.

Effects of hunting on population dynamics

In Chapter 8, I emphasized three primary factors determining whether a predator was likely to control the numbers of its prey: (a) predation rate (a function of predator numerical and functional response); (b) whether the mortality due to predation is compensatory or additive; and (c) which age, stage, or sex of prey is killed. Hunting by humans is of course a form of predation, so we can use that same framework to ask how each of these factors might affect population dynamics for a particular harvest scenario.

Harvest level: numerical and functional responses of hunters

The number of hunters in any season (numerical response) and how many animals any hunter kills (functional response) collectively determine the total number of animals killed by hunters. Laws and regulations are (we hope) a primary determinant of numerical and functional responses, because the number of hunters and the number of animals taken per hunter are affected by bag limits, season lengths, special licenses through drawings, and hunting zones. Other factors like weather and hunter interest, which can in turn be affected by economics and politics, can also be nearly as important.

Less obvious factors affecting the numerical and functional response of hunters on a hunted population are poaching, crippling losses, and incidental take. **Poaching losses** can be extreme, especially where enforcement is weak and for species of high commercial value. Although I will not go into the many ways that poaching can occur and the consequences it can have, Box 14.1 describes important new insights provided by genetic tools.

Crippling losses include animals that must be accounted for as animals killed because they are mortally wounded but not found by the hunter. For deer in the USA,

---

Box 14.1 Genetic tools and poaching

Many of the forensic tools mentioned in Chapter 3 and elsewhere have some of their most important applications in providing law-enforcement agents with insights into poaching. Matches of microsatellite profiles can definitively connect a crime scene (say, the gut pile of a poached deer) to an individual poacher (say, a piece of venison in someone's freezer). The assignment test may be used to determine the birthplace or origin of a poached individual, facilitating both detection of illegal harvest and trade routes (Manel et al. 2002). For example, African elephants are being decimated, in large part due to the illegal trade in ivory. But where exactly do poachers operate and how do they move ivory out of Africa? A range-wide data-base of genotype frequencies has been developed based on both tissues and non-invasively sampled feces; when an elephant tusk is seized, DNA may be extracted from its tusk and its geographic origin determined (Wasser et al. 2004). In fact, even the threat of prosecution based on the assignment test can lead to confessions: an assignment test indicated that an unusually large 5.5 kg salmon originated not from the location of a fishing contest in Finland but rather from a nearby area that supplied fish markets; when confronted with this genetic evidence the fisherman confessed to buying the salmon at a local fish shop and sneaking it into the competition (Primmer et al. 2000).

---

crippling losses may exceed 20% of the legal firearms kill and nearly 40% of the legal archery kill (Connelly et al. 2005).

Incidental take, called by-catch in fisheries, refers to animals of one species taken accidentally in the process of harvesting another species. It is easy to imagine how this happens when a net is drawn through the ocean for, say, shrimp, and other species such as sea turtles get caught and killed in the net. Incidental harvest can also occur in terrestrial harvests, as in Serengeti National Park in Tanzania, where game-meat hunting using snares has led to substantial harvest of nontarget carnivores, including 11% of the spotted hyaena population during 1991 (Hofer et al. 1996).

Is hunting mortality additive or compensatory?

As you saw in Chapters 6 and 8, compensatory mortality via predation (or human harvesters) occurs when survival, reproduction, or movement into the population increases, thereby ameliorating the effects of harvest mortality on population growth. On the other hand, predator or harvest mortality could be additive, increasing the death rate and decreasing population growth without compensatory effects.

Consider for the moment compensation operating only through survival rates (Nichols et al. 1984, Williams et al. 2002). Figure 14.1 shows how annual survival rate would be affected by the extreme cases of completely compensatory versus totally additive hunting mortality. Hunting mortality is H_t, annual survival rate in the absence of hunting is S_0, and realized annual survival in the presence of hunting is S_t (of course, survival is $1 -$ mortality). If hunting mortality is additive, then harvest mortality and

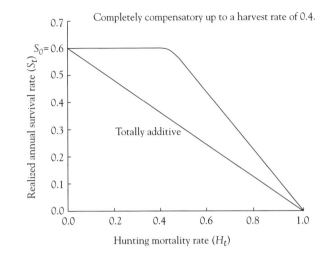

Fig. 14.1 How additive and compensatory mortality affect annual survival rate. With complete compensation, a harvest rate as large as 40% does not decrease the annual nonhunting survival rate of 0.6; however survival declines when harvest exceeds the compensation threshold of 40%. For totally additive mortality, annual survival declines linearly as hunting mortality rate increases.

nonharvest mortality are independent competing risks. Animals have to survive the hunting season with probability $(1-H_t)$ and they must survive everything else (S_0). Combining these two probabilities gives[1]

$$\text{Realized survival under additive mortality} = S_t = S_0(1-H_t) = S_0 - S_0H_t \qquad (14.1)$$

If, on the other hand, hunting mortality were completely compensatory, an increase in hunting mortality (H_t) prompts an equivalent decrease in nonharvest mortality, so realized annual survival in the presence of hunting (S_t) is the same as the background annual survival rate in the absence of hunting (S_0):

$$\text{Realized survival under complete compensation} = S_t = S_0 \qquad (14.2)$$

Obviously, mortality from other causes can only compensate for harvest levels up to a point. Specifically, if hunting mortality (H_t) exceeds nonhunting mortality $(1-S_0)$, then harvest becomes additive and survival will decline, even if hunting mortality is compensatory at less intense levels.[2] Typically, the threshold where mortality can no longer compensate for harvest will be variable but well below $1-S_0$.

---

[1]Technically, we are making the assumption that all hunting mortality (H_t) and all nonhunting mortality $(1-S_0)$ occur at separate times; however, this approach also adequately captures additivity in more realistic situations where nonhunting mortality also occurs during the hunting season (Nichols et al. 1984:537–40).

[2]Annual survival above the threshold of perfect compensation $(C) = S_0[(1-H_t)/(1-C)]$ (Williams et al. 2002:228).

Thus the maximum harvest rate threshold where hunting mortality can be completely compensated for via survival alone will be higher for species that have higher background mortality (or lower background survival) in the absence of exploitation. A general extension of this prediction is that species characterized by short life times and high reproductive rates will tend to sustain relatively high hunting rates, while species characterized by longer life times (high yearly survival) and low reproductive rates will be less resilient to hunting (Leopold 1933, Cardillo et al. 2005).

Even under additive mortality there is some degree of numerical compensation that arises not from the density dependence of compensatory mortality but rather because hunting takes out some animals that would have died due to nonhunting mortality (Nichols et al. 1984). Consider for example 100 quail in August, and totally additive mortality due to hunting. During the fall hunting season, 30% are killed ($H_t=0.3$), and nonhunting annual mortality is 20% ($S_0=1-0.2=0.8$). Without hunting, 20 quail would die ($100*0.2$). With hunting, 30 die due to hunting ($100*0.3$), but only 14 would die due to nonhunting mortalities [$N(1-H_t)(1-S_0)=100*0.7*0.2=14$]. Thus, of the 30 birds killed by hunting, six of them ($20-14=6$) would have died anyway without hunting. These are not compensatory mortality rate changes, but rather just observation of the mathematical fact that dying from one factor (hunting) makes you unavailable to die from something else.

Of course, a population will almost never experience either perfect compensation or additivity in mortality rates. Nevertheless, the extremes are useful to bracket how compensation might affect a wildlife population. And remember: so far we have only considered compensatory versus additive mortality. But the more fundamental question of interest is whether population growth is affected by hunting. Because λ is a function of age-specific survival and reproduction – and movement among populations – even totally additive hunting mortality for certain age classes may be entirely compensated for by increased survival in other age classes, by reproduction, or by immigration, leaving abundance in the presence of hunting unchanged or even increased.

Coyotes are a spectacular example of compensation of harvest mortality via a number of vital rates, as heavy exploitation leads to increased litter sizes of surviving females, increased reproduction in yearlings, higher juvenile survival, and higher immigration into the population, all of which frustrate efforts to reduce coyote populations in relatively small areas (Knowlton et al. 1999). Similarly, upland game birds such as bobwhite quail can compensate mortality due to sport hunting by increasing reproduction (Roseberry & Klimstra 1984). As an example of adjusted movement rates among populations affecting harvest compensation, hunting mortality on ruffed grouse in Wisconsin became increasingly more additive as immigration from adjacent areas with lower hunting mortality became restricted by habitat fragmentation, thereby exposing the heavily harvested populations to declines (Small et al. 1991; see also Labonté et al. 1998 for moose). The phenomenon of spatial compensation of mortalities in harvested populations by animals from nearby reserves underlies the **spatial harvest control method** of harvest management whereby a certain amount of area is closed to harvest to ensure that harvest in other areas – connected to the refuges – remains sustainable (McCullough 1996).

All of which brings us back to the empirical question: is hunting compensatory or additive? As you might expect, the correct answer is – without shame – "sometimes" or "it depends." In waterfowl, hunting mortality is mostly additive for geese, and mostly mixed for most duck species (Nichols et al. 1995, Nichols 2000). Compensation is similarly variable in ungulates, where current levels of sport harvest are sometimes compensatory (e.g. deer harvest in Ontario, Canada; Giles & Findlay 2004) and sometimes additive (e.g. elk harvest in Yellowstone National Park; Vucetich et al. 2005).

In short, the degree to which hunting mortality can be compensated by other mortality, or by reproduction or movement, is a species- and context-specific piece helping to determine how harvest affects a population. If nothing is known about compensation, the most cautious approach would assume none, and set harvest models (to be discussed below) as if the harvest mortality were completely additive. Next we'll consider how the effect of harvest depends on which individuals are harvested.

Which ages and sexes are harvested

By now, it should be a fundamental truism for you that harvesting different ages or sexes will have different impacts on population growth. First, animals of different ages and sexes are harvested at different rates. In many ungulate harvests, for example, hunters seek out older males because they are bigger and have larger antlers or horns (other hunters may prefer smaller animals or females because they taste better). Furthermore, population growth will be influenced by sex- or age-specific harvests because different ages (and sexes) contribute differently to population growth (Chapter 7). All other things being equal, harvesting a given number of animals of low reproductive value will affect the population less than killing the same number having high reproductive value (Goodman 1980). For example, hunters in Yellowstone's northern range tend to harvest cow elk with high reproductive value whereas wolves tend to harvest older cow elk and calves, both of which have low reproductive value, implying that the elk population could sustain higher wolf predation than human harvest (Wright et al. 2006).

Just as the effects of harvesting particular age or size classes can be evaluated, so too can the effects of preferentially harvesting males or females. For most bird and mammal species, management has traditionally skewed harvest toward males. In part, male-biased harvests come from preferences by hunters: males tend to have flashy ornaments or plumage, are often bigger, and in many cases are more vulnerable to harvest. The more biologically based rationale for harvesting primarily males arises from the demographic advantage; if the population has a fixed carrying capacity (and polygynous mating), then harvesting males to skew the sex ratio toward proportionately more breeding females will increase overall reproduction, thereby increasing the sustained yield (Caughley 1977). This general premise – to harvest primarily males so that the remaining population has a higher proportion of reproducing females – has been shown to be reasonable in a number of cases. After all, for polygynous species, where males do not substantially contribute to raising young and one male can fertilize several females (e.g. perhaps up to 50 for reindeer; Mysterud et al. 2002), males may be less relevant to population growth than females.

However, when carried to an extreme, harvesting primarily males to skew the sex ratio toward females can negatively affect physiology, behavior, and genetic effective population size, mitigating or even reversing the positive demographic benefit of more females (Ginsberg & Milner-Gulland 1994). In ungulates, the normally polygynous breeding sex ratio may be pushed to extremes where males become unable to inseminate females, as appears to have occurred for critically endangered saiga antelope subject to heavy poaching in central Asia (especially for the horns that are highly valued in traditional Chinese medicine); when the proportion of adult males dropped below one per 36 females, the proportion of females reproducing decreased from about 80% to less than 20% (Milner-Gulland et al. 2003b).

Similarly, strong preferential harvest of particular stage or sex classes can disrupt social structure or have other behavioral effects, cascading into effects on other vital rates and population growth (R.B. Harris & W. Wall, personal communication). A classic case occurs when excessive harvests of dominant older males makes younger males become the primary breeders; because females may be hesitant to breed with younger males, calving may be delayed or become less synchronous (Singer & Zeigenfuss 2002, Mysterud et al. 2002). A delay in breeding or synchrony is particularly problematic for species that rely on synchronous calving for predator swamping, or whose birth dates must correspond to optimal forage quality (for an example with elk; see Box 14.2).

Another potential effect of strongly biasing the sex ratio via harvest is reduction of the genetic effective population size (N_e; Chapter 9). If populations are relatively small, or locally structured because of strong site fidelity, the reduction in N_e due to a skewed sex ratio could exacerbate inbreeding depression (Harris et al. 2002, Peek et al. 2002).

Overall, then, the fact that all sex and age classes are not equal must mean that preferential harvest of particular sexes and ages can affect vital rates and population growth through a variety of mechanisms. Of course, age- or sex-specific harvest will not always be problematic, as there are plenty of examples where skewing of sex ratios or age structure does not have negative effects (White et al. 2001, Mysterud et al. 2002). The consequences of sex-ratio skew under harvest will depend on case-specific factors including density, weather, and female nutritional condition (Connelly et al. 2005); also, the more strongly polygynous a population is (which will again vary due to many factors), the more it will be able to withstand male-biased harvest[3]. The possible effects of stage or sex-structured harvest should be considered – along with the other factors such as harvest rate and the degree to which harvest is compensatory – when evaluating how a particular harvest program might affect population growth.

---

[3]Trophy hunting usually focuses on species with strong sexual dimorphism, which in turn is correlated with strong polygyny (that's what drives the big horns and antlers). Therefore, a relatively low offtake of highly polygynous species via trophy hunting will often be sustainable. Genetic consequences of selecting trophies will be considered next.

> **Box 14.2** An example of how a strong skew in sex ratio could affect reproduction
>
> North American elk in unhunted populations would be expected to have roughly 25 mature males (more than 5 years old) per 100 females. In most hunted populations there is a strong tendency to harvest males, leading to mature male/female ratios of less than 10:100, and often less than 5:100. Research on North American elk and their close relative, the European red deer, indicate that as the proportion of males becomes very small, conception and birth dates are later and less compressed. Juvenile mortality may increase by 1% for each day a calf is born past the median birth date, both due to the loss of the benefits of predator swamping with birth synchrony and because of breakdown in the timing of birth relative to highest forage quality, which affects both optimal lactation of females and highest growth of calves before seasonal forage quality declines (Wisdom & Cook 2000).

Long-term effects: hunting as a selective force

The first strong empirical hints that harvest can change the very nature of a population came from the fisheries world (Law 2000, Ashley et al. 2003). Fishermen prefer larger fish, and regulatory mechanisms (for example, minimum net mesh size) tend to capture larger fish while smaller fish go free. Thus, heavy harvesting has led to strong selection in favor of smaller fish, with few large or older individuals. Selection on body size can cascade into demographic consequences, because favored genotypes that grow slower and reproduce earlier are exposed to higher mortality due to predation and produce fewer or smaller eggs (Conover & Munch 2002). The result is smaller fish with slower population growth. Northern cod off the coast of northeast North America underwent one of the worst fisheries collapses ever experienced following intensive harvests up to the early 1990s, when only 4% of 1-year-olds survived to the age of first reproduction (4 years) in this long-lived species (Cook et al. 1997). Sure enough, cod evolved to mature at earlier ages and smaller sizes (Olsen et al. 2004). Following a moratorium on fishing in 1992, the evolutionary shifts halted and even began to reverse, but abundance has not yet recovered.

These findings from the fisheries world are relevant to some harvests in wildlife populations. For example, African elephants, a species in deep decline due to illegal ivory hunting, are exhibiting an increasing proportion of tuskless females, a trait that appears to be sex-linked and heritable (Jachmann et al. 1995); the effects of this selection against tusks is as yet unknown. Similarly, elk harvests that strongly remove males, especially large males, can greatly decrease the effective population size (N_e), which could have a modest negative effect on a number of correlated traits affecting fitness, such as male reproductive success and number of antler points (Hard et al. 2006).

In a detailed study linking field observation to genetic analyses (Coltman et al. 2003), a population of bighorn sheep in Alberta, Canada, experienced a 30-year trophy-hunting regime whereby any ram that reached a minimum legal horn size could be harvested during the hunting season. As a result, hunters preferentially harvested rams

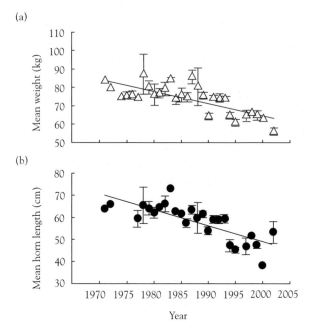

Fig. 14.2 Preferential harvesting of male bighorn sheep that rapidly put on weight and grew large horns led to an inherited decrease in both traits over 30 years. The graphs show (a) weight and (b) horn length for 4-year-old bighorn sheep rams. From Coltman et al. (2003; reproduced with permission of Nature Publishing Group).

that quickly grew big horns and body size. Based on reconstruction of pedigrees from field observation, coupled with paternity analysis using DNA (Chapter 3), the characteristics of horn and body size were found to be highly heritable, meaning that their phenotypic expression has a large genetic component. Because the preferentially harvested rams were removed between ages 4 and 8, at the beginning of their prime reproductive years, rams of higher breeding value for both horn size and body weight had their expected lifespan and lifelong reproductive output cut short by selection imposed by trophy hunting. Thus the genes for most rapid and impressive horn and body growth – the very traits desired for trophy management – were being removed from the population before they could be passed on, so sheep in this population became lighter and carried smaller horns (Fig. 14.2).

How to deal with this paradox that harvest of trophy bighorn rams contributed to the decline in the very traits (big horns and big bodies) most sought after by trophy hunters? A simple solution was to institute so-called full-curl restrictions, limiting harvest to rams with horns whose tips extend beyond the tip of the ram's nose, so that the males carrying genes for large horns and body size are able to reproduce for several years before being harvested. Such a restriction was implemented in parts of Alberta in 1996, and although it decreases the number of animals legally harvested it no doubt increases quality. If the trophy phenotype is mostly independent of age, then a related solution might be to emphasize harvest based on age and not trophy characteristics.

For example, the age of African lions can be reliably distinguished based on nose pigmentation up until about age 9; a harvest strategy targeting only male lions older than age 5 (whose noses are more than half black) is sustainable and allows the animals with trophy phenotypes to reproduce (Whitman et al. 2004)[4].

Overall, then, selective harvest of certain classes can lead to genetic changes that permanently affect morphology. Extending the findings from the commercial fishery world, there could also be unexpected cascading effects on genetically correlated traits (e.g. female body weight, calving rate, and disease resistance) that could actually lead to the evolution of a lower population growth rate. Again, there is no reason to think that selective harvest will inevitably lead to these changes, but the possibilities should be considered in evaluating harvest regimes.

Models to guide sustainable harvest

A long-term sustainable harvest, or sustained yield, is one that does not lead to extinction or unacceptable decline in the harvested population. We have seen that the sustainability of a population in the face of harvest depends on the number of animals harvested, who gets harvested, and how much the mortality imposed by harvest can be compensated. The simplest harvest models ignore age structure and try to capture the harvest rate and benefits of compensation by comparing expected population yield under density dependence against possible harvest levels. Although the so-called **maximum sustained yield** (MSY) – the highest amount that can be harvested sustainably – is a slippery and potentially dangerous concept, I will introduce MSY with simple logistic growth and then explain how more complex population models can help direct sustainable harvest levels for the long term.

Fixed-quota harvests

In Chapter 6 I presented the logistic growth equation, exhorting a healthy skepticism about applications that relied too heavily on this one form of density dependence. Retaining that skepticism, we can still take home an important applied point: under logistic growth, a population at roughly half of its carrying capacity would be putting the most individuals into the population each time step via recruitment (recall dN/dt from Chapters 5 and 6), so this would be the population size where we could harvest the most animals each year.

The intuitive reason for recruitment, or yield of the population, being maximized at $0.5K$ under logistic growth is that at very small numbers no negative density dependence operates but not many individuals are available to contribute offspring;

---

[4]Although we might not typically think of lions as an important species for sport harvest, they are (Creel & Creel 1997). In Tanzania, tourist hunting in 1992 generated $14 million to the government and hunting outfitters, with much of that money being returned to conservation activities. Lions were one of the top three species harvested.

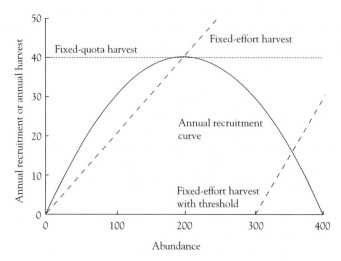

Fig. 14.3 Three different harvest approaches (dotted and dashed lines) to setting annual harvests. Each harvest type is plotted against the annual recruitment parabola (solid line; otherwise known as dN/dt; see Fig. 6.4) assuming logistic growth and $r=0.4$. Whenever the annual harvest is greater than annual recruitment, the harvested population will decline to the abundance where the harvest and recruitment lines intersect, or to extinction, whichever comes first. Thus, for the fixed-quota harvest (dotted line), the MSY (40 animals/year) is achieved when the abundance is held at 200, but if abundance were less than 200 this harvest level would lead to extinction. A fixed-effort harvest would give the same MSY but is more conservative against overharvest because if abundance were less than 200, recruitment exceeds harvest and abundance increases. The most cautious method uses fixed effort and also imposes a threshold (here arbitrarily set at 300 animals), below which no harvest occurs.

conversely, at K many individuals could reproduce, but negative density dependence has its maximal effect, quashing reproduction and/or survival. In the middle zone of population size, dN/dt (the recruitment or population yield) is maximized.

Consider the outcome of a hypothetical harvest strategy based on a fixed annual quota in a population with logistic growth (Fig. 14.3). Harvesting the population from its carrying capacity of 400 animals down to 0.5K (200 animals) would lead to an MSY of 40 animals per year without causing the population to decline[5]. But such a strategy dances on the knife edge of overharvest in the real world where uncertainty exists in knowing the current abundance, carrying capacity, or the true density-dependence function.

Suppose we imposed our harvest of 40 animals per year while holding the population at what we think is 200 animals, but in fact there were actually only 150 animals due to stochasticity or sample variance in the estimated abundance. On Fig. 14.3, draw a line from the x axis at $N=150$ up to the recruitment line. Notice that recruitment

---

[5]Under logistic growth the MSY (yield at the peak of the recruitment parabola) is $rK/4$.

per year of about 35 animals is less than the fixed quota of 40 per year, leading towards overharvest if this fixed-quota harvest continued.

Furthermore, locking in a fixed quota based on a single density-dependent compensation form such as the logistic curve is also dangerous. As preached in Chapter 6, there are many different ways that density dependence can be expressed in wild populations. If, for example, negative density dependence were lighter at low densities but strong closer to the carrying capacity, instead of the linear decline assumed by logistic growth (McCullough 1992) the recruitment curve hump in Fig. 14.3 would shift to the right, requiring a population size considerably greater than $K/2$ to sustain the harvest. The dangers become even greater with Allee effects where vital rates and population growth actually decline at smaller population sizes. Indeed, the fact that many exploited fish species live in schools whose size determines successful defense against predators implies that fisheries collapses may often result from harvests based on models that ignore positive density dependence (Courchamp et al. 1999, Stephens & Sutherland 1999). As Caughley and Sinclair (1994:287) note: "The trick with managing a population for sustained yield is to play it safe. You estimate the MSY on what information is available to you . . . , refine that estimate of the MSY as often as you can or at least as often as your monitoring system allows, but keep the harvest well below the MSY."

Fixed-effort and threshold harvests

Instead of a fixed-quota or constant harvest, with no safety valve to reduce harvest if stochasticity or uncertainty causes recruitment to be less than we expect, an alternative method takes advantage of the fact that the number of animals harvested depends on both abundance and effort. Fixing effort causes the number of animals harvested to change as abundance changes, providing a self-correcting feedback under all but the most extreme harvest rates. In other words, a drop in abundance leads to a drop in the number harvested. Fixing harvest effort (also called the **proportional harvest method**) requires tracking (and control) of many factors, including the number of hunters and their allowable bag limit, season limits and timing, which age classes and sexes get harvested, and the techniques and technology used to make the harvest. This approach does make for a much more cautious harvest approach. Aanes et al. (2002) found that for willow ptarmigan in Fennoscandia, a proportional harvest model with an upper harvest limit (i.e. a bag limit) was the best way to determine the MSY while minimizing potentially dangerous population fluctuations.

Using logistic growth as a simple example, notice that in contrast to a fixed quota, a fixed-effort harvest would not likely lead to extinction even if stochasticity or sampling variance made the population smaller than $0.5K$ (Fig. 14.3). When stochasticity and uncertainty in population estimation are fairly high – as they often are in wildlife populations – a prudent extension of the fixed-effort method prohibits any harvest below a threshold abundance (Fig. 14.3). Both the appropriate threshold and the fraction harvested above the threshold depend on the harvest goal. Increased stochasticity and uncertainty in population dynamics increases should lead to an increase in the threshold abundance and a decrease in the proportion harvested or fixed effort (Lande et al. 2001).

Again, remember that Fig. 14.3 uses logistic growth as just one possible recruitment curve that could be drawn. Different recruitment curves would lead to different harvests. For example, 1–3 million large kangaroos are harvested each year in Australia. The harvest is sustainable (with 16–38 million kangaroos persisting) in part due to the thoughtful use of harvest models. Kangaroo population growth is largely determined by their intake of plants, and plant biomass is driven by rainfall and plant density. A mechanistic model of kangaroo numbers driven by stochastic variation in rainfall and plant biomass indicated that a harvest rate of 10–15% of the population per year would provide the MSY; this rate is at least 25% lower than the harvest rate that might be predicted from a simple deterministic logistic growth model (Grigg & Pople 2001).

Finally, before leaving these simple harvest models we must note the implications of seasonality in harvest and density dependence (Boyce et al. 1999). On their face, it seems that the harvest models of Fig. 14.3 must assume that harvest mortality is not compensated for, because the population size inevitably declines from K down to where harvest rate meets recruitment rate. However, harvest can occur in one part of the year and compensation in another part of the year, so average population size across the year declines under harvest while population size from spring to spring does not decline. In other words, when we talk about reducing the population size to medium densities to exact the greatest yield, the reductions may be seasonal, not permanent.

Adding age structure to harvest models

For some populations, especially those where it is difficult to distinguish ages or sexes (e.g. many birds and carnivores), the harvest models described above can be perfectly sufficient. However, in cases where managers need to know the consequences of harvesting certain age or stage (or sex) classes, one can turn to more realistic stage-structured harvest models. Any of the ideas in the simple models discussed previously can be incorporated into more complex stage-structured models. Building on stage-structured models described in Chapter 7, and basic ideas behind harvesting in this chapter, I'll jump right into some examples applying structured models to harvest predictions.

Sex, size, and age structure have long been a part of setting harvest strategies for ungulates, both due to recognition of how different stages and sexes affect population dynamics and because there is great interest in managing particular stages and sexes. The effects of specific harvest regimes – perhaps targeting primarily big bucks or spike-antlered males as opposed to does or fawns – can be tracked with structured models. As just one example, Langvatn and Loison (1999) used structured models and detailed vital rates from the field for red deer in Norway to conclude that current harvest of males could increase by perhaps 10% without decreasing population growth.

Several colleagues and I performed a similar analysis for North American elk in Washington (Peek et al. 2002). Our model needed to track stages that were biologi-cally meaningful, readily identifiable in the field, and of interest to hunters and managers: male and female calves, yearlings, young adults, adults, and old adults. We built a two-sex (male and female), five-stage (calf, yearling, young adult, adult, and old adult) matrix-projection model. Because elk vital rates and abundances are typically estimated in late winter (March), just before the birth of calves, we used a pre-birth-pulse projection matrix. The base matrix is a little complicated and so is shown in Box 14.3.

With the base matrix in hand, how does harvest of different ages or stages affect population dynamics? Of particular interest was how harvest would affect stages distinguishable in the field: spike bulls (yearling males), so-called raghorn bulls (young adult males), mature bulls (adult and old adult males), all antlered bulls (yearling, young adult, and adult), calves (males and females), and antlerless elk (all females, plus

---

Box 14.3 The base elk model (without harvest) used to explore effects of harvest on particular ages and stages

This is a pre-birth-pulse matrix model, where subscript numbers refer to stage (0, calf; 1, yearling; 2, young adult; 3, adult; 4, old adult) and letters refer to sex (f, male; m, female). Odd-numbered columns and rows refer to female vital rates and even numbered to males. So, the top two rows include both fecundity (m) and the survival of newborns to be counted as a female (row 1) or male (row 2) calf about to become a yearling 1 year later (P_0). The G terms represent survival and transition to the next stage. Harvest was added to the model either as a simple additive effect ($1 - H$) or by a more complex function where bull harvest led to lower reproductive output of females (see text, and remember that a variety of other approaches are possible; that's the beauty of exploring what-if-type scenarios with structured models).

The stages ...	Calf female	Calf male	Yearling female	Yearling male	Young adult female	Young adult male	Adult female	Adult male	Old adult female	Old adult male
	0	0	$P_{0f}m_{1f}$	0	$P_{0f}m_{2f}$	0	$P_{0f}m_{3f}$	0	$P_{0f}m_{4f}$	0
	0	0	$P_{0m}m_{1m}$	0	$P_{0m}m_{2m}$	0	$P_{0m}m_{3m}$	0	$P_{0m}m_{4m}$	0
	G_{1f}	0	0	0	0	0	0	0	0	0
	0	G_{1m}	0	0	0	0	0	0	0	0
	0	0	G_{2f}	0	P_{2f}	0	0	0	0	0
	0	0	0	G_{2m}	0	P_{2m}	0	0	0	0
	0	0	0	0	G_{3f}	0	P_{3f}	0	0	0
	0	0	0	0	0	G_{3m}	0	P_{3m}	0	0
	0	0	0	0	0	0	G_{4f}	0	P_{4f}	0
	0	0	0	0	0	0	0	G_{4m}	0	P_{4m}

A pre-birth-pulse matrix model used to explore effects of harvest on particular ages and stages.

male calves). As input to these sensitivity analysis models, we primarily relied on relatively high-quality data from the Blue Mountains (Washington) elk herd.

The simplest way to incorporate harvest rate per year (H) on a given sex or stage class is to multiply survival terms (P or G in Box 14.3) by ($1-H$) for that stage, assuming totally additive mortality; compensation for harvest mortality on any stage or stages would reduce the effect of that harvest. A complication involves how reductions in mature bulls might reduce reproductive performance by females and/or reduce survival of calves. To address this possibility we considered a scenario where calf production was reduced in areas of high mature bull harvest[6].

To evaluate the effects of different harvest strategies on population dynamics, we used the ratio of abundance 50 years into the future with a particular harvest strategy divided by the baseline abundance at the same time but without the imposed harvest strategy. We also considered how harvest would affect sex ratio, determined as the overall male/female ratio, as well as the proportion of mature bulls.

Current harvest rates are quite low on calves and females ($H=0.019$), relatively high on raghorn and mature bulls ($H=0.14$), and very high on spike males ($H=0.6$). Using our model with both additive mortality and a negative feedback between mature-bull harvest and recruitment, we asked how a 25% increase over current harvest rates would affect future dynamics and sex ratios. The population could sustain a 25% increase in harvest of raghorns or calves quite readily (Fig. 14.4a); these scenarios would also result in sex ratios with plenty of males, including mature bulls (Fig. 14.4b). By contrast, a 25% increase in harvest of only antlered bulls would both limit population growth and skew the sex ratio, a bad outcome on both fronts. Figure 14.4 also shows that increased harvest of antlerless elk (cow and calf) reduces population size, but retains a high male sex ratio.

Many other scenarios can and should be included in an exercise such as this one, including some with compensatory mortality and other density dependence, some without the negative feedback between male bull harvest and recruitment, and a range of harvest levels and vital rates. With a palate of what-if scenarios, one can embrace uncertainty by bracketing the zone where true dynamics, population responses, and vital rates are likely to be. Notice also that we used the structured models to present to managers how both future relative abundance and sex ratio may be affected under different harvest regimes. Population size is of obvious interest, but sex ratio may be of equal or greater importance for ungulates in terms of hunter satisfaction (e.g. availability of bulls to harvest), evolutionary outcomes, and population dynamics through feedback effects.

Finally, the effects of harvest on population growth are in the context of natural variation in vital rates (Chapters 5 and 7). Calf elk survival varies much more than any other vital rate – due to predation, habitat, and climate changes – thereby acting as a

---

[6]For interested readers who want to know the details, terms in the top two rows of the matrix (Box 14.3) were decremented based on $1-(0.27*[(\text{harvest rate of adult bulls})+(\text{harvest rate of old bulls})/2])$. Of course, this is just one possible approach based on data available to us; as for any uncertain input, one should consider several different possibilities.

(a)

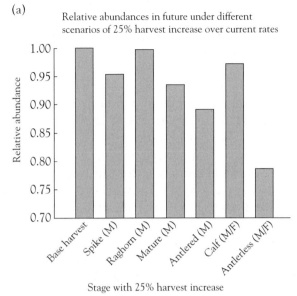

Relative abundances in future under different
scenarios of 25% harvest increase over current rates

Stage with 25% harvest increase

(b)

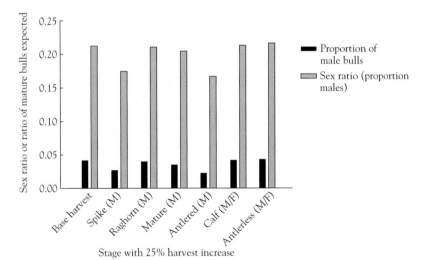

Expected sex ratio and proportion of mature bulls under
different scenarios of 25% harvest increase

Stage with 25% harvest increase

Fig. 14.4 Relative effects of harvesting different stages and sexes of elk in Blue Mountain, Washington. The structured model contained additive harvest mortality and a feedback between bull harvest rate and reproductive contribution by females (discussed in the text). Each stage experienced a 25% increase over its current harvest rate (with other stages held at the current baseline rate). (a) Relative abundance, calculated as predicted abundance at time step 50 for a harvest scenario divided by the predicted abundance at time step 50 under the current harvest rates. (b) Future expected sex ratio. M, male; F, female.

driver of elk population growth rate; any harvest-management strategy for elk must therefore recognize that yearly changes in population size will typically depend on swings in calf survival (Raithel et al. 2006).

So, structured models can often pay off in big dividends for management insights. I will close this section with comments on often-used simple alternatives to structured models in ungulate harvest management: **age ratios** (e.g. fawn/doe) or **sex ratios** (e.g. buck/doe). These are often used to index population growth or general population health because they are relatively easy and cheap to obtain (Caughley 1974, Bonenfant et al. 2005).

Higher age ratios (higher fawn/doe ratio) are typically interpreted to indicate healthier populations, with better recruitment of young. But there are multiple reasons why age ratios alone say very little about population dynamics. First, the ratio could be biased by differential detectability among sex and age classes (Bonenfant et al. 2005). Second, annual reproduction is only one vital rate that contributes to population change. A change in survival, for instance, that affects all stages equally could lead to a screaming population increase or a terrifying decrease but have no effect on the age ratio. Conversely, very different age-ratio changes can be produced by identical changes in population growth rate, depending on which vital rates are altered. Another problem is that the tipping point between a good and bad ratio cannot be discerned from the age ratios alone; for example, tweaking cow elk harvest rates from 0.01 to 0.15 could cause a population to go from being stationary ($\lambda=1.0$) to a 7% decline ($\lambda=0.93$), even though the changes in calf/cow ratios from $27:100$ to $34:100$ would probably be indistinguishable with field data (Connelly et al. 2005)[7]. Finally, even if age-ratio changes did relate to population growth, a poor ratio (few fawns relative to does) could be interpreted as either a population that exceeded average carrying capacity – in which case the interpretation would be to increase harvest – or a population confronting a lowered carrying capacity, in which case the appropriate action would be reduce harvesting (R.B. Harris and W. Wall, personal communication).

The bottom line is that you need more than just age ratios to be able to say anything about the status of a population, as Caughley (1974:562) pointed out more than 30 years ago.

> The interpretation of age ratios is obviously a hazardous undertaking. Of themselves they reveal little about the demography of a population, and their unsupported use can lead to serious blunders of interpretation.

Likewise, sex ratios by themselves are not a great indicator of the things they are often used for, including male survival, population health, or where the population size is relative to carrying capacity (McCullough 1994). An increase in the ratio of females to males could occur via any one of five ways: (i) females increase while males remain constant; (ii) females remain constant while males decrease; (iii) both females and males increase, but females more so; (iv) both females and males decrease, but

---

[7]Notice also in this example that the healthier population had the lower calf/cow ratio.

males more so; or (v) females increase while males decrease. Thus, sex ratios alone are not coupled to changes in population size. Whereas sex-ratio data may help interpret whether harvest is skewing sex ratios too much, there is no reason to think that a high female/male ratio necessarily says anything about the population's size or relationship to carrying capacity.

Waterfowl harvest and adaptive harvest management

In this chapter we have seen that we can better understand the effects of harvest using insights into population sex and age structure, density dependence and compensation, and the role of stochasticity from weather and other factors. Overlaying the biological interactions, of course, is a huge human dimension, whereby harvest regulations are influenced by socioeconomic considerations. One way to bring these pieces together into a transparent, cohesive whole is the **adaptive harvest management** (AHM) framework being used to regulate the sport harvest of waterfowl in the USA. Although it is still evolving, AHM represents a shining example where data collection, monitoring, and population modeling have been married to management goals in a way that explicitly and proudly embraces uncertainty. First I will give a bit of history of the management of waterfowl, particularly in the USA, and use that as a platform to briefly explain the AHM framework (Nichols et al. 1995, Nichols 2000, Williams et al. 2002).

A brief history of waterfowl management in the USA

Waterfowl refers to members of the bird family Anatidae – ducks, geese, and swans – with 45 native species in North America. Before 1900, virtually all waterfowl were considered nearly infinite in number, and harvested without regulation. That attitude was problematic, and led to calls for more regulation. The Migratory Bird Treaty Act of 1918 established management mandates that still hold today: to protect migratory bird populations (including waterfowl) with the secondary objective of providing hunting opportunity compatible with protection. By the 1950s waterfowl banding and winter survey data were being methodically collected and were coupled with population models to determine how duck populations responded to hunting mortality. Politics swirled, to be sure, but the science was sound, and by the 1970s waterfowl data collection and population modeling was maturing into one of the best monitoring systems in the world for a widespread group of organisms, including winter surveys, harvest surveys, banding programs, and aerial ground surveys for numbers and reproduction.

Currently, roughly 13 million waterfowl are harvested annually in the USA by 1.5 million sport hunters, with a total annual economic output of $1.6 billion (Williams et al. 2002). This is a harvest success story; most populations are healthy, harvest is carefully monitored and regulated, and habitat is being protected. The keys to success lie in successful legislation and curtailment of commercial hunting, combined with a management philosophy that eschewed the maximum harvest idea and supported the development of excellent science (Nichols 2000). Thus a key component of the regulatory process includes data collected each year on population status, habitat conditions, production, and harvest levels, coupled with population models. These data and models, and the uncertainty that goes with them, form the heart of AHM.

AHM

For a long time it has been recognized that experiments on wildlife populations at appropriate temporal and spatial scales are horrifically expensive and logistically daunting, so it makes sense to harness widespread management actions as rigorous scientific experiments (Macnab 1983). Carl Walters (1986) and others have formalized systematic approaches to adaptive resource management, providing a way to use information from management actions to learn about population dynamics and to inform future decision-making, even in the face of considerable uncertainty[8].

A form of adaptive management for waterfowl harvest –AHM – uses suites of predictive models expressing different hypotheses about population dynamics to explicitly incorporate uncertainty about wildlife populations into the formulation of management decisions (Williams et al. 1996). Following a period of political intervention in harvest regulations by US Senators and others, AHM was adopted by the US Fish and Wildlife Service in 1995 to provide a transparent process for establishing hunting regulations for mid-continent mallards (Williams et al. 1996, Nichols 2000, Johnson et al. 2002). AHM embraces three forms of uncertainty in addition to environmental and demographic stochasticity, as follows.

1 **Structural uncertainty** in underlying biological mechanisms, including both biological mechanisms and the effects of management on the population; for example, is harvest mortality compensatory or additive?
2 **Partial observability** about the population status and trend; this is sampling variance, and must be quantified if we want to understand the real biological processes.
3 **Partial controllability** in how the proposed regulations will feed back to population-level effects, recognizing that harvest regulations do not lead to precise changes in harvest rates. For example, doubling bag limits will not necessarily double the harvest, because hunter preferences and abilities come into play. Partial controllability makes it hard to predictably change harvest to reduce structural uncertainty (for example, modifying harvest rates to explore whether harvest mortality is additive).

The old-fashioned way to deal with these forms of uncertainty is to be risk-averse, making the best prediction possible and then regulating harvest more conservatively than the prediction. Such an approach sacrifices hunter opportunity, but more importantly it fails to teach us anything. AHM makes objective regulatory decisions in the face of uncertainty, and also reduces uncertainty about harvest impacts by following

---

[8]For years I have been chanting the "embrace uncertainty" mantra to students, thinking I was the first to make it up. But as I wrote this section I noticed that a classic adaptive resource management paper by Walters and Holling (1990:2061) uses "Embracing Uncertainty" as a section heading.

a systematic process of cycling through monitoring, assessment, and decision-making steps as follows (Johnson et al. 2002).

- List objectives, which may include multiple components and constraints. Often, an objective might be to maximize cumulative harvest over the long term, subject to constraints such as not going below a minimum population size.
- Determine an array of realistic management or regulatory options, including both political palatability and feasibility. For AHM, these may range from restrictive to liberal harvests, and from being constant to varying from year to year. The options would account for desires of hunters, who are, after all, imposing the experimental pulse.
- Build a set of population models that predict population responses to management as functions of various management actions, population characteristics, and environmental conditions. These models depend on a collection of alternative hypotheses about how harvest affects waterfowl populations (e.g. is mortality compensatory or additive?). Because each model represents a different view of how the world works, models differ in predictions and the management strategies they support.
- Establish a monitoring program to provide data to compare with predictions to learn which hypotheses are best supported, to determine the degree to which objectives are being met, and to feed into the harvest models.
- Evaluate which models best approximate the monitored system, accounting for uncertainties in likelihoods among models and over time. Often this step is accomplished through Bayesian analysis (Chapter 2). Here is where you learn which models best capture the system by observing how well the data fit each model.

Each year, you cycle back through the process, iterating regulatory decisions based on resource status (from the monitoring), management needs, and your view of how the world works (from the models).

Management of overabundant and pest populations

Although most of this chapter has focused on population-biology principles to facilitate harvest without causing long-term decline, in many instances the management goal is to intentionally reduce population numbers, as with exotic species or native species reaching numbers large enough to cause environmental and economic damage (e.g. whitetail deer in the eastern USA and Canada geese in many parts of North America). Where the goal is to decrease population numbers through harvest, one simply turns upside down the usual goal of sustainable harvest and tries to drive the population below an ecological or socially derived threshold. Phenomena that are sources of caution for sustainable harvest become swords to wield when trying to reduce numbers. For example, positive density dependence (Chapter 6) could be harnessed by pushing numbers low enough that Allee effects, or other factors in the extinction vortex (Chapter 12), cause numbers to spiral down even more.

Because control efforts are by definition manipulations that alter vital rates and shift densities, heavy harvest of pest species can be informative as to how harvest affects populations. As an example, mountain goats were introduced to Olympic National

Park in Washington state during the 1920s and increased to over 1000 by the 1980s, when park biologists began to remove goats via live capture (see Box 5.2). Although the goat population did show density-dependent responses in litter size and age of first breeding as numbers were decreased, the responses were insufficient to compensate for the removals, and goat numbers declined. So, additive harvest mortality (in this case, live removal to somewhere else) accomplished the goal of decreasing the population in the park, at least for the short term (Houston & Stevens 1988).

Of course, age-structured models can help to develop efficient control approaches (see also examples in Chapter 7). Red fox have become a major exotic pest in Australia, both killing farm animals and affecting the persistence of several native species. Fox control has included shooting, poisoning, and immunocontraception, with little thought toward how much any of these methods would actually reduce population growth. However, McLeod and Saunders (2001) conducted a sensitivity analysis of red fox and concluded that a proportionate change in juvenile survival (and to a lesser extent, reproduction) would decrease population growth more than the same reductions in adult survival. They therefore suggested that targeting reproduction and juvenile survival would be more efficacious in reducing fox numbers than broad-based poisoning of all age classes[9].

Summary

Harvest can be a deterministic driver of population decline and even extinction, especially when a commercial profit stands to be made from the harvest. In fact, the fisheries world has grappled with case after case of horrific overexploitation (e.g. Myers & Worm 2003), leading some to conclude that ". . . there is remarkable consistency in the history of resource exploitation: resources are inevitably overexploited, often to the point of collapse or extinction." (Ludwig et al. 1993:17).

Others are more optimistic, even in the fisheries world (e.g. Rosenberg et al. 1993). For many wildlife species, harvest continues to provide sustainable and profoundly important opportunities for subsistence and recreation. Although setting harvest regulations has a critical human dimension, wildlife population-biology principles are pivotal to determine the number and characteristics of animals that could be removed without causing population decline.

The effects of harvest in any particular instance are informed by knowledge of the harvest rate, which age or stages are being killed, and the degree to which the harvest can be compensated for by changes in mortality or other vital rates. Compensation might occur seasonally for survival (where high harvest mortality is compensated by lower nonharvest mortality), or it may occur via increases in other survival rates, reproduction, or even immigration into the population. Although compensation means that animals can be removed without decreasing population growth, there will always be an upper limit to the amount of harvest that can be compensated.

---

[9]Assuming that reproduction and juvenile survival can be changed, of course.

Simple harvest models based on negative density dependence (such as the logistic model) are informative in showing that both recruitment of new individuals into the population and harvest out of the population can be maximized when the population is reduced to a medium density at roughly half of the carrying capacity. However, there are plenty of reasons to avoid trying to maximize harvest based on the simple logistic model. One avenue for appropriate caution is to harvest fewer individuals than might be expected under MSY using logistic growth. Another way would be to fix effort – so that harvest fluctuates with abundance – instead of fixing a set quota at the MSY expected with logistic density dependence.

Incorporating population structure improves harvest models simply because all age and sex classes are not created equal in either their interest to hunters or in their effects on population dynamics. Some ages can be harvested with little effect on population growth, whereas others will have large effects; the consequences of specific harvest strategies are not necessarily intuitive, but can be clarified using population models. Similarly, some harvest-initiated skew toward females dominating the population can increase population growth, but too strong a skew in sex ratio can decrease reproduction and genetic effective size and increase mortality. Also, strong selection on animals whose phenotype tends to be expressed at younger ages (as with many trophy-hunted species) could cause evolutionary shifts that decrease the trophy characteristics. Overall, models that specifically track different age and sex structures are more informative than relying solely on simple metrics like doe/fawn or doe/buck ratios.

All of the pieces of population biology join with management constraints and political feasibility in the framework of Adaptive Harvest Management (AHM). AHM is not simply doing management and calling it science, but rather is a formal process that evaluates management options against expected population responses. Uncertainty in underlying biological mechanisms (structural uncertainty), population status and trend (partial observability) and in the ability to achieve particular outcomes (partial controllability) are all explicitly accounted for in AHM.

If the goal is to reduce the numbers of a particular population through harvest, one can grasp the golden opportunity to use the manipulation to learn about how populations respond to harvest. We are at an unprecedented place in human history where biological insights and models, coupled with ever-improving data from the field, can inform decisions into how harvest can fill the objective of either sustainability or an intentional decrease in numbers of a pest species.

Further reading

Reynolds, J.D., Mace, G.M., Redford, K.H., and Robinson, J.G. (eds.) (2001) *Conservation of Exploited Species*. Cambridge University Press, Cambridge. An edited volume that spans from the biological to the sociopolitical and socioeconomic aspects of harvest at the global scale.

Epilogue

We've come full circle from Chapter 1, where we confronted a human population size of over 6 billion, growing by about 200,000 per day and interacting strongly in both numerical dominance and per-capita effect on other species. Although the resolution of human needs and those of wildlife must occur on many fronts, one piece of the puzzle includes applied wildlife population biology. Throughout the book, the line I have been walking is this: on the one hand, common sense and intuition will only get you so far in resolving wildlife conflicts. Application of ecological science – including seemingly esoteric pieces from math and invisible DNA molecules – will often expose non-intuitive and surprising paths to effective management. I've rained examples on you, and emphasized the mantra that all ecological processes (e.g. age classes, vital rates, interactions, harvest rates, responses to management, and so on) are not equal. But on the other hand, these paths toward solutions are not inscrutable, not so complex that they are irresolvable, and not an exercise in obfuscation. Rather, we learn from the broad field of population biology, then apply the tools and concepts to particular problems, embrace uncertainty, and move forward.

A danger in covering tough and sometimes sad outcomes that are part of wildlife and conservation biology is the onset of despair, causing aspiring professionals to walk away so they don't have to think about it. I hope you fight that feeling with the conviction that knowledge forms a foundation for change. Learning this subject is not the end game, not the final resolution, but it is surely one essential part.

As you prepare to go forth to meet the challenges of integrating wildlife population biology into conservation management decision-making, I will leave you with four short thoughts.

First, whether you do management, education, research, or something else for your career, always be true to yourself and true to your profession. To do that does mean, of course, relying on the facts and concepts that you have learned. It also means remembering how to gain knowledge of your own, obtaining reliable facts and holding them up against innovative and thoughtful ideas of your mind (recall $idea_{mind}$ from Chapter 2). Being true to your profession also includes interacting with others in a respectful, trustworthy, and ethical manner. At times, you will supplement your knowledge with the less tangible signals that come from your sense of what is right.

Second, have confidence in yourself, and your hunches. Trust yourself and question absurdity. Sometimes things will seem huge and scary, like you will never be able to take into account all those complex factors and interactions and contradictions. But if you're honest and open about what you know, you'll find that you actually know a lot. Michael Soulé dedicated one of the first books on conservation biology (Soulé 1986:12) to "the students who will come after, who will witness the worst and accomplish the most." That's you.

The corollary of this thought is that you not only know things, but also have a good basis to understand complexity and to embrace uncertainty. Barry Lopez (1986:413) wrote: "There are simply no answers to some of the great pressing questions. You continue to live them out, making your life a worthy expression of a leaning into the light." So, lean into the light.

Third, acknowledge complications and politics, but don't let them freeze or discourage you. Of course you will encounter constraints in applying population biology to wildlife conservation and management issues. Politics, economics, bureaucracy, and apathy; all of these and more will play a role, and complicate decisions far beyond the guidance of applied population biology. But that doesn't take away from the importance of seeking to know how biological systems work, and how different processes – including human perturbations – will lead to different outcomes. As I've stated in several contexts throughout the book, an inability to deal with complexity and uncertainty leads to a paralysis of indecision, which is itself a decision to do nothing.

I have an old friend, Terry Shrader, who has spent his career as a fisheries-management biologist. A decade or so ago, as we sat by the campfire the week before I began teaching my first wildlife population biology course, I asked Terry what he thought were the most important things to tell my students. He answered:

> Give them the quantitative tools and rigorous philosophy so that they can convincingly implement changes to business-as-usual in management. I've seen it time and time again – real bright people come in with the knowledge to make research and management better, and then they hit the first paycheck and everything they learned in college goes right out their ear . . . In management, your time is so limited by social and political constraints that you have to really work to change things for the better, to keep from just doing things because that's the way they've been done for years.

And finally, be aware of your surroundings, and enjoy them! I hope this is not a shocker to you: your studies and commitment to wildlife will not get you rich, or make you a society power-broker. Rather, your interest in this field probably comes from a connection that you have with the land and the life on the land. You chose it because you want to have contact – either directly through your day-to-day job or indirectly through the analysis, writing, management, and outreach that you do – with wildlife. These are the things that you will look back on, as testament to a life well lived and a profession well chosen. Aldo Leopold (1966:20) recognized the value of this sort of awareness:

> I once knew an educated lady . . . who told me that she had never heard or seen the geese that twice a year proclaim the revolving seasons to her well-insulated roof. Is education possibly a process of trading awareness for things of lesser worth? The goose who trades his is soon a pile of feathers.

And in another great statement of the merits of broad-thinking awareness, the science fiction writer Robert Heinlein (1988:248) wrote:

> A human being should be able to change a diaper, plan an invasion, butcher a hog, conn a ship, design a building, write a sonnet, balance accounts, build a wall, set a bone, comfort the dying, take orders, give orders, cooperate, act alone, solve equations, analyze a new problem, pitch manure, program a computer, cook a tasty meal, fight efficiently, die gallantly. Specialization is for insects.

And so, let your education and awareness meet your commitment, and have fun. Margaret Murie, who with her husband Olaus were two of the most remarkable field biologists and natural historians ever to have lived, shared one of her favorite thoughts from a tombstone in England:

> The wonder of the world, the beauty and the power, the shapes of things, their colours, lights, and shades – these I saw. Look ye also while life lasts.

I hope you can use population biology – the remarkable insights from invisible DNA, non-intuitive ecological concepts, and amazing matrix-projection models – to sharpen your vision as you marvel at the world.

References

Aanes, S., Engen, S., Sæther, B.-E., Willebrand, T., and Marcström, V. (2002) Sustainable harvesting strategies of willow ptarmigan in a fluctuating environment. *Ecological Applications* **12**, 281–90.

Abrams, P.A. and Ginzburg, L.R. (2000) The nature of predation: prey dependent, ratio dependent or neither. *Trends in Ecology and Evolution* **15**, 337–41.

Acevedo-Whitehouse, K., Gulland, F., Greig, D., and Amos, W. (2003) Disease susceptibility in California sea lions. *Nature* **422**, 35.

Adams, J.R., Kelly, B.T., and Waits, L.P. (2003) Using faecal DNA sampling and GIS to monitor hybridization between red wolves (*Canis rufus*) and coyotes (*Canis latrans*). *Molecular Ecology* **12**, 2175–86.

Aguilar, A., Roemer, G., Debenham, S., Binns, M., Gercelon, D., and Wayne, R.K. (2004) High MHC diversity maintained by balancing selection in an otherwise genetically monomorphic mammal. *Proceedings of the National Academy of Sciences USA* **101**, 3490–4.

Akçakaya, H.R. (2000). Population viability analyses with demographically and spatially structured models. *Ecological Bulletins* **48**, 23–38.

Akçakaya, H.R. and Raphael, M.G. (1998) Assessing human impact despite uncertainty: viability of the northern spotted owl metapopulation in the northwestern USA. *Biodiversity & Conservation* **7**, 875–94.

Akçakaya, H.R. and Sjögren-Gulve, P. (2000) Population viability analysis in conservation planning: an overview. *Ecological Bulletins* **48**, 9–21.

Akçakaya, H.R., Ferson, S., Burgman, M.A., Keith, D.A., Mace, G.M., and Todd, C.R. (2000) Making consistent IUCN classifications under uncertainty. *Conservation Biology* **14**, 1001–13.

Akçakaya, H.R., Burgman, M.A., Kindvall, O. et al. (2004) *Species Conservation and Management.* Oxford University Press, Oxford.

Aldridge, C.L., Oyler-McCance, S.J., and Brigham, R.M. (2001) Occurrence of greater sage-grouse x sharp-tailed grouse hybrids in Alberta. *The Condor* **103**, 657–60.

Allee, W.C. (1931) *Animal Aggregations: a Study in General Sociology.* University of Chicago Press, Chicago.

Allee W.C., Park, O., Emerson, A.E., Park, T., and Schmidt, K.P. (1949) *Principles of Animal Ecology.* W.B. Saunders Company, Philadelphia.

Allendorf, F.W. and Leary, R.F. (1986) Heterozygosity and fitness in natural populations of animals. In: *Conservation Biology: the Science of Scarcity and Diversity* (ed. M.E. Soulé), pp. 57–76. Sinauer Associates, Sunderland, MA.

Allendorf, F.W. and Ryman, N. (2002) The role of genetics in population viability analysis. In: *Population Viability Analysis* (eds. S.R. Beissinger and D.R. McCullough), pp. 50–85. University of Chicago Press, Chicago.

Allendorf, F.W. and Luikart, G. (2006) *Conservation and the Genetics of Populations.* Blackwell Publishing, Cambridge, MA.

Allendorf, F.W., Leary, R.F., Spruell, P., and Wenburg, J.K. (2001) The problems with hybrids: setting conservation guidelines. *Trends in Ecology and Evolution* **16**, 613–22.

Andelman, S.J. and Fagan, W.F. (2000) Umbrellas and flagships: efficient conservation surrogates or expensive mistakes? *Proceedings of the National Academy of Sciences USA* **97**, 5954–9.

Anderson, D.R., Burnham, K.P., and Thompson, W.L. (2000) Null hypothesis testing: problems, prevalence, and an alternative. *Journal of Wildlife Management* **64**, 912–23.

Ankney, C.D. (1996) Why did the ducks come back in 1994 and 1995: was Johnny Lynch right? *Proceedings of the Seventh International Waterfowl Symposium* **7**, 40–4.

Aoki, K. (2005) Avoidance and prohibition of brother-sister sex in humans. *Population Ecology* **47**, 13–19.

Ashley, M.V., Willson, M.F., Pergams, O.R.W., O'Dowd, D.J., Gende, S.M., and Brown, J.S. (2003) Evolutionarily enlightened management. *Biological Conservation* **111**, 115–23.

Avise, J.C. (1994) *Molecular Markers, Natural History, and Evolution.* Chapman and Hall, New York.

Avise, J.C. and Nelson, W.S. (1989) Molecular genetic relationships of the extinct dusky seaside sparrow. *Science* **243**, 646–8.

Bailey, L.L., Simons, T.R., and Pollock, K.H. (2004a) Estimating detection probability parameters for pethodon salamanders using the robust capture-recapture design. *Journal of Wildlife Management* **68**, 1–13.

Bailey, L.L., Simons, T.R., and Pollock, K.H. (2004b) Spatial and temporal variation in detection probability of plethodon salamanders using the robost capture-recapture design. *Journal of Wildlife Management* **68**, 14–24.

Baker, C.S. and Palumbi, S.R. (1994) Which whales are hunted? A molecular genetic approach to monitoring whale hunting. *Science* **265**, 1538–9.

Bakker, V.J. and Van Vuren, D.H. (2004) Gap-crossing decisions by the red squirrel, a forest-dependent small mammal. *Conservation Biology* **18**, 689–97.

Ballou, J.D. (1997) Ancestral inbreeding only minimally affects inbreeding depression in mammalian populations. *Journal of Heredity* **88**, 169–78.

Banks, P.B. (1999) Predation by introduced foxes on native bush rats in Australia: do foxes take the doomed surplus? *Journal of Applied Ecology* **36**, 1063–71.

Banks, R. (1985). *Continental Drift.* HarperCollins Publishers, New York.

Bart, J., Droege, S., Geissler, P., Peterjohn, B., and Ralph, C.J. (2004) Density estimation in wildlife surveys. *Wildlife Society Bulletin* **32**, 1242–7.

Batáry, P. and Baldi, A. (2004) Evidence of an edge effect on avian nest success. *Conservation Biology* **18**, 389–400.

Battin, J. (2004) When good animals love bad habitats: ecological traps and the conservation of animal populations. *Conservation Biology* **18**, 1482–91.

Baum, K.A., Haynes, K.J., Dillemuth, F.P., and Cronin, J.T. (2004) The matrix enhances the effectiveness of corridors and stepping stones. *Ecology* **85**, 2671–6.

Bayes, T. (1763) An essay towards solving a problem in the doctrine of chances. *The Philosophical Transactions* **53**, 370–418.

Bean, M.J. and Rowland, M.J. (1997) *The Evolution of National Wildlife Law,* 3rd edn. Praeger, Westport, CT.

Becker, J. (1996) *Hungry Ghosts: Mao's Secret Famine.* The Free Press, New York.

Beebee, T.J.C. (1995) Amphibian breeding and climate. *Nature* **374**, 219–20.

Beedlow, P.A., Tingey, D.T., Phillips, D.L., Hogsett, W.E., and Olszyk, D.M. (2004) Rising atmospheric CO_2 and carbon sequestration in forests. *Frontiers in Ecology and the Environment* **2**, 315–22.

Beets, G. (1999) Education and age at first birth. *Demos* **15**, 5–8.

Beier, P. (1996) Metapopulation models, tenacious tracking, and cougar conservation. In: (ed. D.R. McCullough) *Metapopulations and Wildlife Conservation*, pp. 293–323. Island Press, Washington DC.

Beier, P. and Noss, R.F. (1998) Do habitat corridors provide connectivity. *Conservation Biology* **12**, 1241–52.

Beissinger, S.R. (1995) Modeling extinction in periodic environments: everglades water levels and snail kite population viability. *Ecological Applications* **5**, 618–31.

Beissinger, S.R. and Snyder, N.F.R. (2002) Water levels affect nest sucess of the snail kite in Florida: AIC and the omission of relevant candidate models. *The Condor* **102**, 208–15.

Beissinger, S.R. and Westphal, M.I. (1998) On the use of demographic models of population viability in endangered species management. *Journal of Wildlife Management* **62**, 821–41.

Bekoff, M. and Wells, M.C. (1986) Social ecology and behavior of coyotes. *Advances in the Study of Behavior* **16**, 251–338.

Belovsky, G.E., Bissonette, J.A., Dueser, R.D. et al. (1994) Management of small populations: concepts affecting the recovery of endangered species. *Wildlife Society Bulletin* **22**, 307–16.

Belovsky, G.E., Mellison, C., Larson, C., and Van Zandt, P.A. (2002) How good are PVA models? Testing their practices with experimental data on the brine shrimp. In: *Population Viability Analysis* (eds. S.R. Beissinger and D.R. McCullough), pp. 257–83. University of Chicago Press, Chicago.

Bennetts, R.E., Nichols, J.D., Lebreton, J.-D., Pradel, R., Hines, J.E., and Kitchens, W.M. (2001) Methods for estimating dispersal probabilities and related parameters using marked animals. In: *Dispersal* (eds. J. Clobert, E. Danchin, A.A. Dhondt, and J.D. Nichols), pp. 3–17. Oxford University Press, Oxford.

Berger, J. (1997) Population constraints associated with the use of black rhinos as an umbrella species for desert herbivores. *Conservation Biology* **11**, 69–78.

Berger, J., Testa, J.W., Roffe, T., and Monforts, S.L. (1999) Conservation endocrinology: a noninvasive tool to understand relationships between carnivore colonization and ecological carrying capacity. *Conservation Biology* **13**, 980–9.

Berry, O., Tocher, M.D., and Sarre, S.D. (2004) Can assignment tests measure dispersal? *Molecular Ecology* **13**, 551–61.

Berry, O., Tocher, M.D., Gleeson, D.M., and Sarre, S.D. (2005) Effect of vegetation matrix on animal dispersal: genetic evidence from a study of endangered skinks. *Conservation Biology* **19**, 855–64.

Berteaux, D., Réale, D., McAdam, A.G., and Boutin, S. (2004) Keeping pace with fast climate change: can arctic life count of evolution? *Integrative and Comparative Biology* **44**, 140–51.

Biek, R., Funk, W.C., Maxell, B.A., and Mills, L.S. (2002) What is missing in amphibian decline research: insights from ecological sensitivity analysis. *Conservation Biology* **16**, 728–34.

Bierzychudek, P. (1982) The demography of jack-in-the-pulpit, a forest perennial that changes sex. *Ecological Monographs* **52**, 335–51.

Bijlsma, R., Bundgaard, J., and Boerema, A.C. (2000) Does inbreeding affect the extinction risk of small populations? predictions from *Drosophila*. *Journal of Evolutionary Biology* **13**, 502–14.

Bjornstad, O.N., Ims, R.A., and Lambin, X. (1999) Spatial population dynamics: analyzing patterns and processes of population synchrony. *Trends in Ecology and Evolution* **14**, 427–32.

Blackburn, T.M., Cassey, P., Duncan, R.P., Evans, K.L., and Gaston, K.J. (2004) Avian extinction and mammalian introductions on oceanic islands. *Science* **305**, 1955–8.

Blanchong, J.A., Scribner, K.T., and Winterstein, S.R. (2002) Assignment of individuals to populations: Bayesian methods and multi-locus genotypes. *Journal of Wildlife Management* **66**, 321–9.

Blejwas, K.M., Williams, C.L., Shin, G.T., Mccullough, G.R., and Jaeger, M.M. (2006) Salivary DNA evidence convicts breeding male coyotes of killing sheep. *Journal of Wildlife Management* **70** (in press).

Blumstein, D.T. and Daniel, J.C. (2005) The loss of anti-predator behavior following isolation on islands. *Proceedings of the Royal Society of London Series B* **272**, 1663–8.

Boatman, N.D., Brickle, N.W., Hart, J.D. et al. (2004) Evidence for the indirect effects of pesticides on farmland birds. *Ibis* **146**, 131–43.

Bolen, E.G., and Robinson, W.L. (2003) *Wildlife Ecology and Management*, 5th edn. Prentice Hall, Upper Saddle River, NJ.

Bonenfant, C., Gaillard, J.-M., Klein, F., and Hamann, J.-L. (2005) Can we use the young: female ratio to infer ungulate population dynamics? An empirical test using red deer *Cervus elaphus* as a model. *Journal of Applied Ecology* **42**, 361–70.

Bongaarts, J. (1994) Population policy options in the developing world. *Science* **263**, 771–6.

Bouzat, J.L., Lewin, H.A., and Paige, K.N. (1998a). The ghost of genetic diversity past: historical DNA analysis of the greater prairie chicken. *American Naturalist* **152**, 1–6.

Bouzat, J.L., Cheng, H.H., Lewin, H.A., Westemeier, R.L., Brawn, J.D., and Paige, K.N. (1998b) Genetic evaluation of a demographic bottleneck in the greater prairie chicken. *Conservation Biology* **12**, 836–43.

Box, G.E.P. (1979) Robustness in the strategy of scientific model building. In: *Robustness in Statistics* (eds. R.L. Launer and G.N. Wilkinson), pp. 201–36. Academic Press, New York.

Boyce, M.S. (1977) Population growth with stochastic fluctuations in the life table. *Theoretical Population Biology* **12**, 366–73.

Boyce, M.S. (1992) Population viability analysis. *Annual Review of Ecology and Systematics* **23**, 481–506.

Boyce, M.S., Sinclair, A.R.E., and White, G.C. (1999) Seasonal compensation of predation and harvesting. *Oikos* **87**, 419–26.

Brashares, J.S., Arcese, P., Sam, M.K., Coppolillo, P.B., Sinclair, A.R.E., and Balmford, A. (2004) Bushmeat hunting, wildlife declines, and fish supply in West Africa. *Science* **306**, 1180–3.

Brault, S. and Caswell, H. (1993) Pod-specific demography of killer whales (*Orcinus orca*). *Ecology* **74**, 1444–54.

Britten, H.B. (1996) Meta-analyses of the association between multilocus heterozygosity and fitness. *Evolution* **50**, 2158–64.

Brook, B.W. (2004) Australasian bird invasions: accidents of history? *Ornithological Science* **3**, 33–42.

Brook, B.W. and Bowman, D.M.J.S. (2002) Explaining the Pleistocene megafaunal extinctions: models, chronologies, and assumptions. *Proceedings of the National Academy of Sciences USA* **99**, 14624–7.

Brook, B.W. and Bowman, D.M.J.S. (2004). The uncertain blitzkrieg of Pleistocene megafauna. *Journal of Biogeography* **31**, 517–23.

Brook, B.W. and Bowman, D.M.J.S. (2005) One equation fits overkill: why allometry underpins both prehistoric and modern body size-based extinctions. *Population Ecology* **47**, 137–41.

Brook, B.W. and Whitehead, P.J. (2005) Sustainable harvest regimes for magpie geese (*Anseranas semipalmata*) under spatial and temporal heterogeneity. *Wildlife Research* **32**, 459–64.

Brook, B.W. and Bradshaw, C.J.A. (2006) Strength of evidence for density dependence in abundance time series of 1198 species. *Ecology* **87**, 1445–51.

Brook, B.W., O'Grady, J.J., Chapman, A.P., Burgman, M.A., Akçakaya, H.R., and Frankham, R. (2000) Predictive accuracy of population viability analysis in conservation biology. *Nature* **404**, 385–7.

Brook, B.W., Burgman, M.A., Akçakaya, H.R., O'Grady, J.J., and Frankham, R. (2002) Critiques of PVA ask the wrong questions: throwing the jeuristic baby out with the numerical bath water. *Conservation Biology* **16**(1), 262–3.

Brook, B.W., Sodhi, N.S., and Ng, P.K.L. (2003) Catastrophic extinctions follow deforestation in Singapore. *Nature* **424**, 420–3.

Brooker, L. and Brooker, M. (2002) Dispersal and population dynamics of the blue-breasted fairywren, *Malurus pulcherrimus*, in fragmented habitat in the Western Australian wheatbelt. *Wildlife Research* **29**, 225–33.

Broquet, T. and Petit, E. (2004) Quantifying genotyping errors in noninvasive population genetics. *Molecular Ecology* **13**, 3601–8.

Brown, J.H. (1995) *Macroecology*. University of Chicago Press, Chicago.

Brown, J.H. and Kodric-Brown, A. (1977) Turnover rates in insular biogeography: effect of immigration on extinction. *Ecology* **58**, 445–9.

Brown, J.L., Li, S., and Bhagabati, N. (1999) Long-term trend toward earlier breeding in an American bird: a response to global warming? *Proceedings of the National Academy of Sciences USA* **96**, 5565–9.

Bruford, M.W., Cheesman, D.J., Coote, T. et al. (1996) Microsatellites and their application to conservation genetics. In: *Molecular Genetic Approaches in Conservation* (eds. T.B. Smith and R.K. Wayne), pp. 278–97. Oxford University Press, New York.

Buckland, S.T., Anderson, D.R., Burnham, K.P., Laake, J.L., Borchers, D.L., and Thomas, L. (2001) *Introduction to Distance Sampling: Estimating Abundance of Biological Populations.* Oxford University Press, New York.

Buckman, R. (2003) *Human Wildlife: The Life That Lives On Us.* Johns Hopkins University Press, Baltimore, MD.

Burgman, M. and Possingham, H. (2000) Population viability analysis for conservation: the good, the bad and the undescribed. In: *Genetics, Demography and Viability of Fragmented Populations* (eds. A.G. Young and G.M. Clarke), pp. 97–112. Cambridge University Press, Cambridge.

Burgman, M.A., Ferson, S., and Akçakaya, H.R. (1993) *Risk Assessment in Conservation Biology.* Chapman and Hall, London.

Burke, T., Hanotte, O., and Van Pijlen, I. (1996) Minisatellite analysis in conservation genetics. In: *Molecular Genetic Approaches in Conservation* (eds. T.B. Smith and R.K. Wayne), pp. 251–77. Oxford University Press. New York.

Burnham, K.P. and Anderson, D.R. (2002) *Model Selection and Multimodel Inference,* 2nd edn. Springer, New York.

Byers, D.L. and Waller, D.M. (1999) Do plant populations purge their genetic load? Effects of population size and mating history on inbreeding depression. *Annual Review of Ecology and Systematics* **30**, 479–513.

Byers, J.A. (1997) *American Pronghorn: Social Adaptations and the Ghosts of Predators Past.* University of Chicago Press, Chicago.

Calaby, J.H. and Grigg, G.C. (1989) Changes in macropodoid communities and populations in the past 200 years, and the future. In: *Kangaroos, Wallabies and Rat Kangaroos* (eds. G. Grigg, P. Jarman, and I. Hume), pp. 813–20. Surrey Beatty and Sons, Sydney.

Calambokidis, J. and Barlow, J. (2004) Abundance of blue and humpback whales in the eastern North Pacific estimated by capture-recapture and line-transect methods. *Marine Mammal Science* **20**, 63–85.

Cam, E., Lougheed, L., Bradley, R., and Cooke, F. (2003) Demographic assessment of a marbled murrelet population from capture-recapture data. *Conservation Biology* **17**, 1118–26.

Cano, R.J., Poinar, H.N., Pieniazek, N.J., Acra, A., and Poinar, Jr., G.O. (1993) Amplification and sequencing of DNA from a 120–135-million-year-old weevil. *Nature* **363**, 536–8.

Cardillo, M., Mace, G.M., Jones, K.E. et al. (2005) Multiple causes of high extinction risk in large mammal species. *Science* **309**, 1239–41.

Caro, T.M. (2003) Umbrella species: critique and lessons from East Africa. *Animal Conservation* **6**, 171–81.

Caro, T.M. and O'Doherty, G. (1999) On the use of surrogate species in conservation biology. *Conservation Biology* **13**, 805–14.

Carson, R. (1962) *Silent Spring.* Houghton Mifflin Company, Boston, MA.

Case, T.J. (2000) *An Illustrated Guide to Theoretical Ecology.* Oxford University Press, New York.

Case, T.J. and Bolger, D.T. (1991) The role of introduced species in shaping the distribution and abundance of island reptiles. *Evolutionary Ecology* **5**, 272–90.

Caswell, H. (1989) *Matrix Population Models: Construction, Analysis, and Interpretation.* Sinauer Associates, Sunderland, MA.

Caswell, H. (2001) *Matrix Population Models: Construction, Analysis, and Interpretation,* 2nd edn. Sinauer Associates, Sunderland, MA.

Caughley, G. (1974) Interpretation of age ratios. *Journal of Wildlife Management* **38**, 557–62.

Caughley, G. (1977) *Analysis of Vertebrate Populations.* John Wiley and Sons, New York.

Caughley, G. (1985) Harvesting of wildlife: past, present, and future. In: *Game Harvest Management* (eds. S.L. Beason and Roberson, S.F.), pp. 3–14. Caesar Kleberg Wildlife Research Institute, Kingsville, TX.

Caughley, G. and Gunn, A. (1996) *Conservation Biology in Theory and Practice.* Blackwell Science, Cambridge, MA.

Caughley, G. and Sinclair, A.R.E. (1994) *Wildlife Ecology and Management.* Blackwell Scientific Publications, Oxford.

Chalfoun, A.D., Thompson, F.R.I., and Ratnaswamy, M.J. (2002) Nest predators and fragmentation: a review of meta-analysis. *Conservation Biology* **16**, 306–18.

Chaneton, E.J. and Bonsall, M.B. (2000) Enemy-mediated apparent competition: empirical patterns and the evidence. *Oikos* **88**, 380–94.

Chang, K. (2003) Crunching the market's numbers: risk, yes; reward, maybe. *New York Times,* January 3, p. D3.

Channell, R. and Lomolino, M.V. (2000) Dynamic biogeography and conservation of endangered species. *Nature* **403**, 84–6.

Chappell, F. (1987) *I am One of You Forever: a Novel.* Louisiana State University Press, Baton Rouge, LA.

Charlesworth, B. and Charlesworth, D. (1999) The genetic basis of inbreeding depression. *Genetical Research* **74**, 329–40.

Citta, J.J. (2005) *Population Dynamics of Mountain Bluebirds.* PhD thesis, University of Montana, Missoula, MT.

Citta, J.J. and Mills, L.S. (1999) What do demographic sensitivity analyses tell us about controlling brown-headed cowbirds? *Studies in Avian Biology* **18**, 121–34.

Claridge, A.W. and Lindenmayer, D.B. (1994) The need for a more sophisticated approach toward wildlife corridor design in the multiple-use forests of southeastern Australia: the case for mammals. *Pacific Conservation Biology* **1**, 301–7.

Clark, J.A. and May, R.M. (2002) Taxonomic bias in conservation research. *Science* **297**, 191–2.

Clavero, M. and García-Berthou, E. (2005) Invasive species are a leading cause of animal extinctions. *Trends in Ecology and Evolution* **20**, 110.

Clobert, J., Danchin, E., Dhondt, A.A., and Nichols, J.D. (2001) *Dispersal.* Oxford University Press, Oxford.

Clout, M.N., Graeme, E.P., and Robertson, B.C. (2002) Effects of supplementary feeding on the offspring sex ratio of kakapo: a dilemma for the conservation of a polygynous parrot. *Biological Conservation* **107**, 13–18.

Cohen, J.E. 1995. *How Many People can the Earth Support?* W.W. Norton & Co., New York.

Cole, L.C. (1954) The population consequences of life history phenomena. *Quarterly Review of Biology* **29**, 103–37.

Cole, L.C. (1957) Sketches of general and comparative demography. *Cold Spring Harbor Symposia on Quantitative Biology* **22**, 1–15.

Coltman, D.W., Pilkington, J.G., Smith, J.A., and Pemberton, J.M. (1999) Parasite-mediated selection against inbred Soay sheep in a free-living, island population. *Evolution* **53**, 1259–67.

Coltman, D.W., O'Donoghue, P., Jorgenson, J.T., Hogg, J.T., Strobeck, C., and Festa-Bianchet, M. (2003) Undesirable evolutionary consequences of trophy hunting. *Nature* **426**, 655–8.

Compton, B.B., Zager, P., and Servheen, G. (1995) Survival and mortality of translocated woodland caribou. *Wildlife Society Bulletin* **23**, 490–6.

Connelly, J.W., Reese, K.P., Garton, E.O., and Commons-Kemner, M.L. (2003) Response of greater sage-grouse *Centrocercus urophasianus* populations to different levels of exploitation in Idaho, USA. *Wildlife Biology* **9**, 335–40.

Connelly J.W., Gammonley, J.H., and Peek, J.M. (2005) Harvest management. In: *Techniques For Wildlife Investigations and Management* (ed. C. Braun), pp. 658–90. The Wildlife Society, Washington DC.

Connor, E.F., Yoder, J.M., and May, J.A. (1999) Density-related predation by the Carolina chickadee, *Poecile carolinensis*, on the leaf-mining moth, *Cameraria hamadryadella* at three spatial scales. *Oikos* **87**, 105–12.

Conover, D.O. and Munch, S.B. (2002) Sustaining fisheries yields over evolutionary time scales. *Science* **297**, 94–6.

Cooch, E. (2001) First steps with Program MARK: linear models. In: *Wildlife, Land and People: Priorities for the 21st Century* (eds. R. Field, R.J. Warren, H. Okarma, and P.R. Sievert), pp. 343–9. The Wildlife Society, Bethesda, MD.

Cook, R.M., Sinclair, A., and Stefansson, G. (1997) Potential collapse of North Sea cod stocks. *Nature* **385**, 521–2.

Côté, I.M. and Sutherland, W.J. (1997) The effectiveness of removing predators to protect bird populations. *Conservation Biology* **11**, 395–405.

Coulson, T., Catchpole, E.A., Albon, S.D. et al. (2001) Age, sex, density, winter weather, and population crashes in Soay Sheep. *Science* **292**, 1528–31.

Courchamp, F., Clutton-Brock, T., and Grenfell, B. (1999) Inverse density dependence and the Allee effect. *Trends in Ecology and Evolution* **14**, 405–10.

Courchamp, F., Langlais, M., and Sugihara, G. (2000) Rabbits killing birds: modeling the hyperpredation process. *Journal of Animal Ecology* **69**, 154–64.

Courchamp, F., Woodroffe, R., and Roemer, G. (2003) Removing protected populations to save endangered species. *Science* **302**, 1532.

Cox, Jr., R.R., Eadie, J.M., Otis, D.L., and Reinecke, K.J. (2004) *Science-based allocation of resources for waterfowl conservation in the contiguous U.S.: implications for the Migratory Bird Conservation Fund*. Technical Report to the Migratory Bird Conservation Fund Allocation Process Team, US Fish and Wildlife Service.

Craig, J., Anderson, S., Clout, M. et al. (2000) Conservation issues in New Zealand. *Annual Review of Ecology and Systematics* **31**, 61–78.

Crandall, K.A., Bininda-Emonds, O.R.P., Mace, G.M., and Wayne, R.K. (2000) Considering evolutionary processes in conservation biology. *Trends in Ecology and Evolution* **15**, 290–5.

Cree, A., Thompson, M.B., and Daugherty, C.H. (1995) Tuatara sex determination. *Nature* **375**, 543.

Creel, S. and Creel, N.M. (1997) Lion density and population structure in the Selous Game Reserve: evaluation of hunting quotas and offtake. *African Journal of Ecology* **35**, 83–93.

Creel S. and Creel, N.M. (2002) The African wild dog: behavior, ecology, and conservation. *Monographs in behavior and ecology*. Princeton University Press, Princeton, NJ.

Creel, S., Spong, G., Sands, J.L. et al. (2003) Population size estimation in Yellowstone wolves with error-prone noninvasive microsatellite genotypes. *Molecular Ecology* **12**, 2003–9.

Cristancho, S. and Vining, J. (2004) Culturally-defined keystone species. *Human Ecology Review* **11**, 153–64.

Crnokrak, P. and Roff, D.A. (1999) Inbreeding depression in the wild. *Heredity* **83**, 260–70.

Crone, E.E. (2001) Is survivorship a better fitness surrogate than fecundity? *Evolution* **55**, 2611–14.

Crooks, J.A. and Soulé, M.E. (1999) Lag times in population explosions of invasive species: causes and implications. In: *Invasive Species and Biodiversity Management* (eds. O.T. Sandlund, P.J. Schei, and Å. Viken), pp. 103–25. Kluwer Academic Publishers, Dordrecht.

Crooks, K.R., Sanjayan, M.A., and Doak, D.F. (1998) Reassessment of the importance of cub predation for cheetah conservation. *Conservation Biology* **12**, 889–95.

Cross, P.C. and Beissinger, S.R. (2001) Using logistic regression to analyze the sensitivity of PVA models: a comparison of methods based on African wild dog models. *Conservation Biology* **15**, 1335–46.

Crouse, D.T., Crowder, L.B., and Caswell, H. (1987) A stage-based population model for loggerhead sea turtles and implications for conservation. *Ecology* **68**, 1412–23.

Crowder, L.B., Crouse, D.T., Heppell, S.S., and Martin, T.H. (1994) Predicting the impact of turtle excluder devices on loggerhead sea turtle populations. *Ecological Applications* **4**, 437–45.

Crowley, T.J. (2000) Causes of climate change over the past 1000 years. *Science* **289**, 270–7.

Crozier, L. (2003) Winter warming facilitates range expansion: cold tolerance of the butterfly *Atalopedes campestris*. *Oecologia* **135**, 648–56.

Cunningham, A.A. (1996) Disease risks of wildlife translocations. *Conservation Biology* **10**, 349–53.

Cunningham, M. and Moritz, C. (1998) Genetic effects of forest fragmentation on a rainforest restricted lizard (Scincidae: *Gnypetoscincus queenslandiae*). *Biological Conservation* **83**, 19–30.

Cuthbert, R. and Davis, L.S. (2002) The impact of predation by introduced stoats on Hutton's shearwaters, New Zealand. *Biological Conservation* **108**, 79–92.

Cuthbert, R., Fletcher, D., and Davis, L.S. (2001) A sensitivity analysis of Hutton's shearwater: prioritizing conservation research and management. *Biological Conservation* **100**, 163–72.

Czech, B., Krausman, P.R., and Borkhataria, R. (1998) Social construction, political power, and the allocation of benefits to endangered species. *Conservation Biology* **12**, 1103–12.

Daniels, S.J. and Walters, J.R. (2000) Inbreeding depression and its effects on natal dispersal in red-cockaded woodpeckers. *The Condor* **102**, 482–91.

Daniels, S.J., Priddy, J.A., and Walters, J.R. (2000) Inbreeding in small populations of red-cockaded woodpeckers: insights from a spatially explicit individual-based model. In: *Genetics, Demography and Viability of Fragmented Populations* (eds. A.G. Young and G.M. Clarke), pp. 129–48. Cambridge University Press, Cambridge.

Darwin, C. (1859) *On the Origin of Species by Means of Natural Selection, or the Preservation of Favoured Races in the Struggle for Life*. Penguin Books, Harmondsworth.

Darwin, C. (1896) *The Variation of Animals and Plants under Domestication*, D. Appleton and Co., New York.

Daszak, P., Cunningham, A.A., and Hyatt, A.D. (2000) Emerging infectious diseases of wildlife-threats to biodiversity and human health. *Science* **287**, 443–9.

Daszak, P., Cunningham, A.A., and Hyatt, A.D. (2003) Infectious disease and amphibian population declines. *Diversity and Distributions* **9**, 141–50.

Daugherty, C.H., Cree, A., Hay, J.M., and Thompson, M.B. (1990) Neglected taxonomy and continuing extinctions of tuatara (Sphenodon). *Nature* **347**, 177–9.

Dawkins, R. and Krebs, J.R. (1979) Arms races between and within species. *Proceedings of the Royal Society of London Series B* **205**, 489–511.

Debinski, D.M. and Holt, R.D. (2000) A survey and overview of habitat fragmentation experiments. *Conservation Biology* **14**, 342–55.

deMaynadier, P. and Hunter, M.L. (1994) Keystone support. *BioScience* **44**, 2.

den Boer, P.J. (1981) On the survival of populations in a heterogeneous and variable environment. *Oecologia* **50**, 39–53.

Dennis, B. (1996) Discussion: should ecologists become Bayesians? *Ecological Applications* **6**, 1095–1103.

Dennis, B. and Otten, M.R.M. (2000) Joint effects of density dependence and rainfall on abundance of San Joaquin kit fox. *Journal of Wildlife Management* **64**, 388–400.

Dennis, B., Munholland, P.L., and Scott, J.M. (1991) Estimation of growth and extinction parameters for endangered species. *Ecological Monographs* **61**, 115–43.

Derocher, A.E., Lunn, N.J., and Stirling, I. (2004) Polar bears in a warming climate. *Integrative and Comparative Biology* **44**, 163–76.

Desrochers, A. and Hannon, S.J. (1997) Gap crossing decisions by forest songbirds during the post-fledging period. *Conservation Biology* **11**, 1204–10.

Diamond, J. (1999) *Guns, Germs, and Steel: the Fates of Human Societies.* W.W. Norton & Co., New York.

Diamond, J. (2005) *Collapse: How Societies Choose to Fail or Succeed.* Viking, New York.

Didham, R.K., Ewers, R.M., and Gemmell, N.J. (2005) Comment on "Avian extinction and mammalian introductions on oceanic islands". *Science* **307**, 1412a.

Dietz, J.M., Baker, A.J., and Ballou, J.D. (2000) Demographic evidence of inbreeding depression in wild golden lion tamarins. In: *Genetics, Demography and Viability of Fragmented Populations* (eds. A.G. Young and G.M. Clarke), pp. 203–11. Cambridge University Press, Cambridge.

Doak, D.F. and Mills, L.S. (1994) A useful role for theory in conservation. *Ecology* **75**, 615–26.

Dobson, A., Ralls, K., Foster, M. et al. (1999) Connectivity: maintaining flows in fragmented landscapes. In: *Continental Conservation: Scientific Foundations of Regional Reserve Networks* (eds. M.E. Soulé and J. Terborgh), pp. 129–71. Island Press, Washington DC.

Dowling, T.E., Minckley, W.L., Douglas, M.E., Marsh, P.C., and DeMarais, B.D. (1992) Response to Wayne, Nowak, and Phillips and Henry: use of molecular characters in conservation biology. *Conservation Biology* **6**, 600–3.

Dreitz, V.J., Bennetts, R.E., Toland, B., Kitchens, W.M., and Collopy, M.W. (2002) Snail kite nest success and water levels: a reply to Beissinger and Snyder. *The Condor* **104**, 216–21.

Driscoll, M.J.L. and Donovan, T.M. (2004) Landscape context moderates edge effects: nesting success of wood thrushes in central New York. *Conservation Biology* **18**, 1330–8.

Dudash, M.R. and Fenster, C.B. (2000) Inbreeding and outbreeding depression in fragmented populations. In: *Genetics, Demography and Viability of Fragmented Populations* (eds. A.G. Young and G.M. Clarke), pp. 35–53. Cambridge University Press, Cambridge.

Dunn, P.O. and Winkler, D.W. (1999) Climate change has affected the breeding date of tree swallows throughout North America. *Proceedings of the Royal Society of London Series B* **266**, 2487–90.

Easterling, D.R., Meehl, G.A., Parmesan, C., Changnon, S.A., Karl, T.R., and Mearns, L.O. (2000) Climate extremes: observations, modeling, and impacts. *Science* **289**, 2068–74.

Eberhardt, L.L. and Simmons, M.A. (1992) Assessing rates of increase from trend data. *Journal of Wildlife Management* **56**, 603–10.

Edmands, S. (1999) Heterosis and outbreeding depression in interpopulation crosses spanning a wide range of divergence. *Evolution* **53**, 1757–68.

Eldridge, M.D.B., King, J.M., Loupis, A.K. et al. (1999) Unprecedented low levels of genetic variation and inbreeding depression in an island population of the black-footed rock-wallaby. *Conservation Biology* **13**, 531–41.

Ellison, A.M. (1996) An introduction to Bayesian inference for ecological research and environmental decision-making. *Ecological Applications* **6**, 1036–46.

Ellison, A.M. (2004) Bayesian inference in ecology. *Ecology Letters* **7**, 509–20.

Ellner, S.P., Fieberg, J., Ludwig, D., and Wilcox, C. (2002) Precision of population viability analysis. *Conservation Biology* **16**, 258–61.

Elton, C.S. (1958) *The Ecology of Invasions by Animals and Plants.* Wiley, New York.

Emlen, S.T. (1990) White-fronted bee-eaters: helping in a colonially nesting species. In: *Cooperative Breeding in Birds: Long-Term Studies of Ecology and Behavior* (eds. R.B. Stacey and W.D. Koenig), pp. 489–526. Cambridge University Press, Cambridge.

Erdrick, L. (1993) Skunk dreams. *The Georgia Review* **XLVII**, 85–94.

Erman, D.C. and Pister, E.P. (1989) Ethics and the environmental biologist. *Fisheries* **14**, 4–7.

Errington, P.L. (1946) Predation and vertebrate populations. *Quarterly Review of Biology* **21**, 144–77; 221–45.

Errington, P.L. (1956) Factors limiting higher vertebrate populations. *Science* **124**, 304–7.

Erwin, T.L. (1982) Tropical forests: their richness in coleoptera and other arthropod species. *The Coleopterists Bulletin* **36**, 74–5.

Estes, J.A. and Duggins, D.O. (1995) Sea otters and kelp forests in Alaska: generality and variation in a community ecological paradigm. *Ecological Monographs* **65**, 75–100.

Estes, J., Crooks, K., and Holt, R. (2001) Predators, Ecological role of. *Encyclopedia of Biodiversity*, vol. 4. Academic Press, New York.

Ezenwa, V.O. (2004) Parasite infection rates of impala (*Aepyceros melampus*) in fenced game reserves in relation to reserve characteristics. *Biological Conservation* **118**, 397–401.

Fa, J.E. and Peres, C.A. (2001) Game vertebrate extraction in Africa and Neotropical forests: an intercontinental comparison. In: *Conservation of Exploited Species* (eds. J.D. Reynolds, G.M. Mace, K.H. Redford, and J.G. Robinson), pp. 203–41. Cambridge University Press, Cambridge.

Fagan, W.F., Meir, E., Prendergast, J., Folarin, A., and Karieva, P. (2001) Characterizing population vulnerability for 758 species. *Ecology Letters* **4**, 132–8.

Fahrig, L. (2003) Effects of habitat fragmentation on biodiversity. *Annual Review of Ecology, Evolution, and Systematics* **34**, 487–515.

Farrell, L.E., Roman, J., and Sunquist, M.E. (2000) Dietary separation of sympatric carnivores identified by molecular analysis of scats. *Molecular Ecology* **9**, 1583–90.

Fauth, J.E. (1999) Identifying potential keystone species from field data – an example from temporary ponds. *Ecology Letters* **2**, 36–43.

Favre, L., Balloux, F., Goudet, J., and Perrin, N. (1997) Female-biased dispersal in the monogamous mammal *Crocidura russula*: evidence from field data and microsatellite patterns. *Proceedings of the Royal Society of London Series B* **264**, 127–32.

Ferson, S. and Burgman, M.A. (1995) Correlations, dependency bounds and extinction risks. *Biological Conservation* **73**, 101–5.

Fieberg, J. and Ellner, S.P. (2000) When is it meaningful to estimate an extinction probability? *Ecology* **81**, 2040–7.

Fisher, D.O., Blomberg, S.P., and Owens, I.P.F. (2003) Extrinsic versus intrinsic factors in the decline and extinction of Australian marsupials. *Proceedings of the Royal Society of London Series B* **270**, 1801–8.

Fitzsimmons, N.N., Buskirk, S.W., and Smith, M.H. (1997) Genetic changes in reintroduced rocky mountain bighorn sheep populations. *Journal of Wildlife Management* **61**, 863–72.

Fleishman, E., Murphy, D.D., and Brussard, P.F. (2000) A new method for selection of umbrella species for conservation planning. *Ecological Applications* **10**, 569–79.

Fleishman, E., Blair, R.B., and Murphy, D.D. (2001) Empirical validation of a method for umbrella species selection. *Ecological Applications* **11**, 1489–1501.

Forman, R.T.T. (2000) Estimate of the area affected ecologically by the road system in the United States. *Conservation Biology* **14**, 31–5.

Fox, G.A. and Gurevitch, J. (2000) Population numbers count: tools for near-term demographic analysis. *American Naturalist* **156**, 242–56.

Frankel, O.H. and Soulé, M.E. (1981) *Conservation and Evolution*. Cambridge University Press, Cambridge.

Frankham, R. (1995) Conservation genetics. *Annual Review of Genetics* **29**, 305–27.

Frankham, R. (1996) Relationship of genetic variation to population size in wildlife. *Conservation Biology* **10**, 1500–8.

Frankham, R. (1997) Do island populations have less genetic variation than mainland populations? *Heredity* **78**, 311–27.

Frankham, R., Ballou, J.D., and Briscoe, D.A. (2002) *Introduction to Conservation Genetics*. Cambridge University Press, Cambridge.

Franklin, A.B., Gutierrez, R.J., Nichols, J.D. et al. (2004) Population dynamics of the California spotted owl (*Strix occidentalis occidentalis*): a meta-analysis. *Ornithological Monographs* **54**, 1–55.

Franklin, I.R. (1980) Evolutionary change in small populations. In: *Conservation Biology: an Evolutionary-Ecological Perspective* (eds. M.E. Soulé and B.A. Wilcox), pp.135–49. Sinauer Associates, Sunderland, MA.

Franklin, J.F. (1993) Preserving biodiversity: species, ecosystems, or landscapes? *Ecological Applications* **3**, 202–5.

Fraser, D.J. and Bernatchez, L. (2001) Adaptive evolutionary conservation: towards a unified concept for defining conservation units. *Molecular Ecology* **10**, 2741–52.

Freddy, D.J., White, G.C., Kneeland, M.C. et al. (2004) How many mule deer are there? Challenges of credibility in Colorado. *Wildlife Society Bulletin* **32**, 916–27.

Fritts, T.H. and Rodda, G.H. (1998) The role of introduced species in the degradation of island ecosystems: a case history of Guam. *Annual Review of Ecology and Systematics* **29**, 113–40.

Fujiwara, M. and Caswell, H. (2001) Demography of the endangered North Atlantic right whale. *Nature* **414**, 537–41.

Fuller, M.R., Millspaugh, J.J., Church, K.E., and Kenward, R.E. (2005) Wildlife radiotelemetry. In: *Techniques for Wildlife Investigations and Management*, 6th edn (ed. C.E. Braun), pp. 377–417. The Wildlife Society, Bethesda, MD.

Funk, W.C., Greene, A.E., Corn, P.S., and Allendorf, F.W. (2005) High dispersal in a frog species suggests that it is vulnerable to habitat fragmentation. *Biology Letters* **1**, 13–16.

Gaillard, J.-M., Festa-Gianchet, M., and Yoccoz, N.G. (1998) Population dynamics of large herbivores: variable recruitment with constant adult survival. *Trends in Ecology and Evolution* **13**, 58–63.

Gaillard, J.-M., Festa-Blanchet, M., and Yoccoz, N.G. (2001) Not all sheep are equal. *Science* **292**, 1499–1531.

Gaines, M.S., Diffendorfer, J.E., Tamarin, R.H., and Whittam, T.S. (1997) The effects of habitat fragmentation on the genetic structure of small mammal populations. *Journal of Heredity* **88**, 294–304.

Gaona, P., Ferreras, P., and Delibes, M. (1998) Dynamics and viability of a metapopulation of the endangered iberian lynx (*Lynx pardinus*). *Ecological Monographs* **68**, 349–70.

Gärdenfors, U. (2000) Population viability analysis in the classification of threatened species: problems and potentials. *Ecological Bulletins* **48**, 181–90.

Gärdenfors, U., Hilton-Taylor, C., Mace, G.M., and Rodríguez, J.P. (2001) The application of IUCN Red List criteria at regional levels. *Conservation Biology* **15**, 1206–12.

Garnier, J.N., Bruford, M.W., and Goossens, B. (2001) Mating system and reproductive skew in the black rhinoceros. *Molecular Ecology* **10**, 2031–41.

Gascoigne, J.C. and Lipcius, R.N. (2004) Allee effects driven by predation. *Journal of Applied Ecology* **41**, 801–10.

Gaston, K.J. (1994) *Rarity*. Chapman and Hall, London.

Gates, J.E. and Gysel, L.W. (1978) Avian nest dispersion and fledging success in field-forest ecotones. *Ecology* **59**, 871–83.

Gemmell, N.J. and Allendorf, F.W. (2001) Mitochondrial mutations may decrease population viability. *Trends in Ecology and Evolution* **16**, 115–17.

Gende, S.M., Quinn, T.P., and Willson, M.F. (2001) Consumption choice by bears feeding on salmon. *Oecologia* **127**, 372–82.

Gerard, P.D., Smith, D.R., and Weerakkody, G. (1998) Limits of retrospective power analysis. *Journal of Wildlife Management* **62**(2), 801–7.

Gerber, L.R., DeMaster, D.P., and Kareiva, P.M. (1999) Gray whales and the value of monitoring data in implementing the U.S Endangered Species Act. *Conservation Biology* **13**, 1215–19.

Gerlach, G. and Musolf, K. (2000) Fragmentation of landscape as a cause for genetic subdivision in Bank voles. *Conservation Biology* **14**, 1066–74.

Gibbs, J.P. and Breisch, A.R. (2001) Climate warming and calling phenology of frogs near Ithaca, New York, 1900–1999. *Conservation Biology* **15**, 1175–8.

Gibbs, J.P., Droege, S., and Eagle, P. (1998) Monitoring populations of plants and animals. *Bioscience* **48**, 935–40.

Gilbert, D.A., Packer, C., Pusey, A.E., Stephens, J.C., and O'Brien, S.J. (1991) Analytical DNA fingerprinting in lions: parentage, genetic diversity, and kinship. *Journal of Heredity* **82**, 378–86.

Gilbert, F.F. and Dodds, D.G. (2001) *The Philosophy and Practice of Wildlife Management.* Krieger Publishing Company, Malabar, FL.

Giles, B.G. and Findlay, C.S. (2004) Effectiveness of a selective harvest system in regulating deer populations in Ontario. *Journal of Wildlife Management* **68**(2), 266–77.

Gillis, E.A. and Krebs, C.J. (2000) Survival of dispersing versus philopatric juvenile snowshoe hares: do dispersers die? *Oikos* **90**, 343–6.

Gilpin, M.E. and Soulé, M.E. (1986) Minimum viable populations: processes of species extinction. In: *Conservation Biology: the Science of Scarcity and Diversity* (ed. M.E. Soulé), pp. 18–34. Sinauer Associates, Sunderland, MA.

Gilson, A., Syvanen, M., Levine, K., and Banks, J. (1998) Deer gender determination by polymerase chain reaction: validation study and applicatoin to tissues, bloodstains, and hair forensic samples from California. *California Fish and Game* **84**, 159–69.

Ginsberg, J.R. and Milner-Gulland, E.J. (1994) Sex-biased harvesting and population dynamics in ungulates: implications for conservation and sustainable use. *Conservation Biology* **8**, 157–66.

Ginzburg, L.R., Ferson, S., and Akçakaya, H.R. (1990) Reconstructability of density dependence and the conservative assessment of extinction risks. *Conservation Biology* **4**, 63–70.

Goodman, D. (1980) Demographic intervention for closely managed populations. In: *Conservation Biology: an Evolutionary-Ecological Perspective* (eds. M.E. Soulé and B.A. Wilcox), pp. 171–95. Sinauer Associates, Sunderland, MA.

Goodman, D. (2002) Predictive Bayesian population viability analysis: a logic for listing criteria, delisting criteria, and recovery plans. In: *Population Viability Analysis* (eds. S.R. Beissinger and D.R. McCullough), pp. 447–69. University of Chicago Press, Chicago.

Gotelli, N.J. (2001) *A Primer of Ecology*, 3rd edn. Sinauer Associates, Sunderland, MA.

Goudet, J., Perrin, N., and Waser, P. (2002) Tests for sex-biased dispersal using bi-parentally inherited genetic markers. *Molecular Ecology* **11**, 1103–14.

Gove, N.E., Skalski, J.R., Zager, P., and Townshend, R.L. (2002) Stistical models for population reconstruction using age-at-harvest data. *Journal of wildlife Management* **66**, 310–20.

Grant, A. and Benton, T.G. (2000) Elasticity analysis for density-dependent populations in stochastic environments. *Ecology* **81**, 680–93.

Griffin, A.S., Blaumstein, D.T., and Evans, C.S. (2000) Training captive-bred or translocated animals to avoid predators. *Conservation Biology* **14**, 1317–26.

Griffin, P. (2004) *Landscape Ecology of Snowshoe Hares in Montana.* PhD thesis, University of Montana, Missoula, MT.

Griffin, P.C., Bienen, L., Gillin, C.M., and Mills, L.S. (2003) Estimating pregnancy rates and litter size in snowshoe hares using ultrasound. *Wildlife Society Bulletin* **31**, 1066–72.

Griffiths, R., Double, M.C., Orr, K., and Dawson, R.J.G. (1998) A DNA test to sex most birds. *Molecular Ecology* **7**, 1071–5.

Grigg, G.C. and Pople, A.R. (2001) Sustainable use and pest control in conservation: Kangaroos as a case study. In: *Conservation of Exploited Species* (eds. J.D. Reynolds, G.M. Mace, K.H. Redford, and J.G. Robinson), pp. 403–23. Cambridge University Press, Cambridge.

Groom, M.J. and Pascual, M. (1998) The analysis of population persistence: an outlook on the practice of viability analysis. In: *Conservation Biology for the Coming Decade*, 2nd edn (eds. P. Fielder and P. Kareiva), pp. 4–27. Chapman and Hall, New York.

Groombridge, B. and Jenkins, M.D. (2002) *World Atlas of Biodiversity*. UNEP World Conservation Monitoring Centre. University of California Press, Berkeley, CA.

Groves, C.R., Jensen, D.B., Valutism, L.L. et al. (2002) Planning for biodiversity conservation: putting conservation science into practice. *BioScience* **52**, 499–512.

Guthery, F.S., Brennan, L.A., Peterson, M.J. and Lusk, J.J. (2005) Information theory in wildlife science: critique and viewpoint. *Journal of Wildlife Management* **69**(2), 457–65.

Haddad, N.M., Bowne, D.R., Cunningham, A. et al. (2003) Corridor use by diverse taxa. *Ecology* **84**, 609–15.

Hagan, J.M., Haegen, W.M.V., and Mckinley, P.S. (1996) The early development of forest fragmentation effects on birds. *Conservation Biology* **10**, 188–202.

Haig, S.M. and Ballou, J.D. (2002) Pedigree analyses in wild populations. In: *Population Viability Analysis* (eds. S.R. Beissinger and D.R. McCullough), pp. 388–405. University of Chicago Press, Chicago.

Haig, S.M., Mullins, T.D., Forsman, E.D., Trail, P.W., and Wennerberg, L. (2004) Genetic identification of spotted owls, barred owls, and their hybrids: legal implications of hybrid identity. *Conservation Biology* **18**, 1347–57.

Hallerman, E.M. (2003) *Population Genetics: Principles and Applications for Fisheries Scientists.* American Fisheries Society, Bethesda, MD.

Hames, R.S., Rosenberg, K.V., Lowe, J.D., and Dhondt, A.A. (2001) Site reoccupation in fragmented landscapes: testing predictions of metapopulation theory. *Journal of Animal Ecology* **70**, 182–90.

Hanski, I. (1996) Metapopulation ecology. In: *Population Dynamics in Ecological Space and Time* (eds. O.E. Rhodes, Jr., R.K. Chesser, and M.H. Smith), pp. 13–43. University of Chicago Press, Chicago.

Hanski, I. (2002) Metapopulations of animals in highly fragmented landscapes and population viability analysis. In: *Population Viability Analysis* (eds. S.R. Beissinger and D.R. McCullough), pp. 86–108. University of Chicago Press, Chicago.

Hanski, I. and Simberloff, D. (1997) The metapopulation approach, its history, conceptual domain, and application to conservation. In: *Metapopulation Biology: Ecology, Genetics, and Evolution* (eds. I. Hanski and M.E. Gilpin), pp. 5–26. Academic Press, San Diego.

Hanski, I. and Gaggiotti, O.E. (2004) *Ecology, Genetics, and Evolution of Metapopulations.* Elsevier Academic Press, Burlington, MA.

Hard, J.J., Mills, L.S., and Peek, J.M. (2006) Genetic implications of reduced survival of male red deer. *Wildlife Biology* **12** (in press).

Harder, J.D. and Kirkpatrick, R.L. (1994) Physiological methods in wildlife research. In: *Research and Management Techniques for Wildlife and Habitats* (ed. T. Bookhout), pp. 275–306. The Wildlife Society, Bethesda, MD.

Harding, E.K., Doak, D.F., and Albertson, J.D. (2001) Evaluating the effectiveness of predator control: the non-native red fox as a case study. *Conservation Biology* **15**, 1114–22.

Hardy, I.C.W. (2002) *Sex Ratios: Concepts and Research Methods.* Cambridge University Press, Cambridge.

Harper, K.A., Macdonald, S.E., Burton, P.J. et al. (2005) Edge influence on forest structure and composition in fragmented landscapes. *Conservation Biology* **19**, 768–82.

Harris, R.B. and Allendorf, F.W. (1989) Genetically effective population size of large mammals: an assessment of estimators. *Conservation Biology* **3**, 181–91.

Harris, R.B., Wall, W.A., and Allendorf, F.W. (2002) Genetic consequences of hunting: what do we know and what should we do? *Wildlife Society Bulletin* **30**, 634–43.

Harrison, S. (1994) Metapopulations and conservation. In: *Large-Scale Ecology and Conservation Ecology* (eds. P.J. Edwards, R.M. May, and N.R. Webb), pp. 111–28. Blackwell Science Publications, Oxford.

Harrison, S. and Fahrig, L. (1995) Landscape pattern and population conservation. In: *Mosaic Landscapes and Ecological Processes* (eds. L. Hansson, L. Fahrig, and G. Merriam), pp. 293–308. Chapman and Hall, London.

Hartl, D.L. and Clark, A.G. (1997) *Principles of Population Genetics*, 3rd edn. Sinauer Associates, Sunderland, MA.

Hastings, A., Hom, C.L., Ellner, S., Turchin, P., and Godfray, H.C.J. (1993) Chaos in ecology: is Mother Nature a strange attractor? *Annual Review of Ecology and Systematics* **24**, 1–33.

Haub, C. (2002) How many people have ever lived on earth? *Population Today* Nov/Dec 3–4.

Hebblewhite, M., Paquet, P.C., Pletscher, D.H., Lessard, R.B., and Callaghan, C.J. (2003) Development and application of a ratio estimator to estimate wolf kill rates and variance in a multiple-prey system. *Wildlife Society Bulletin* **31**, 933–46.

Hedrick, P.W. (1995) Gene flow and genetic restoration: the Florida panther as a case study. *Conservation Biology* **9**, 996–1007.

Hedrick, P. (2003) The major histocompatibility complex (MHC) in declining populations: an example of adaptive variation. In: *Reproduction Science and Integrated Conservation* (eds. W.V. Holt, A.R. Pickard, J.C. Rodger, and D.E. Wildt), pp. 97–113. Cambridge University Press, Cambridge.

Hedrick, P.W. and Kalinowski, S.T. (2000) Inbreeding depression in conservation biology. *Annual Review of Ecology and Systematics* **31**, 139–62.

Heinlein, R.A. (1988) *Time Enough For Love.* Penguin Putnam, New York.

Heppell, S.S., Walters, J.R., and Crowder, L.B. (1994) Evaluating management alternatives for red-cockaded woodpeckers: a modeling approach. *Journal of Wildlife Management* **58**, 479–87.

Heppell, S.S., Caswell, H., and Crowder, L.B. (2000) Life histories and elasticity patterns: perturbation analysis for species with minimal demographic data. *Ecology* **81**, 654–65.

Hersteinsson, P. and Macdonald, D.W. (1992) Interspecific competition and the geographical distribution of red and arctic foxes Vulpes vulpes and Alopex lagopus. *Oikos* **64**, 505–15.

Hilborn, R. and Mangel, M. (1997) *The Ecological Detective.* Princeton University Press, Princeton, NJ.

Hilty, J. and Merenlender, A. (2000) Faunal indicator taxa selection for monitoring ecosystem health. *Biological Conservation* **92**, 185–97.

Hitchings, S.P. and Beebee, T.J.C. (1997) Genetic substructuring as a result of barriers to gene flow in urban *Rana temporaria* (common frog) populations: implications for biodiversity conservation . *Heredity* **79**, 117–27.

Hixon, M.A., Pacala, S.W., and Sandin, S.A. (2002) Population regulation: historical context and contemporary challenges of open vs. closed systems. *Ecology* **83**, 1490–1508.

Hoekman, S.T., Mills, L.S., Howerter, D.W., Devries, J.H., and Ball, I.J. (2002) Sensitivity analysis of the life cycle of mid-continent mallards. *Journal of Wildlife Management* **66**, 883–900.

Hoekman, S.T., Gabor, T.S., Maher, R., Murkin, H.R., and Lindberg, M.S. (2006) Demographics of breeding female mallards in southern Ontario, Canada. *Journal of Wildlife Management* **70**, 111–20.

Hoelzel, A.R., Halley, J., O'Brien, S.J. et al. (1993) Elephant seal genetic variation and the use of simulation models to investigate historical population bottlenecks. *Journal of Heredity* **84**, 443–9.

Hofer, H., Campbell, K.L.I., East, M.L., and Huish, S.A. (1996) The impact of game meat hunting on target and non-target species in the Serengeti. In: *The Exploitation of Mammal Populations* (eds. V.J. Taylor and N. Dunstone), pp. 117–46. Chapman and Hall, London.

Holdaway, R.N. and Jacomb, C. (2000) Rapid extinction of the Moas (Aves: Dinornithiformes): model, test, and implications. *Science* **287**, 2250–4.

Holden, C. (1987) Why do women live longer than men? *Trends in Ecology and Evolution* **238**, 158–61.

Holling, C.S. (1959) The components of predation as revealed by a study of small-mammal predation of the European pine sawfly. *The Canadian Entomologist* **91**, 293–320.

Holmes, E.E. (2001) Estimating risks in declining populations with poor data. *Proceedings of the National Academy of Sciences USA* **98**, 5072–7.

Holway, D.A., Lach, L., Suarez, A.V., Tsutsui, N.D., and Case, T.J. (2002) The causes and consequences of ant invasions. *Annual Review of Ecology and Systematics* **33**, 181–229.

Hoogland, J.L. (1995) *The Black-Tailed Prairie Dog: Social Life of a Burrowing Mammal.* University of Chicago Press, Chicago.

Horvitz, C., Schemske, D.W., and Caswell, H. (1997) The relative "importance" of life-history stages to population growth: prospective and retrospective analyses. In: *Structured-Population Models in Marine, Terrestrial, and Freshwater Systems* (eds. S. Tuljapurkar and H. Caswell), pp. 247–71. Chapman and Hall, New York.

Houlden, B.A., England, P.R., Taylor, A.C., Greville, W.D., and Sherwin, W.B. (1996) Low genetic variability of the koala *Phascolarctos cinereus* in south-eastern Australia following a severe population bottleneck. *Molecular Ecology* **5**, 269–81.

Houston, D.B. and Stevens, V. (1988) Resource limitation in moutain goats: a test by experimental cropping. *Canadian Journal of Zoology* **66**, 228–38.

Houston, D.B., Schreiner, E.G., and Moorhead, B.B. (1994) *Mountain Goats in Olympic National Park: Biology and Management of an Introduced Species*. Natural Resources Publication Office, National Park Service, Denver, CO.

Hunter, Jr., M.L. (2002) *Fundamentals of Conservation Biology*. Blackwell Publishing, Malden, MA.

Hurlbert, S.H. (1984) Pseudoreplication and the design of ecological field experiments. *Ecological Monographs* **54**, 187–211.

Hurlbert, S.H. (1997) Functional importance vs keystoneness: reformulating some questions in theoretical biocenology. *Australian Journal of Ecology* **22**, 369–82.

IPCC (2001) *Climate Change 2001: the Scientific Basis.* Contribution of Working Group I to the third assessment report of the Intergovernmental Panel on Climate Change. Cambridge University Press, Cambridge.

IUCN (2001a) The application of IUCN Red List criteria at regional levels. *Conservation Biology* **15**, 1206–12.

IUCN (2001b) IUCN red list categories and criteria. Version 3.1. www.redlist.org/info/categories_criteria2001. IUCN, Cambridge.

Jablonski, D. (1991) Extinctions: a paleontological perspective. *Science* **253**, 754–7.

Jachmann, H., Berry, P.S.M., and Imae, H. (1995) Tusklessness in African elephants: a future trend. *African Journal of Ecology* **33**, 230–5.

Janczewski, D.N., Yuhki, N., Gilbert, D.A., Jefferson, G.T., and O'Brien, S.J. (1992) Molecular phylogenetic inference from saber-toothed cat fossils of Rancho La Brea. *Proceedings of the National Academy of Sciences USA* **89**, 9769–73.

Janzen, F.J. (1994) Climate change and temperature-dependent sex determination in reptiles. *Population Biology* **91**, 7487–90.

Jarvis, J.U.M., Bennett, N.C., and Spinks, A. (1998) Food availability and foraging by wild colonies of Damaraland mole-rats (*Cryptomys damarensis*): implications for sociality. *Oecologia* **113**, 290–8.

Jedrzejewska, B. and Jedrzejewski, W. (1989) Seasonal surplus killing as hunting strategy of the weasel *Mustela nivalis* – test of a hypothesis. *Acta Theriologica* **34**, 347–59.

Jenkins, D., Watson, A., and Miller, G.R. (1964) Predation and red grouse populations. *Journal of Applied Ecology* **1**, 183–95.

Jeschke, J.M., Kopp, M., and Tollrian, R. (2002) Predator functional responses: discriminating between handling and digesting prey. *Ecological Monographs* **72**, 95–112.

Johnson, D.H. (1999) The insignificance of statistical significance testing. *Journal of Wildlife Management* **63**(3), 763–72.

Johnson, D.H. (2002) The importance of replication in wildlife research. *Journal of Wildlife Management* **66**, 919–32.

Johnson, F.A., Kendall, W.L., and Dubovsky, J.A. (2002) Conditions and limitations on learning in the adaptive management of mallard harvests. *Wildlife Society Bulletin* **30**, 176–85.

Johnson, J.B. and Omland, K.S. (2004) Model selection in ecology and evolution. *Trends in Ecology and Evolution* **19**(2), 101–8.

Johnson, K.H. and Braun, C.E. (1999) Viability and conservation of an exploited sage grouse population. *Conservation Biology* **13**, 77–84.

Johnson, K.N., Agee, J., Bescha, R. et al. (1999) Sustaining the people's lands: recommendations for stewardship of the national forests and grasslands into the next century. *Journal of Forestry* **97**(5), 6–12.

Johnson, W.C., Millett, B.V., Gilmanov, T., Voldseth, R.A., Guntenspergen, G., and Naugle, D. (2005) Vulnerability of northern prairie wetlands to climate change. *BioScience* **55**, 863–72.

Joly, D.O. and Patterson, B.R. (2003) Use of selection indices to model the functional response of predators. *Ecology* **84**, 1635–9.

Jones, C. (2002) A model for the conservation management of a "secondary" prey: sooty shearwater (*Puffinus griseus*) colonies on mainland New Zealand as a case study. *Biological Conservation* **108**, 1–12.

Jones, A.G. and Ardren, W.R. (2003) Methods of parentage analysis in natural populations. *Molecular Ecology* **12**, 2511–23.

Journal of Wildlife Management (1937) Statement of policy. *Journal of Wildlife Management* **1**, 1–2.

Kalinowski, S.T. and Waples, R.S. (2002) Relationship of effective to census size in fluctuating populations. *Conservation Biology* **16**, 129–36.

Kalinowski, S.T., Sawaya, M., and Taper, M.L. (2006) Individual identification and distributions of genotypic differences between individuals. *Journal of Wildlife Management* **70** (in press).

Karanth, K.U. and Nichols, J.D. (1998) Estimation of tiger densities in India using photographic captures and recaptures. *Ecology* **79**, 2852–62.

Karels, T.J. and Boonstra, R. (2000) Concurrent density dependence and independence in populations of arctic ground squirrels. *Nature* **408**, 460–3.

Kauffman, M.J., Frick, W.F., and Linthicum, J. (2003) Estimation of habitat-specific demography and population growth for peregrine falcons in California. *Ecological Applications* **13**, 1802–16.

Kays, R.W. and DeWan, A.A. (2004) Ecological impact of inside/outside house cats around a suburban nature preserve. *Animal Conservation* **7**, 273–83.

Keilman, N. (2003) The threat of small households. *Nature* **421**, 489–90.

Keller, I. and Largiader, C.R. (2003) Recent habitat fragmentation caused by major roads leads to reduction of gene flow and loss of genetic variability in ground beetles. *Proceedings of the Royal Society of London Series B* **270**, 417–23.

Keller, L.F. (1998) Inbreeding and its fitness effects in an insular population of song sparrows (*Melospiza melodia*). *Evolution* **52**, 240–50.

Keller, L.F. and Waller, D.M. (2002) Inbreeding effects in wild populations. *Trends in Ecology and Evolution* **17**, 230–41.

Keller, L.F., Arcese, P., Smith, J.N.M., Hochachka, W.M., and Stearns, S.C. (1994) Selection against inbred song sparrows during a natural population bottleneck. *Nature* **372**, 356–7.

Keller, L.F., Grant, P.R., Grant, B.R., and Petren, K. (2002) Environmental conditions affect the magnitude of inbreeding depression in survival of Darwin's finches. *Evolution* **56**, 1229–39.

Kelley, Jr., J.R. (1996) Line-transect sampling for estimating breeding wood duck density in forested wetlands. *Wildlife Society Bulletin* **24**, 32–6.

Kelly, M.J. (2001) Lineage loss in Serengeti cheetahs: consequences of high reproductive variance and heritability of fitness on effective population size. *Conservation Biology* **15**, 137–47.

Kelly, M.J. and Durant, S.M. (2000) Viability of the Serengeti cheetah population. *Conservation Biology* **14**, 786–97.

Kelly, S.T. and DeCapita, M.E. (1982) Cowbird control and its effect on Kirtland's Warbler reproductive success. *Wilson Bulletin* **94**, 363–5.

Kendall, W.L. (1999) Robustness of closed capture-recapture methods to violations of the closure assumption. *Ecology* **80**, 2517–25.

Kendall, W.L. and Nichols, J.D. (2004) On the estimation of dispersal and movement of birds. *The Condor* **106**, 720–31.

Kendall, W.L., Nichols, J.D., and Hines, J.E. (1997) Estimating temporary emigration using capture-recapture data with Pollock's robust design. *Ecology* **78**, 563–78.

Kindvall, O. (2000) Comparative precision of three spatially realistic simulation models of metapopulation dynamics. *Ecological Bulletins* **48**, 101–10.

Kingsland, S.E. (1985) *Modeling Nature*. University of Chicago Press, Chicago.

Knowlton, F.F., Gese, E.M., and Jaeger, M.M. (1999) Coyote depredation control: An interface between biology and management. *Journal of Range Management* **52**, 398–412.

Koenig, W.D., Van Vuren, D., and Hooge, P.N. (1996) Detectability, philopatry, and the distribution of dispersal distances in vertebrates. *Trends in Ecology and Evolution* **11**, 514–17.

Kohn, M.H. and Wayne, R.K. (1997) Facts from feces revisited. *Trends in Ecology and Evolution* **12**, 223–7.

Kohn, M.H., York, E.C., Kamradt, D.A., Haught, G., Sauvajot, R.M., and Wayne, R.K. (1999) Estimating population size by genotyping faeces. *Proceedings of the Royal Society of London Series B* **266**, 657–63.

Korpimäki, E. and Norrdahl, K. (1998) Experimental reduction of predators reverses the crash phase of small-rodent cycles. *Ecology* **79**, 2448–55.

Korzukhin, M.D., Porter, S.D., Thompson, L.C., and Wiley, S. (2001) Modeling temperature-dependent range limits for the fire ant *Solenopsis invicta* (Hymenoptera: Formicidae) in the United States. *Environmental Entomology* **30**, 645–55.

Kotliar, N.B. (2000) Application of the new keystone-species concept to prairie dogs: how well does it work? *Conservation Biology* **14**, 1715–21.

Krausman, P.R. (2002) *Introduction to Wildlife Management: the Basics.* Prentice Hall, Upper Saddle River, NJ.

Krebs, C.J. (1999) *Ecological Methodology*, 2nd edn. Benjamin/Cummings, Menlo Park, CA.

Krebs, C.J. (2003) Two complementary paradigms for analyzing population dynamics. In: *Wildlife Population Growth Rates* (eds. R.M. Sibly, J. Hone, and T.H. Clutton-Brock), pp. 110–26. The Royal Society and Cambridge University Press, Cambridge.

Krebs, C.J., Boutin, S., Boonstra, R. et al. (1995) Impact of food and predation on the snowshoe hare cycle. *Science* **269**, 1112–15.

Kruuk, H. (1972) Surplus killing by carnivores. *Journal of Zoology* **166**, 233–44.

Kuussaari, M., Saccheri, I., Camara, M., and Hanski, I. (1998) Allee effect and population dynamics in the Glanville fritillary butterfly. *Oikos* **82**, 384–92.

Labonté, J., Ouellet, J.-P., Courtois, R., and Belisle, F. (1998) Moose dispersal and its role in the maintenance of harvested populations. *Journal of Wildlife Management* **62**, 225–35.

Lacy, R.C. (1997) Importance of genetic variation to the viability of mammalian populations. *Journal of Mammalogy* **78**, 320–35.

Lacy, R.C. (2000) Structure of the VORTEX simulation model for population viability analysis. *Ecological Bulletins* **48**, 191–203.

Lacy, R.C. and Miller, P.S. (2002) Incorporating human populations and activities into population viability analysis. In: *Population Viability Analysis* (eds. S.R. Beissinger and D.R. McCullough), pp. 490–510. University of Chicago Press, Chicago.

Lambeck, R.J. (1997) Focal species: a multi-species umbrella for nature conservation. *Conservation Biology* **11**, 849–56.

Lancia, R.A., Nichols, J.D., Pollock, K.H., and Kendall, W.L. (2005) Estimating the number of animals in wildlife populations. In: *Techniques for Wildlife Investigations and Management*, 6th edn (ed. C.E. Braun), pp. 106–53. The Wildlife Society, Bethesda, MD.

Lanciani, C.A. (1998) A simple equation for presenting reproductive value to introductory biology and ecology classes. *Bulletin of the Ecological Society of America* **79**, 192–3.

Land, E.D. and Lacy, R.C. (2000) Introgression level achieved through Florida panther genetic restoration. *Endangered Species Update* **17**, 99–103.

Lande, R. (1988a) Genetics and demography in biological conservation. *Science* **241**, 1455–60.

Lande, R. (1988b) Demographic models of the northern spotted owl (*Strix occidentalis caurina*). *Oecologia* **75**, 601–7.

Lande, R., Sæther, B.-E., and Engen, S. (2001) Sustainable exploitation of fluctuating populations. In: *Conservation of Exploited Species* (eds. J.D. Reynolds, G.M. Mace, K.H. Redford, and J.G. Robinson), pp. 67–86. Cambridge University Press, Cambridge.

Lande, R., Engen, S., and Sæther, B.-E. (2003) *Stochastic Population Dynamics in Ecology and Conservation*. Oxford University Press, Oxford.

Landeau, L. and Terborgh, J. (1986) Oddity and the "confusion effect" in predation. *Animal Behaviour* **34**, 1372–80.

Landres, P.B., Verner, J., and Thomas, J.W. (1988) Ecological uses of vertebrate indicator species: a critique. *Conservation Biology* **2**, 316–28.

Langtimm, C.A., O'Shea, T.J., Pradel, R., and Beck, C.A. (1998) Estimates of annual survival probabilities for adult Florida manatees (*Trichechus manatus latirostris*). *Ecology* **79**, 981–97.

Langvatn, R. and Loison, A. (1999) Consequences of harvesting on age structure, sex ratio and population dynamics of red deer *Cervus elaphus* in central Norway. *Wildlife Biology* **5**, 213–23.

Larson, G. (1998) *There's a Hair In My Dirt! A Worm's Story*. HarperCollins, New York.

Laurance, S.G.W., Stouffer, P.C., and Laurance, W.F. (2004) Effects of road clearings on movement patterns of understory rainforest birds in central Amazonia. *Conservation Biology* **18**, 1099–1109.

Laurance, W.F. (1990) Comparative responses of five arboreal marsupials to tropical forest fragmentation. *Journal of Mammalogy* **71**, 641–53.

Laurance, W.F. (2004) Forest-climate interactions in fragmented tropical landscapes. *Philosophical Transactions of the Royal Society* **359**, 345–52.

Laurance, W.F. and Cochrane, M.A. (2001) Synergistic effects in fragmented landscapes. *Conservation Biology* **15**, 1488–9.

Laurance, W.F., Lovejoy, T.E., Vasconcelos, H.L. et al. (2002) Ecosystem decay of Amazonian forest fragments: a 22-year investigation. *Conservation Biology* **16**, 605–18.

Law, R. (2000) Fishing, selection, and phenotypic evolution. *International Council for the Exploration of the Sea Journal of Marine Sciences* **57**, 659–68.

Leader-Williams, N. and Dublin, H.T. (2000) Charismatic megafauna as "flagship species". In: *Priorities for the Conservation of Mammalian Diversity: Has the Panda had its Day?* (eds. A. Entwistle and N. Dunstone), pp. 53–81. Cambridge University Press, Cambridge.

Leberg, P.L. (1991) Influence of fragmentation and bottlenecks on the genetic divergence of wild turkey populations. *Conservation Biology* **5**, 522–30.

Leberg, P.L. (1992) Effects of population bottlenecks on genetic diversity as measured by allozyme electrophoresis. *Evolution* **46**, 477–94.

Leberg, P.L. (2002) Estimating allelic richness: effects of sample size and bottlenecks. *Molecular Ecology* **11**, 2445–9.

Lebreton, J.D. and Clobert, J. (1991) Bird population dynamics, management and conservation: the role of mathematical modelling. In: *Bird Population Studies: Relevance to Conservation and Management* (eds. C.M. Perrins, J.D. Lebreton and G.J.M. Hirons), pp. 105–25. Oxford University Press, Oxford.

Lebreton, J.-D., Burnham, K.P., Clobert, J., and Anderson, D.R. (1992) Modeling survival and testing biological hypotheses using marked animals: a unified approach with case studies. *Ecological Monographs* **62**, 67–118.

Lee, D.C. (2000) Assessing land-use impacts on bull trout using Bayesian belief networks. In: *Quantitative Methods in Biology* (eds. S. Ferson and M. Burgman), pp. 127–47. Springer-Verlag, New York.

Lefkovitch, L.P. (1965) The study of population growth in organisms grouped in stages. *Biometrics* **21**, 1–18.

Lehman, N., Eisenhawer, A., Hansen, K. et al. (1991) Introgression of coyote mitochondrial DNA into sympatric North American gray wolf populations. *Evolution* **45**, 104–19.

Lento, G.M., Lavery, S., Funahashi, N., Dalebout, M.L., and Baker, C.S. (2001) Market surveys in Japan and Korea, 2000–2001: implications for boundaries of protected stocks. Unpublished report (SC/53/SD6) to the Scientific Committee of the International Whaling Commission, Cambridge.

Leopold, A. (1933) *Game Management*. Charles Scribner's Sons. New York.

Leopold, A. (1953) *Round River*. Oxford University Press, New York.

Leopold, A. (1966) *A Sand County Almanac*. Ballantine Books, New York.

Leopold, A.C. (2004) Living with the land ethic. *BioScience* **54**, 149–54.

Leslie, P.H. (1945) On the use of matrices in certain population mathematics. *Biometrika* **33**, 183–212.

Leung, L.K.P., Dickman, C.R., and Moore, L.A. (1993) Genetic variation in fragmented populations of an Australian rainforest rodent, *Melomys cervinipes*. *Pacific Conservation Biology* **1**, 58–65.

Levins, R. (1970) Extinction. *Lectures on Mathematics in the Life Sciences* **2**, 75–107.

Lidicker, Jr., W.Z. (1975) The role of dispersal in the demography of small mammals. In: *Small Mammals: their Productivity and Population Dynamics* (eds. F.B. Golley, K. Petrusewicz, and L. Ryszkowski), pp. 103–28. Cambridge University Press, Cambridge.

Liermann, M. and Hilborn, R. (2001) Depensation: evidence, models and implications. *Fish and Fisheries* **2**, 33–58.

Lindenmayer, D.B., Manning, A.D., Smith, P.L. et al. (2002) The focal-species approach and landscape restoration: a critique. *Conservation Biology* **16**, 338–45.

Link, W.A. and Nichols, J.D. (1994) On the importance of sampling variance to investigations of temporal variation in animal population size. *Oikos* **69**, 539–44.

Liu, J., Daily, G.C., Ehrlich, P.R., and Luck, G.W. (2003) Effects of household dynamics on resource consumption and biodiversity. *Nature* **421**, 530–3.

Lodge, D.M. and Shrader-Frechette, K. (2003) Nonindigenous species: ecological explanation, environmental ethics, and public policy. *Conservation Biology* **17**, 31–7.

LoGiudice, K. (2003) Trophically transmitted parasites and the conservation of small populations: raccoon roundworm and the imperiled allegheny woodrat. *Conservation Biology* **17**, 258–66.

Long, E.S., Diefenbach, D.R., Rosenberry, C.S., Wallingford, B.D., and Grund, M.D. (2005) Forest cover influences dispersal distance of white-tailed deer. *Journal of Mammalogy* **86**, 623–9.

Lopez, B. (1986) *Arctic Dreams*. Random House, New York.

Losos, J.B., Schoener, T.W., Warheit, K.I., and Creer, D. (2001) Experimental studies of adaptive differentiation in Bahamian *Anolis* lizards. *Genetics* **112–13**, 399–415.

Lovejoy, T.E., Bierregaard, Jr., R.O., Rylands, A.B. et al. (1986) Edge and other effects of isolation on amazon forest fragments. In: *Conservation Biology: the Science of Scarcity and Diversity* (ed. M.E. Soulé), pp. 257–85. Sinauer Associates, Sunderland, MA.

Lowe, W.H. (2003) Linking dispersal to local population dynamics: a case study using a headwater salamander system. *Ecology* **84**, 2145–54.

Ludwig, D. (1999) Is it meaningful to estimate a probability of extinction? *Ecology* **80**, 293–310.

Ludwig, D. and Walters, C.J. (2002) Fitting population viability analysis into adaptive management. In: *Population Viability Analysis* (eds. S.R. Beissinger and D.R. McCullough), pp. 511–20. University of Chicago Press, Chicago.

Ludwig, D., Hilborn, R., and Walters, C. (1993) Uncertainty, resource exploitation, and conservation: lessons from history. *Science* **260**, 17, 36.

Luikart, G. and England, P.R. (1999) Statistical analysis of microsatellite DNA data. *Trends in Ecology and Evolution* **14**, 253–6.

Lukacs, P.M. and Burnham, K.P. (2005) Estimating population size from DNA-based closed capture-recapture data incorporating genotyping error. *Journal of Wildlife Management* **69**, 396–403.

Lutz, W., Sanderson, W., and Scherbov, S. (2001) The end of world population growth. *Nature* **412**, 543–5.

Lyver, P.O., Moller, H., and Robertson, C.J.R. (2000) Predation at sooty shearwater *Puffinus griseus* colonies on the New Zealand mainland: is there safety in numbers? *Pacific Conservation Biology* **5**, 347–57.

MacArthur, R.H. and Wilson, E.O. (1967) *The Theory of Island Biogeography*. Princeton University Press, Princeton, NJ.

Macdonald, D.W., Mace, G.M., and Barretto, G.R. (1999) The effects of predators on fragmented prey populations: a case study for the conservation of endangered prey. *Journal of Zoology* **247**, 487–506.

Mace, G.M. (1994) An investigation into methods for categorizing the conservation status of species. In: *Large-Scale Ecology and Conservation Ecology* (eds. P.J. Edwards, R.M. May, and N.R. Webb), pp. 293–312. Blackwell Scientific Publications, Oxford.

Mace, G.M. (1995) Classification of threatened species and its role in conservation planning. In: *Extinction Rates* (eds. J.H. Lawton and R.M. May), pp. 197–213. Oxford University Press, New York.

Mace, G.M. and Lande, R. (1991) Assessing extinction threats: towards a reevaluation of IUCN threatened species categories. *Conservation Biology* **5**, 148–57.

Mace, G.M. and Kershaw, M. (1997) Extinction risk and rarity on an ecological timescale. In: *The Biology of Rarity* (eds. W.E. Kunin and K.J. Gaston), pp. 130–49. Chapman and Hall, London.

Mace, G.M. and Hudson, E.J. (1999) Attitudes toward sustainability and extinction. *Conservation Biology* **13**, 242–6.

Mack, R.N., Simberloff, D., Lonsdale, W.M., Evans, H., Clout, M., and Bazzaz, F.A. (2000) Biotic invasions: causes, epidemiology, global consequences, and control. *Ecological Applications* **10**, 689–710.

MacKenzie, D.I., and Kendall, W.L. (2002) How should detection probability be incorporated into estimates of relative abundance? *Ecology* **83**, 2387–93.

MacKenzie, D.I., Nichols, J.D., Sutton, N., Kawanishi, K., and Bailey, L.L. (2005) Improving inferences in population studies of rare species that are detected imperfectly. *Ecology* **86**, 1101–13.

Macnab, J. (1983) Wildlife management as scientific experimentation. *Wildlife Society Bulletin* **11**, 397–401.

Madsen, T., Stille, B., and Shine, R. (1996) Inbreeding depression in an isolated population of adders *Vipera berus*. *Biological Conservation* **75**, 113–18.

Madsen, T., Shine, R., Olsson, M., and Wittzell, H. (1999) Restoration of an inbred adder population. *Nature* **402**, 34–5.

Maehr, D.S. and Lacy, R.C. (2002) Avoiding the lurking pitfalls in Florida panther recovery. *Wildlife Society Bulletin* **30**, 971–8.

Maguire, L.A. (1991) Risk analysis for conservation biologists. *Conservation Biology* **5**, 123–5.

Malcolm, J.R. (1994) Edge effects in central Amazonian forest fragments. *Ecology* **75**, 2438–45.

Malthus, T.R. (1798) *An Essay on the Principle of Population*. Reprinted and edited by G. Gilbert (1999), Oxford University Press, Oxford.

Manel, S., Berthier, P., and Luikart, G. (2002) Detecting wildlife poaching: identifying the origin of individuals with Bayesian assignment tests and multilocus genotypes. *Conservation Biology* **16**, 650–9.

Manel, S., Gaggiotti, O.E., and Waples, R.S. (2005) Assignment methods: matching biological questions with appropriate techniques. *Trends in Ecology and Evolution* **20**, 136–42.

Mangel, M. and Tier, C. (1994) Four facts every conservation biologist should know about persistence. *Ecology* **75**, 607–14.

Marcot, B.G., Castellano, M.A., Christy, J.A. et al. (1997) Terrestrial ecology assessment. In: *An Assessment of Ecosystem Components in the Interior Columbia Basin and Portions of the Klamath and Great Basins*. USDA Forest Service General Technical Report PNW-GTR-405, vol. III (eds. T.M. Quigley and S.J. Arbelbide), pp. 1497–1713. USDA Forest Service Pacific Northwest Research Station, Portland, OR.

Marcot, B.G., Holthausen, R.S., Raphael, M.G., Rowland, M.M., and Wisdom, M.J. (2001) Using Bayesian belief networks to evaluate fish and wildlife population viability under land management alternatives from an environmental impact statement. *Forest Ecology and Management* **153**, 29–42.

Markham, B. (1983) *West with the Night*. North Point Press, New York.

Marzluff, J.M. and Restani, M. (1999) The effects of forest fragmentation on avian nest predation. In: *Forest Fragmentation: Wildlife and Management Implications* (eds. J.A. Rochelle, L.A. Lehmann and J. Wisniewski), pp. 155–69. Brill Academic Publishing, Leiden.

Master, L.L., Stein, B.A., Kutner, L.S., and Hammerson, G.A. (2000) Vanishing assets: conservation status of U.S. species. In: *Precious Heritage: the Status of Biodiversity in the United States* (eds. B.A. Stein, L.S. Kutner, and J.S. Adams), pp. 93–118. Oxford University Press, New York.

May, R.M. (1974) Biological populations with nonoverlapping generations: stable points, stable cycles, and chaos. *Science* **186**, 645–7.

May, R.M. (1994) Biological diversity: differences between land and sea. *Philosophical Transactions of the Royal Society of London Series B* **343**, 105–11.

May, R.M. (1997) The dimensions of life on earth. In: *Nature and Human Society: the Quest for a Sustainable World* (eds. P.H. Raven), pp. 30–45. National Academy Press, Washington DC.

May, R.M., Lawton, J.H., and Stork, N.E. (1995) Assessing extinction rates. In: *Extinction Rates* (eds. J.H. Lawton and R.M. May), pp. 1–24. Oxford University Press, Oxford.

McCallum, H. (2000) *Population Parameters: Estimation for Ecological Models*. Blackwell Publishing, Oxford.

McCarthy, M.A., Possingham, H.P., Day, J.R., and Tyre, A.J. (2001) Testing the accuracy of population viability analysis. *Conservation Biology* **15**, 1030–8.

McCarthy, M.A., Keith, D., Tietjen, J. et al. (2004) Comparing predictions of extinction risk using models and subjective judgement. *Acta Oecologica* **26**, 67–74.

McCullough, D.R. (1992) Concepts of large herbivore population dynamics. In: *Wildlife 2001: Populations* (eds. D.R. McCullough and R.H. Barrett), pp. 967–84. Elsevier Applied Science, London.

McCullough, D.R. (1994) What do herd composition counts tell us? *Wildlife Society Bulletin* **22**, 295–300.

McCullough, D.R. (1996) Spatially structured populations and harvest theory. *Journal of Wildlife Management* **60**(1), 1–9.

McDaniel, G.W., McKelvey, K.S., Squires, J.R., and Ruggiero, L.F. (2000) Efficacy of lures and hair snares to detect lynx. *Wildlife Society Bulletin* **28**, 119–23.

McGraw, J.B. and Furedi, M.A. (2005) Deer browsing and population viability of a forest understory plant. *Science* **307**, 920–2.

McGuane, T. (1997) The heart of the game. In: *A Hunter's Heart: Honest Essays on Blood Sport* (ed. D. Petersen), pp. 162–73. Henry Holt and Co, New York.

McLean, I.G., Lundie-Jenkins, G., and Jarman, P.J. (1996) Teaching an endangered mammal to recognise predators. *Biological Conservation* **75**, 51–62.

McLeod, S.R. and Saunders, G.R. (2001) Improving management strategies for the red fox by using projection matrix analysis. *Wildlife Research* **28**, 333–40.

Meagher, S. (1999) Genetic diversity and *Capillaria hepatica* (Nematoda) prevalence in Michigan deer mouse populations. *Evolution* **53**, 1318–24.

Meagher, S., Penn, D.J., and Potts, W.K. (2000) Male-male competition magnifies inbreeding depression in wild house mice. *Proceedings of the National Academy of Sciences USA* **97**, 3324–9.

Meir, E. and Fagan, W.F. (2000) Will observation error and biases ruin the use of simple extinction models? *Conservation Biology* **14**, 148–54.

Meslow, E.C., Holthausen, R.S., and Cleaves, D.A. (1994) Assessment of terrestrial species and ecosystems. *Journal of Forestry* **92**, 24–7.

Messier, F. (1994) Ungulate population models with predation: a case study with the North American moose. *Ecology* **75**, 478–88.

Messierm, F. (1995) On the functional and numerical responses of wolves to changing prey density. In: *Ecology and Conservation of Wolves in a Changing World* (eds. L.N. Carbyn, S.H. Fritts, and D.R. Seip), pp. 187–97. Occasional publication no. 35, Canadian Circumpolar Institute, Edmonton.

Mildenstein, T.L., Stier, S.C., Nuevo-Diego, C.E., and Mills, L.S. (2005) Habitat selection of endangered and endemic large flying-foxes in Subic Bay, Philippines. *Biological Conservation* **126**, 93–102.

Miller, B., Biggins, D., Wemmer, C. et al. (1990) Development of survival skills in captive raised Siberian polecats (*Mustela eversmanni*) II: predator avoidance. *Journal of Ethology* **8**, 95–104.

Miller, B., Ralls, K., Reading, R.P., Scott, J.M., and Estes, J. (1999) Biological and technical considerations of carnivore translocation: a review. *Animal Conservation* **2**, 59–68.

Miller, C.R. and Waits, L.P. (2003) The history of effective population size and genetice diversity in the Yellowstone grizzly (*Ursus arctos*): implications for conservation. *Proceedings of the National Academy of Sciences USA* **100**, 4334–9.

Miller, H.C. and Lambert, D.M. (2004) Genetic drift outweighs balancing selection in shaping post-bottleneck major histocompatibility complex variation in New Zealand robins (Petroicidae). *Molecular Ecology* **13**, 3709–21.

Mills, L.S. (1995) Edge effects and isolation: red-backed voles on forest remnants. *Conservation Biology* **9**, 395–403.

Mills, L.S. (1996) Fragmentation of a natural area: dynamics of isolation for small mammals on forest remnants. In: *National Parks and Protected Areas: their Role in Environmental Protection* (eds. G. Wright), pp. 199–219. Blackwell Press, Cambridge, MA.

Mills, L.S. (2002) False samples are not the same as blind controls. *Nature* **415**, 471.

Mills, L.S. and Allendorf, F.W. (1996) The one-migrant-per-generation rule in conservation and management. *Conservation Biology* **10**, 1509–18.

Mills, L.S. and Knowlton, F.F. (1989) Observer performance in known and blind radio-telemetry accuracy tests. *Journal of Wildlife Management* **53**, 340–2.

Mills, L.S. and Smouse, P.E. (1994) Demographic consequences of inbreeding in remnant populations. *American Naturalist* **144**, 412–31.

Mills, L.S. and Tallmon, D.A. (1999) The role of genetics in understanding forest fragmentation. In: *Forest Fragmentation: Wildlife and Management Implications* (eds. J.A. Rochelle, L.A. Lehmann, and J. Wisniewski), pp. 171–84. Brill Academic Publishing, Leiden.

Mills, L.S. and Lindberg, M.S. (2002) Sensitivity analysis to evaluate the consequences of conservation actions. In: *Population Viability Analysis* (eds. S.R. Beissinger and D.R. McCullough), pp. 338–66. University of Chicago Press, Chicago.

Mills, L.S., Soulé, M.E., and Doak, D.F. (1993) The keystone-species concept in ecology and conservation. *BioScience* **43**, 219–24.

Mills, L.S., Hayes, S.G., Baldwin, C. et al. (1996) Factors leading to different viability predictions for a grizzly bear data set. *Conservation Biology* **10**, 863–73.

Mills, L.S., Doak, D.F., and Wisdom, M.J. (1999) The reliability of conservation actions based on sensitivity analysis of matrix models. *Conservation Biology* **13**, 815–29.

Mills, L.S., Pilgrim, K.L., Schwartz, M.K., and McKelvey, K. (2000a) Identifying lynx and other North American felids based on mtDNA analysis. *Conservation Genetics* **1**, 285–8.

Mills, L.S., Citta, J.J., Lair, K.P., Schwartz, M.K., and Tallmon, D.A. (2000b) Estimating animal abundance using noninvasive DNA sampling: promise and pitfalls. *Ecological Applications* **10**, 283–94.

Mills, L.S., Doak, D.F., and Wisdom, M.J. (2001) Elasticity analysis for conservation decision-making: reply to Ehrlen et al., *Conservation Biology* **15**, 281–3.

Mills, L.S., Schwartz, M.K., Tallmon, D.A., and Lair, K.P. (2003) Measuring and interpreting connectivity for mammals in coniferous forests. In: *Mammal Community Dynamics: Management and Conservation in the Coniferous Forests of Western North America* (eds. C.J. Zabel and R.G. Anthony), pp. 587–613. Cambridge University Press, Cambridge.

Mills, L.S., Scott, J.M., Strickler, K.M., and Temple, S.A. (2005) Ecology and management of small populations. In: *Techniques for Wildlife Investigations and Management*, 6th edn. (eds. C.E. Braun), pp. 691–713. The Wildlife Society, Bethesda, MD.

Milly, P.C.D., Dunne, K.A., and Vecchia, A.V. (2005) Global pattern of trends in streamflow and water availability in a changing climate. *Nature* **438**, 347–50.

Milner-Gulland, E.J., Bennett, E.L., and the SCB 2002 Annual Meeting Wild Meat Group (2003a) Wild meat: the bigger picture. *Trends in Ecology and Evolution* **18**(7), 351–7.

Milner-Gulland, E.J., Bukreeva, O.M., Coulson, T. et al. (2003b) Reproductive collapse in saiga antelope harems. *Nature* **422**, 135.

Mitchell, C.E. and Power, A.G. (2003) Release of invasive plants from fungal and viral pathogens. *Nature* **421**, 625–7.

Modi, W.S. and Crews, D. (2005) Sex chromosomes and sex determination in reptiles. *Current Opinion in Genetics and Development* **15**, 660–5.

Moilanen, A. (2000) The equilibrium assumption in estimating the parameters of metapopulation models. *Journal of Animal Ecology* **69**, 143–53.

Mooney, H.A. and Cleland, E.E. (2001) The evolutionary impact of invasive species. *Proceedings of the National Academy of Sciences USA* **98**, 5446–51.

Moore, B. (1996) *The Lochsa Story: Land Ethics in the Bitterroot Mountains.* Mountain Press, Missoula, MT.

Moore, R.D., Griffiths, R.A., O'Brien, C.M., Murphy, A., and Jay, D. (2004) Induced defences in an endangered amphibian in response to an introduced snake predator. *Oecologia* **141**, 139–47.

Moore, S.L. and Wilson, K. (2002) Parasites as a viability cost of sexual selection in natural populations of mammals. *Science* **297**, 2015–17.

Morin, P.A. and Woodruff, D.S. (1996) Noninvasive genotyping for vertebrate conservation. In: *Molecular Genetic Approaches in Conservation* (eds. T.B. Smith and R.K. Wayne), pp. 298–313. Oxford University Press, New York.

Moritz, C. (1994) Defining "Evolutionary Significant Units" for conservation. *Trends in Ecology and Evolution* **9**, 373–5.

Moritz, C. (1995) Uses of molecular phylogenies for conservation. *Philosophical Transactions of the Royal Society of London Series B* **349**, 113–18.

Morris, E. (2002) *Theodore Rex.* Modern Library Paperbacks. New York.

Morris, W.F. and Doak, D.F. (2002) *Quantitative Conservation Biology: Theory and Practice of Population Viability Analysis.* Sinauer Associates, Sunderland, MA.

Morrison, M.L. and Hahn, D.C. (2002) Geographic variation in cowbird distribution, abundance, and parasitism. *Studies in Avian Biology* **25**, 65–72.

Moulton, M.P. and Sanderson, J. (1997) *Wildlife Issues in a Changing World.* St. Lucie Press, Delray Beach, FL.

Mowat, G. and Strobeck, C. (2000) Estimating population size of grizzly bears using hair capture, DNA profiling, and mark-recapture analysis. *Journal of Wildlife Management* **64**, 183–93.

Mrosovsky, N. (1982) Sex ratio bias in hatchling sea turtles from artificially incubated eggs. *Biological Conservation* **23**, 309–14.

Murphy, M.A., Waits, L.P., Kendall, K.C., Wasser, S.K., Higbee, J.A., and Bogden, R. (2002) An evaluation of long-term preservation methods for brown bear (*Ursus arctos*) faecal DNA samples. *Conservation Genetics* **3**, 435–40.

Myers, R.A. and Worm, B. (2003) Rapid worldwide depletion of predatory fish communities. *Nature* **423**, 280–3.

Myers, J., Simberloff, J., Kuris, A., and Carey, J. (2000) Reply from J. Myers, D. Simberloff, A. Kuris and J. Carey. *Trends in Ecology and Evolution* **15**, 515–16.

Mysterud, A., Coulson, T., and Stenseth, N.C. (2002) The role of males in the dynamics of ungulate populations. *Journal of Animal Ecology* **71**, 907–15.

Nathan, R., Perry, G., Cronin, J.T., Strand, A.E., and Cain, M.L. (2003) Methods for estimating long-distance dispersal. *Oikos* **103**, 261–73.

Nei, M., Maruyama, T., and Chakraborty, R. (1975) The bottleneck effect and genetic variability in populations. *Evolution* **29**, 1–10.

Newton, I. (1998) Pollutants and pesticides. In: *Conservation Science and Action*, 2nd edn. (ed. W.J. Sutherland), pp. 66–89. Blackwell Science, Malden, MA.

Nichols, J.D. (2000) Evolution of harvest management for North American waterfowl: Selective pressures and preadaptations for adaptive harvest management. *Transactions of the North American Wildlife and Natural Resources Conference* **65**, 65–77.

Nichols, J.D. and Coffman, C.J. (1999) Demographic parameter estimation for experimental landscape studies on small mammal populations. In: *Landscape Ecology of Small Mammals* (eds. G.W. Barrett and J.D. Peles), pp. 287–309. Springer, New York.

Nichols, J.D. and Hines, J.E. (2002) Approaches for the direct estimation of lambda, and demographic contributions to lambda, using capture-recapture data. *Journal of Applied Statistics* **29**, 539–68.

Nichols, J.D. and Pollock, K.H. (1983) Estimation methodology in contemporary small mammal capture-recapture studies. *Journal of Mammalogy* **64**, 253–60.

Nichols, J.D. and Pollock, K.H. (1990) Estimation of recruitment from immigration versus in situ reproduction using Pollock's robust design. *Ecology* **71**, 21–6.

Nichols, J.D., Conroy, M.J., Anderson, D.R., and Burnham, K.P. (1984) Compensatory mortality in waterfowl populations: a review of the evidence and implications for research and management. *Transactions of the North American Wildlife and Natural Resources Conference* **49**, 535–54.

Nichols, J.D., Johnson, F.A., and Williams, B.K. (1995) Managing north American waterfowl in the face of uncertainty. *Annual Review of Ecology and Systematics* **26**, 177–99.

Nichols, J.D., Hines, J.E., Sauer, J.R., Fallon, F.W., Fallon, J.E., and Heglund, P.J. (2000a) A double-observer approach for estimating detection probability and abundance from point counts. *The Auk* **117**, 393–408.

Nichols, J.D., Hines, J.E., Lebreton, J.-D., and Pradel, R. (2000b) Estimation of contributions to population growth: a reverse-time capture-recapture approach. *Ecology* **81**, 3362–76.

Niemi, G.J. and McDonald, M.E. (2004) Application of ecological indicators. *Annual Review of Ecology, Evolution, and Systematics* **35**, 89–111.

Niemi, G.J., Hanowski, J.M., Lima, A.R., Nicholls, T., and Weiland, N. (1997) A critical analysis on the use of indicator species in management. *Journal of Wildlife Management* **61**, 1240–52.

Nisbet, E.G. and Sleep, N.H. (2001) The habitat and nature of early life. *Nature* **409**, 1083–91.

Noon, B.R. (2003) Conceptual issues in monitoring ecological resources. In: *Monitoring Ecosystems: Interdisciplinary Approaches for Evaluating Ecoregional Initiatives* (eds. D.E. Busch and J.C. Trexler), pp. 27–71. Island Press, Washington DC.

Noon, B.R. and Sauer, J.R. (1992) Population models for passerine birds: structure, parameterization, and analysis. In: *Wildlife 2001: Populations* (eds. D.R. McCullough and R.H. Barrett), pp. 441–64. Elsevier Applied Science, London.

Noon, B.R. and McKelvey, K.S. (1996) Management of the spotted owl: a case history in conservation biology. *Annual Review of Ecology and Systematics* **27**, 135–62.

Norbury, G. (2001) Conserving dryland lizards by reducing predator-mediated apparent competition and direct competition with introduced rabbits. *Journal of Applied Ecology* **38**, 1350–61.

Norris, P.W. (1881) *Annual Report of the Superintendent of the Yellowstone National Park to the Secretary of the Interior for the Year 1880*. USGPO, Washington.

Noss, R.F. (1999) Assessing and monitoring forest biodiversity: a suggested framework and indicators. *Forest Ecology and Management* **115**, 135–46.

Oaks, J.L., Gilbert, M., Virani, M.Z. et al. (2004) Diclofenac residues as the cause of vulture population decline in Pakistan. *Nature* **427**, 630–3.

O'Brien, S.J. (2000) Adaptive cycles: parasites selectively reduce inbreeding in soay sheep. *Trends in Ecology and Evolution* **15**, 7–9.

O'Brien, S.J. and Mayr, E. (1991) Bureaucratic mischief: recognizing endangered species and subspecies. *Science* **251**, 1187–8.

O'Donoghue, M. (1994) Early survival of juvenile snowshoe hares. *Ecology* **75**, 1582–92.

O'Donoghue, M., Boutin, S., Krebs, C.J., Zuleta, G., Murray, D.L., and Hofer, E.J. (1998) Functional responses of coyotes and lynx to the snowshoe hare cycle. *Ecology* **79**, 1193–208.

Oerlemans, J. (2005) Extracting a climate signal from 169 glacier records. *Science* **308**, 675–7.

O'Grady, J.J., Brook, B.W., Reed, D.H., Ballou, J.D., Tonkyn, D.W., and Frankham, R. (2006) Realistic levels of inbreeding depression strongly affect extinction risk in wild populations. *Biological Conservation* (in press).

Olden, J.D., Poff, N.L., Douglas, M.R., Douglas, M.E., and Fausch, K.D. (2004) Ecological and evolutionary consequences of biotic homogenization. *Trends in Ecology and Evolution* **19**, 18–24.

Oli, M.K. and Dobson, F.S. (2003) The relative importance of life-history variables to population growth rate in mammals: Cole's prediction revisited. *American Naturalist* **161**, 422–40.

Oli, M.K., Holler, N.R., and Wooten, M.C. (2001) Viability analysis of endangered Gulf Coast beach mice (*Peromyscus polionotus*) populations. *Biological Conservation* **97**, 107–18.

Olsen, E.M., Heino, M., Lilly, G.R. et al. (2004) Maturation trends indicative of rapid evolution preceded the collapse of Northern cod. *Nature* **428**, 932–5.

Orrock, J.L. and Damschen, E.I. (2005) Corridors cause differential seed predation. *Ecological Applications* **15**, 793–8.

Ortega, Y.K., Pearson, D.E., and McKelvey, K.S. (2004) Effects of biological control agents and exotic plant invasion on deer mouse populations. *Ecological Applications* **14**, 241–53.

Ostfeld, R.S. (1994) The fence effect reconsidered. *Oikos* **70**, 340–8.

Otis, D.L., Burnham, K.P., White, G.C., and Anderson, D.R. (1978) Statistical inference from capture data on closed animal populations. *Wildlife Monographs* **62**, 1–135.

Owens, I.P.F. (2002) Sex differences in mortality rate. *Science* **297**, 2008–9.

Oyler-McCance, S.J. and Leberg, P.L. (2005) Conservation genetics in wildlife management. In: *Techniques for Wildlife Investigations and Management*, 6th edn (ed. C.E. Braun), pp. 632–57. The Wildlife Society, Bethesda, MD.

Paetkau, D., Calvert, W., Stirling, I., and Strobeck, C. (1995) Microsatellite analysis of population structure in Canadian polar bears. *Molecular Ecology* **4**, 347–54.

Paetkau, D., Slade, R., Burden, M., and Estoup, A. (2004) Genetic assignment methods for the direct, real-time estimation of migration rate: a simulation-based exploration of accuracy and power. *Molecular Ecology* **13**, 55–65.

Paine, R.T. (1969) A note on trophic complexity and community stability. *American Naturalist* **103**, 91–3.

Palomares, F. and Caro, T.M. (1999) Interspecific killing among mammalian carnivores. *American Naturalist* **153**, 492–508.

Palumbi, S.R. and Cipriano, F. (1998) Species identification using genetic tools: the value of nuclear and mitochondrial gene sequences in whale conservation. *Journal of Heredity* **89**, 459–64.

Parmesan, C. (1996) Climate and species' range. *Nature* **382**, 765–6.

Parmesan, C. and Galbraith, H. (2004) *Observed Impacts of Global Climate Change in the U.S.* Pew Center on Global Climate Change, Arlington, VA.

Parmesan, C. and Yohe, G. (2003) A globally coherent fingerprint of climate change impacts across natural systems. *Nature* **421**, 37–42.

Pascual, M.A., Kareiva, P., and Hilborn, R. (1997) The influence of model structure on conclusions about the viability and harvesting of Serengeti wildebeest. *Conservation Biology* **11**, 966–76.

Paterson, S., Wilson, K., and Pemberton, J.M. (1998) Major histocompatibility complex variation associated with juvenile survival and parasite resistance in a large unmanaged ungulate population (*Ovis aries L.*). *Proceedings of the National Academy of Sciences USA* **95**, 3714–19.

Paulos, J.A. (2003) *A Mathematician Plays the Stock Market*. Basic Books, New York.

Paxinos, E., McIntosh, C., Ralls, K., and Fleischer, R. (1997) A noninvasive method for distinguishing among canid species: amplification and enzyme restriction of DNA from dung. *Molecular Ecology* **6**, 483–6.

Peacock, M.M. and Ray, C. (2001) Dispersal in pikas (*Ochotona princeps*): combining genetic and demographic approaches to reveal spatial and temporal patterns. In: *Dispersal* (eds. J. Clobert, E. Danchin, A.A. Dhondt, and J.D. Nichols), pp. 43–56. Oxford University Press, Oxford.

Pearse, D.E. and Crandall, K.A. (2004) Beyond F_{ST}: analysis of population genetic data for conservation. *Conservation Genetics* **5**, 585–602.

Peek, J.M., Boyce, M.S., Garton, E.O., Hard, J.J., and Mills, L.S. (2002) *Risks Involved in Current Management of Elk in Washington*. Washington Department of Fish and Wildlife, Olympia, WA.

Perry, J.N., Smith, R.H., Woiwod, I.P., and Morse, D.R. (2000) *Chaos in Real Data: Analysis of Nonlinear Dynamics from Short Ecological Time Series*. Kluwer Academic Publishers, Norwell, MA.

Petit, J.R., Jouzel, J., Raynaud, D. et al. (1999) Climate and atmospheric history of the past 420,000 years from the Vostok ice core, Antarctica. *Nature* **399**, 429–36.

Phillips, B.L. and Shine, R. (2004) Adapting to an invasive species: toxic cane toads induce morphological change in Australian snakes. *Proceedings of the National Academy of Sciences USA* **101**, 17150–5.

Phillips, B.L., Brown, G.P., and Shine, R. (2003) Assessing the potential impact of cane toads on Australian snakes. *Conservation Biology* **17**(6), 1738–47.

Phillips, L.D. (1973) *Bayesian Statistics for Social Scientists*. Thomas Y. Crowell Company, New York.

Phillips, M.K., Henry, V.G., and Kelly, B.T. (2003) Restoration of the red wolf. In: *Wolves: Behavior, Ecology and Conservation* (eds. L.D. Mech and L. Boitani), pp. 272–88. University of Chicago Press, Chicago.

Piggott, M.P. and Taylor, A.C. (2003) Remote collection of animal DNA and its applications in conservation management and understanding the population biology of rare and cryptic species. *Wildlife Research* **30**, 1–13.

Pilgrim, K.L., McKelvey, K.S., Riddle, A.E., and Schwartz, M.K. (2005) Felid sex identification based on noninvasive genetic samples. *Molecular Ecology Notes* **5**, 60–1.

Pimm, S.L., Russell, G.J., Gittleman, J.L., and Brooks, T.M. (1995) The future of biodiversity. *Science* **269**, 347–50.

Pirsig, R. (1992) *Lila*. Bantam Books, New York.

Piry, S., Alapetite, A., Cornuet, J.M., Paetkau, D., Baudouin, L., and Estoup, A. (2004) GENECLASS2: software for genetic assignment and first-generation migrant detection. *Journal of Heredity* **95**, 536–9.

Pister, E.P. (1999) Professional obligations in the conservation of fishes. *Environmental Biology of Fishes* **55**, 13–20.

Platt, J.R. (1964) Strong Inference. *Science* **146**, 347–53.

Pletscher, D.H. and Schwartz, M.K. (2000) The tyranny of population growth. *Conservation Biology* **14**, 1918–19.

Poiani, K.A. and Johnson, W.C. (1991) Global warming and prairie wetlands: potential consequences for waterfowl habitat. *BioScience* **41**, 611–18.

Pollock, K.H., Winterstein, S.R., Bunck, C.M., and Curtis, P.D. (1989) Survival analysis in telemetry studies: the staggered entry design. *Journal of Wildlife Management* **53**, 7–15.

Pollock, K.H., Nichols, J.D., Brownie, C., and Hines, J.E. (1990) Statistical inference for capture-recapture experiments. *Wildlife Monographs* **107**, 1–97.

Pollock, M.M., Naiman, R.J., Erickson, H.E., Johnston, C.A., Pastor, J., and Pinay, G. (1995) Beaver as engineers: influences on biotic and abiotic characteristics of drainage basins. In: *Linking Species and Ecosystems* (eds. C.G. Jones and J.H. Lawton), pp. 117–26. Chapman and Hall, New York.

Possingham, H.P., Lindenmayer, D.B., and Tuck, G.N. (2002) Decision theory for population viability analysis. In: *Population Viability Analysis* (eds. S.R. Beissinger and D.R. McCullough), pp. 470–89. University of Chicago Press, Chicago.

Potts, G.R. and Aebischer, N.J. (1994) Population dynamics of the grey partridge *Perdix perdix* 1793–1993: monitoring, modelling and management. *Ibis* **137**, 29–37.

Pounds, J.A., Bustamante, M.R., Coloma, L.A. et al. (2006) Widespread amphibian extinctions from epidemic disease driven by global warming. *Nature* **439**, 161–7.

Powell, L.A., Conroy, M.J., Hines, J.E., Nichols, J.D., and Krements, D.G. (2000) Simultaneous use of mark-recapture and radiotelemetry to estimate survival, movement, and capture rates. *Journal of Wildlife Management* **64**, 302–13.

Powell, R.A. (1982) Evolution of black-tipped tails in weasels: predator confusion. *American Naturalist* **119**, 126–31.

Power, M.E. and Mills, L.S. (1995) The keystone cops meet in Hilo. *Trends in Ecology and Evolution* **10**, 182–4.

Power, M.E., Tilman, D., Estes, J.A. et al. (1996) Challenges in the quest for keystones. *BioScience* **46**, 609–20.

Pray, L.A., Schwartz, J.M., Goodnight, C.J., and Stevens, L. (1994) Environmental dependency of inbreeding depression: implications for conservation biology. *Conservation Biology* **8**, 562–8.

Prendergast, J.R., Quinn, R.M., Lawton, J.H., Eversham, B.C., and Gibbons, D.W. (1993) Rare species, the coincidence of diversity hotspots and conservation strategies. *Nature* **365**, 335–7.

Prenter, J., MacNeil, C., Dick, J.T.A., and Dunn, A.M. (2004) Roles of parasites in animal invasions. *Trends in Ecology and Evolution* **19**, 385–90.

Primmer, C.R., Koskinen, M.T., and Piironen, J. (2000) The one that did not get away: individual assignment using microsatellite data detects a case of fishing competition fraud. *Proceedings of the Royal Society of London Series B* **267**, 1699–1704.

Prugnolle, F. and de Meeus, T. (2002) Inferring sex-biased dispersal from population genetic tools: a review. *Heredity* **88**, 161–5.

Pulliam, H.R. (1988) Sources, sinks, and population regulation. *American Naturalist* **132**, 652–61.

Purvis, A., Gittleman, J.L., Cowlishaw, G., and Mace, G.M. (2000) Predicting extinction risk in declining species. *Proceedings of the Royal Society of London Series B* **267**, 1947–52.

Quammen, D. (1996) *The Song of the Dodo: Island Biogeography in an Age of Extinctions.* Touchstone, New York.

Rabinowitz, D., Cairns, S., and Dillon, T. (1986) Seven forms of rarity and their frequency in the flora of the British Isles. In: *Conservation Biology: the Science of Scarcity and Diversity* (ed. M.E. Soulé), pp. 182–204. Sinauer Associates, Sunderland, MA.

Raithel, J.D., Kauffman, M.J., and Pletscher, D.H. (2006) Impact of spatial and temporal variation in calf elk survival on population growth rate. *Journal of Wildlife Management* (in press).

Ralls, K., Harvey, P.H., and Lyles, A.M. (1986) Inbreeding in natural populations of birds and mammals. In: *Conservation Biology: the Science of Scarcity and Diversity* (ed. M.E. Soulé), pp. 35–56. Sinauer Associates, Sunderland, MA.

Ralls, K., Ballou, J.D., and Templeton, A. (1988) Estimates of lethal equivalents and the cost of inbreeding in mammals. *Conservation Biology* **2**, 185–93.

Rand, D.M. (2001) The units of selection on mitochondrial DNA. *Annual Review of Ecology and Systematics* **32**, 415–48.

Raphael, M.G., Wisdom, M.J., Rowland, M.M. et al. (2001) Status and trends of habitats of terrestrial vertebrates in relation to land management in the interior Columbia river basin. *Forest Ecology and Management* **153**, 63–88.

Rasa, O.A.E. (1989) The costs and effectiveness of vigilance behavior in the Dwarf Mongoose: implications for fitness and optimal group size. *Ethology Ecology and Evolution* **1**, 265–82.

Raveling, D.G. (1989) Nest-predation rates in relation to colony size of black brant. *Journal of Wildlife Management* **53**, 87–90.

Raybould, A.F., Clarke, R.T., Bond, J.M., Welters, R.E., and Gliddon, C.J. (2001) Inferring patterns of dispersal from allele frequency data. In: *Dispersal Ecology* (eds. J.M. Bullock, R.E. Kenward, and R.S. Hails), pp. 89–110. Blackwell Publishing, Malden, MA.

Reale, D., McAdam, A.G., Boutin, S., and Berteaux, D. (2003) Genetic and plastic responses of a northern mammal to climate change. *Proceedings of the Royal Society of London Series B* **270**, 591–6.

Redford, K.H. (1992) The empty forest. *BioScience* **42**(6), 412–22.

Reed, D.H. and Frankham, R. (2003) Correlation between fitness and genetic diversity. *Conservation Biology* **17**, 230–5.

Reed, D.H., Lowe, E.H., Briscoe, D.A., and Frankham, R. (2003) Fitness and adaptation in a novel environment: effect of inbreeding, prior environment, and lineage. *Evolution* **57**, 1822–8.

Reed, J.M. (1999) The role of behavior in recent avian extinctions and endangerments. *Conservation Biology* **13**, 232–41.

Reed, J.M. (2002) Animal behavior as a tool in conservation biology. In: *Conservation Medicine: Ecological Health in Practice* (eds. A.A. Aquirre, R.S. Ostfeld, C.A. Houle, G.M. Tabor, and M.C. Pearl), pp. 145–63. Oxford University Press, Oxford.

Reed, J.M., Murphy, D.D., and Brussard, P.F. (1998) Efficacy of population viability analysis. *Wildlife Society Bulletin* **26**, 244–51.

Reed, J.M., Mills, L.S., Dunning, Jr., J.B. et al. (2002) Emerging issues in population viability analysis. *Conservation Biology* **16**, 7–19.

Reiger, J.F. (2001) *American Sportsmen and the Origins of Conservation*. Oregon State University Press, Corvallis, OR.

Reilly, S.B. (1992) Population biology and status of eastern Pacific gray whales: recent developments. In: *Wildlife 2001: Populations* (eds. D.R. McCullough and R.H. Barrett), pp. 1062–74. Elsevier Applied Science, London.

Relyea, R.A. (2005) The lethal impact of roundup on aquatic and terrestrial amphibians. *Ecological Applications* **15**, 1118–24.

Rhodes, Jr., O.E. and Smith, M.H. (1992) Genetic perspectives in wildlife management: the case of large herbivores. In: *Wildlife 2001: Populations* (eds. D.R. McCullough and R.H. Barrett), pp. 985–96. Elsevier Applied Science, London.

Rhymer, J.M. and Simberloff, D. (1996) Extinction by hybridization and introgression. *Annual Review of Ecology and Systematics* **27**, 83–109.

Richter, B.D., Braun, D.P., Mendelson, M.A., and Master, L.L. (1997) Threats to imperiled freshwater fauna. *Conservation Biology* **11**, 1081–93.

Riddle, A.E., Pilgrim, K.L., Mills, L.S., McKelvey, K.S., and Ruggiero, L.F. (2003) Identification of mustelids using mitochondrial DNA and non-invasive sampling. *Conservation Genetics* **4**, 241–3.

Risbey, D.A., Calver, M.C., Short, J., Bradley, J.S., and Wright, I.W. (2000) The impact of cats and foxes on the small vertebrate fauna of Heirisson Prong, Western Australia. II. A field experiment. *Wildlife Research* **27**, 223–35.

Ritters, K.H. and Wickham, J.D. (2003) How far to the nearest road? *Frontiers in Ecology and the Environment* **1**, 125–9.

Roach, J.L., Stapp, P., Van Horne, B., and Antolin, M.F. (2001) Genetic structure of a metapopulation of black-tailed prairie dogs. *Journal of Mammalogy* **82**, 946–59.

Robbins, T. (1984) *Jitterbug Perfume*. Bantam Books, New York.

Roberge, J.-M. and Angelstam, P. (2004) Usefulness of the umbrella species concept as a conservation tool. *Conservation Biology* **18**, 76–85.

Robertson, B.C., Elliott, G.P., Eason, D.K., Clout, M.N., and Gemmell, N.J. (2006) Sex allocation theory aids species conservation. *Biology Letters* **2** (in press).

Robinson, D.H. and Wainer, H. (2002) On the past and future of null hypothesis significance testing. *Journal of Wildlife Management* **66**(2), 263–71.

Rodenhouse, N.L., Sillett, T.S., Doran, P.J., and Holmes, R.T. (2003) Multiple density-dependence mechanisms regulate a migratory bird population during the breeding season. *Proceedings of the Royal Society of London Series B* **270**, 2105–10.

Romesburg, H.C. (1981) Wildlife science: gaining reliable knowledge. *Journal of Wildlife Management* **45**, 293–313.

Rooney, T.P., Wiegmann, S.M., Rogers, D.A., and Waller, D.M. (2004) Biotic impoverishment and homogenization in unfragmented forest understory communities. *Conservation Biology* **18**, 787–98.

Root, T.L., Price, J.T., Hall, K.R., Schneider, S.H., Rosenzweig, C., and Pounds, J.A. (2003) Finger-prints of global warming on wild animals and plants. *Nature* **421**, 57–60.

Roseberry, J.L. and Klimstra, W.D. (1984) *Population Ecology of the Bobwhite*. Southern Illinois University Press, Carbondale, IL.

Rosenberg, A.A., Fogarty, M.J., Sissenwine, M.P., Beddington, J.R., and Shepherd, J.G. (1993) Achieving sustainable use of renewable resources. *Science* **262**, 828–9.

Rosenberg, D.K., Noon, B.R., and Meslow, C. (1997) Biological corridors: form, function and efficacy. *BioScience* **47**, 677–87.

Rosenberg, D.M. and Resh, V.H. (1993) *Freshwater Biomonitoring and Benthic Macroinvertebrates*. Chapman and Hall, New York.

Rothschild, L.J. and Mancinelli, R.L. (2001) Life in extreme environments. *Nature* **409**, 1092–101.

Rothstein, S.I. and Cook, T.L. (2000) Introduction. In: *Ecology and Management of Cowbirds and their Hosts: Studies in the Conservation of North American Passerine Birds* (eds. J.N.M. Smith, T.L. Cook, S.I. Rothstein, S.K. Robinson, and S.G. Sealy), pp. 323–32. University of Texas Press, Austin, TX.

Rowland, M.M., Wisdom, M.J., Johnson, D.H., Wales, B.C., Copeland, J.P., and Edelmann, F.B. (2003) Evaluation of landscape models for wolverines in the interior northwest, United States of America. *Journal of Mammalogy* **84**, 92–105.

Ruggiero, L.F., Hayward, G.D., and Squires, J.R. (1994) Viability analysis in biological evaluations: concepts of population viability analysis, biological population, and ecological scale. *Conservation Biology* **8**, 364–72.

Ruggiero, L.F., Schwartz, M.K., Aubry, K.B., Krebs, C.J., Stanley, A., and Buskirk, S.W. (2000) Species conservation and natural variation among populations. In: *Ecology and Conservation of Lynx in the United States* (eds. L.F. Ruggiero, K.B. Aubry, S.W. Buskirk et al.), pp. 101–16. General Technical Report RMRS-GTR-30WWW. USDA-Forest Service Rocky Mountain Research Station, Fort Collins, CO.

Rugh, D.J., Hobbs, R.C., Lerczak, J.A., and Breiwick, J.M. (2005) Estimates of abundance of the eastern North Pacific stock of gray whales (*Eschrichtius robustus*) 1997–2002. *Journal of Cetacean Research and Management* **7**, 1–12.

Runge, J.P., Runge, M.C., and Nichols, J.D. (2006) The role of local populations within a landscape context: defining and classifying sources and sinks. *American Naturalist* **167**, 925–38.

Sabo, J.L., Holmes, E.E., and Kareiva, P. (2004) Efficacy of simple viability models in ecological risk assessment: does density dependence matter? *Ecology* **85**, 328–41.

Saccheri, I., Kuussaari, M., Kankare, M., Vikman, P., Fortelius, W., and Hanski, I. (1998) Inbreeding and extinction in a butterfly metapopulation. *Nature* **392**, 491–4.

Sæther, B.-E. and Bakke, Ø. (2000) Avian life history variation and contribution of demographic traits to the population growth rate. *Ecology* **81**, 642–53.

Sæther, B.-E. and Engen, S. (2002) Including uncertainties in population viability analysis using population prediction intervals. In: *Population Viability Analysis* (eds. S.R. Beissinger and D.R. McCullough), pp. 191–212. University of Chicago Press, Chicago.

Sage, R.D., Heyneman, D., Lim, K.-C., and Wilson, A.C. (1986) Wormy mice in a hybrid zone. *Nature* **324**, 60–2.

Sala, O.E., Chapin, F.S.I., Armesto, J.J. et al. (2000) Global biodiversity scenarios for the year 2100. *Science* **287**, 1770–4.

Samson, F.B. (2002) Population viability analysis, management, and conservation planning at large scales. In: *Population Viability Analysis* (eds. S.R. Beissinger and D.R. McCullough), pp. 425–41. University of Chicago Press, Chicago.

Samuel, M.D., Garton, E.O., Schlegel, M.W., and Carson, R.G. (1987) Visibility bias during aerial surveys of elk in northcentral Idaho. *Journal of Wildlife Management* **51**, 622–30.

Sanjayan, M.A., Crooks, K., Zegers, G., and Foran, D. (1996) Genetic variation and the immune response in natural populations of pocket gophers. *Conservation Biology* **10**, 1519–27.

Sarre, S. (1995) Mitochondrial DNA variation among populations of *Oedura reticulata* (Gekkonidae) in remnant vegetation: Implications for metapopulation structure and population decline. *Molecular Ecology* **4**, 395–405.

Schiegg, K., Pasinelli, G., Walters, J.R., and Daniels, S.J. (2002) Inbreeding and experience affect response to climate change by endangered woodpeckers. *Proceedings of The Royal Society of London Series B* **269**, 1153–9.

Schlaepfer, M.A., Runge, M.C., and Sherman, P.W. (2002) Ecological and evolutionary traps. *Trends in Ecology and Evolution* **17**, 474–80.

Schmitz, O.J., Hambäck, P.A., and Beckerman, A.P. (2000) Trophic cascades in terrestrial systems: A review of the effects of carnivore removals on plants. *American Naturalist* **155**(2), 141–53.

Schwartz, M.K., Tallmon, D.A., and Luikart, G. (1998) Review of DNA-based census and effective population size estimators. *Animal Conservation* **1**, 293–9.

Schwartz, M.K., Mills, L.S., McKelvey, K.S., Ruggiero, L.F., and Allendorf, F.W. (2002) DNA reveals high dispersal synchronizing the population dynamics of Canada lynx. *Nature* **415**, 520–2.

Schwartz, M.K., Pilgrim, K.L., McKelvey, K.S. et al. (2004) Hybridization between Canada lynx and bobcats: genetics results and managment implications. *Conservation Genetics* **5**, 349–55.

Scott, J.M., Tear, T.H., and Mills, L.S. (1995) Socioeconomics and the recovery of endangered species: Biological assessment in a political world. *Conservation Biology* **9**, 214–16.

Scribner, K.T. and Stüwe, M. (1994) Genetic relationships among alpine ibex *Capra ibex* populations rçe-established from a common ancestral source. *Biological Conservation* **69**, 137–43.

Şekercioğlu, Ç.H., Daily, G.C., and Ehrlich, P.R. (2004) Ecosystem consequences of bird declines. *Proceedings of the National Academy of Sciences USA* **101**, 18042–7.

Shaffer, M. (1987) Minimum viable populations: coping with uncertainty. In: *Viable Populations for Conservation* (ed. M.E. Soulé), pp. 69–86. Cambridge University Press, Cambridge.

Shaffer, M., Watchman, L.H., Snape, III, W.J., and Latchis, I.K. (2002) Population viability analysis and conservation policy. In: *Population Viability Analysis* (eds. S.R. Beissinger and D.R. McCullough), pp. 123–42. University of Chicago Press, Chicago.

Shaw, C.N., Wilson, P.J. and White, B.N. (2003) A reliable molecular method of gender determination for mammals. *Journal of Mammalogy* **84**, 123–8.

Sherwin, W.B. and Moritz, C. (2000) Managing and monitoring genetic erosion. In: *Genetics, Demography and Viability of Fragmented Populations* (eds. A.G. Young and G.M. Clarke), pp. 9–34. Cambridge University Press, Cambridge.

Shore, R.F., Malcolm, H.M., McLennan, D. et al. (2006) Did foot-and-mouth disease-control operations affect rodenticide exposure in raptors? *Journal of Wildlife Management* **70**, 588–93.

Short, J., Kinnear, J.E., and Robley, A. (2001) Surplus killing by introduced predators in Australia-evidence for ineffective anti-predator adaptations in native prey species. *Biological Conservation* **103**, 283–301.

Sibly, R.M., Barker, D., Denham, M.C., Hone, J., and Pagel, M. (2005) On the regulation of populations of mammals, birds, fish and insects. *Science* **309**, 607–10.

Silvy, N.J., Lopez, R.R., and Peterson, M.J. (2005) Wildlife marking techniques. In: *Techniques for Wildlife Investigations and Management*, 6th edn. (ed. C.E. Braun), pp. 339–76. The Wildlife Society, Bethesda, MD.

Simberloff, D. (1988) The contribution of population and community biology to conservation science. *Annual Review of Ecology and Systematics* **19**, 473–511.

Simberloff, D. (1998) Flagships, umbrellas, and keystones: is single-species management passé in the landscape era? *Biological Conservation* **83**, 247–57.

Simberloff, D. (2003) How much information on population biology is needed to manage introduced species? *Conservation Biology* **17**, 83–92.

Simberloff, D., Parker, I.M., and Windle, P.N. (2005) Introduced species policy, management, and future research needs. *Frontiers in Ecology and the Environment* **3**, 12–20.

Sinclair, A.R.E. (1989) Population regulation in animals. In: *Ecological Concepts* (ed. J.M. Cherret), pp. 197–241. Blackwell Scientific Publications, Oxford.

Sinclair, A.R.E. (1996) Mammal populations: fluctuation, regulation, life history theory and their implications for conservation. In: *Frontiers of Population Ecology* (eds. R.B. Floyd, A.W. Sheppard, and P.J. De Barro), pp. 127–54. CSIRO Publishing, Melbourne.

Sinclair, A.R.E., Pech, R.P., Dickman, C.R., Hik, D., Mahon, P., and Newsome, A.E. (1998) Predicting effects of predation on conservation of endangered prey. *Conservation Biology* **12**, 564–75.

Singer, F.J. and Zeigenfuss, L.C. (2002) Influence of trophy hunting and horn size on mating behavior and survivorship of mountain sheep. *Journal of Mammalogy* **83**, 682–98.

Singer, M.C., Thomas, C.D., and Parmesan, C. (1993) Rapid human-induced evolution of insect-host associations. *Nature* **366**, 681–3.

Sjögren-Gulve, P. and Hanski, I. (2000) Metapopulation viability analysis using occupancy models. *Ecological Bulletins* **48**, 53–71.

Sloane, M.A., Sunnucks, P., Alpers, D., Beheregaray, L.B., and Taylor, A.C. (2000) Highly reliable genetic identification of individual northern hairy-nosed wombats from single remotely collected hairs: a feasible censusing method. *Molecular Ecology* **9**, 1233–40.

Small, R.J., Jolzwart, J.C., and Rusch, D.H. (1991) Predation and hunting mortality of ruffed grouse in central Wisconsin. *Journal of Wildlife Management* **55**, 512–20.

Smil, V. (1999) How many billions to go: the peaking of the population growth rate deserves wider recognition. *Nature* **401**, 429.

Smith, A.P. and Quin, D.G. (1996) Patterns and causes of extinction and decline in Australian conilurine rodents. *Biological Conservation* **77**, 243–67.

Smith, F.D.M., May, R.M., Pellew, R., Johnson, T.H., and Walter, K.R. (1993) How much do we know about the current extinction rate? *Trends in Ecology and Evolution* **8**, 375–8.

Smith, J.N.M. and Hellmann, J.J. (2002) Population persistence in fragmented landscapes. *Trends in Ecology and Evolution* **17**, 397–9.

Sorenson, L.G., Goldberg, R., Root, T.L., and Anderson, M.G. (1998) Potential effects of global warming on waterfowl populations breeding in the northern great plains. *Climatic Change* **40**, 343–69.

Soulé, M.E. (1980) Thresholds for survival: maintaining fitness and evolutionary potential. In: *Conservation Biology: an Evolutionary-Ecological Perspective* (eds. M.E. Soulé and B.A. Wilcox), pp. 151–69. Sinauer Associates, Sunderland, MA.

Soulé, M.E. (ed.) (1986) Conservation biology and the "real world." In: *Conservation Biology: the Science of Scarcity and Diversity*, pp. 1–12. Sinauer Associates, Sunderland, MA.

Soulé, M.E. (ed.) (1987) *Viable Populations for Conservation*. Cambridge University Press, New York.

Soulé, M.E. and Wilcox, B.A. (eds.) (1980) *Conservation Biology: an Evolutionary-Ecological Perspective*, pp. 55–69. Sinauer Associates, Sunderland, MA.

Soulé, M.E. and Mills, L.S. (1992) Conservation genetics and conservation biology: a troubled marriage. In: *Conservation of Biodiversity For Sustainable Development* (eds. O.T. Sandlund, K. Hindar, and A.H.D. Brown), pp. 55–69. Scandinavian University Press, Oslo.

Soulé, M.E. and Mills, L.S. (1998) No need to isolate genetics. *Science* **282**, 1658–9.

Soulé, M.E., Bolger, D.T., Alberts, A.C., Wright, J., Sorice, M., and Hill, S. (1988) Reconstructed dynamics of rapid extinctions of chaparral-requiring birds in urban habitat islands. *Conservation Biology* **2**, 75–92.

Soulé, M.E., Estes, J.A., Berger, J., and del Rio, C.M. (2003) Ecological effectiveness: Conservation goals for interactive species. *Conservation Biology* **17**, 1238–50.

Soulé, M.E., Estes, J.A., Miller, B., and Honnold, D.L. (2005) Strongly interacting species: Conservation policy, management, and ethics. *BioScience* **55**, 168–76.

Spencer, C.C., Neigel, J.E., and Leberg, P.L. (2000) Experimental evaluation of the usefulness of microsatellite DNA for detecting demographic bottlenecks. *Molecular Ecology* **9**, 1517–28.

Spencer, R.-J. and Thompson, M.B. (2005) Experimental analysis of the impact of foxes on freshwater turtle populations. *Conservation Biology* **19**, 845–54.

Spielman, D. and Frankham, R. (1992) Modeling problems in conservation genetics using captive *Drosophila* populations: improvement of reproductive fitness due to immigration of one individual into small partially inbred populations. *Zoo Biology* **11**, 343–51.

Spielman, D., Brook, B.W., and Frankham, R. (2004) Most species are not driven to extinction before genetic factors impact them. *Proceedings of the National Academy of Sciences USA* **101**, 15261–4.

Spong, G. and Creel, S. (2001) Deriving dispersal distances from genetic data. *Proceedings of the Royal Society London Series B* **268**, 2571–4.

Stegner, W. (1992) A capsule history of conservation. In: *Where the Bluebird Sings to the Lemonade Springs*, pp. 117–34. Random House, New York.

Steidl, R.J., Hayes, J., and Schauber, E. (1997) Statistical power analysis in wildlife research. *Journal of Wildlife Management* **61**, 270–9.

Steidl, R.J., DeStefano, S., and Matter, W.J. (2000). On increasing the quality, reliability, and rigor of wildlife science. *Wildlife Society Bulletin* **28**, 518–21.

Steinbeck, J. (1960) *The Log From the Sea of Cortez*. Pan Books, London.

Stephens, P.A. and Sutherland, W.J. (1999) Consequences of the Allee Effect for behavior, ecology, and conservation. *Trends in Ecology and Evolution* **14**, 401–4.

Stevens, J.E. (1996) It's a jungle in there. *BioScience* **46**, 314–17.

Stiling, P. (2002) *Ecology: Theories and Applications*. Prentice Hall, Upper Saddle River, NJ.

Stork, N. (1999) The magnitude of global biodiversity and its decline. In: *The Living Planet in Crisis: Biodiversity Science and Policy* (eds. J. Cracraft and F.T. Grifo), pp. 3–32. Columbia University Press, New York.

Stow, A.J. and Sunnucks, P. (2004) Inbreeding avoidance in Cunningham's skinks (*Egernia cunninghami*) in natural and fragmented habitat. *Molecular Ecology* **13**, 443–7.

Strode, P.K. (2003) Implications of climate change for North American wood warblers (*Parulidae*). *Global Change Biology* **9**, 1137–44.

Suorsa, P., Helle, H., Huhta, E., Jantti, A., Nikula, A., and Hakkarainen, H. (2003) Forest fragmentation is associated with primary brood sex ratio in the treecreeper (*Certia familiaris*). *Proceeding of the Royal Society of London Series B* **270**, 2215–22.

Sutherland, W.J. (2002) Science, sex and the kakapo. *Nature* **419**, 265–6.

Swanson, B.J. (1998) Autocorrelated rates of change in animal populations and their relationship to precipitation. *Conservation Biology* **12**, 801–8.

Swindell, W.R. and Bouzat, J.L. (2006) Reduced inbreeding depression due to historical inbreeding in *Drosophila melanogaster*: evidence for purging. *Journal of Evolutionary Biology* **19**, 1257–64.

Sword, G.A., Lorch, P.D., and Gwynne, D.T. (2005) Migratory bands give crickets protection. *Nature* **433**, 703.

Taberlet, P. and Fumagalli, L. (1996) Owl pellets as a source of DNA for genetic studies of small mammals. *Molecular Ecology* **5**, 301–5.

Taberlet, P., Waits, L.P., and Luikart, G. (1999) Non-invasive genetic sampling: look before you leap. *Trends in Ecology and Evolution* **14**, 323–7.

Tallmon, D.A. and Mills, L.S. (2004) Edge effects and isolation: red-backed voles revisited. *Conservation Biology* **18**, 1658–64.

Tallmon, D.A., Draheim, H.M., Mills, L.S., and Allendorf, F.W. (2002) Insights into recently fragmented vole populations from combined genetic and demographic data. *Molecular Ecology* **11**, 699–709.

Tallmon, D.A., Jules, E.S., Radke, N.J., and Mills, L.S. (2003) Of mice and men and trillium: cascading effects of forest fragmentation. *Ecological Applications* **13**, 1193–203.

Tallmon, D.A., Luikart, G., and Waples, R.S. (2004) The alluring simplicity and complex reality of genetic rescue. *Trends in Ecology and Evolution* **19**, 489–96.

Taper, M.L. (2004) Model identification from many candidates. In: *The Nature of Scientific Evidence* (eds. M.L. Taper and S.R. Lele), pp. 488–524. University of Chicago Press, Chicago.

Taper, M.L. and Lele, S.R. (eds.) (2004) *The Nature of Scientific Evidence*. University of Chicago Press, Chicago.

Taylor, A.C., Sherwin, W.B., and Wayne, R.K. (1994) Genetic variation of microsattelite loci in a bottlenecked species: the northern hairy-nosed wombat *Lasiorhinus krefftii*. *Molecular Ecology* **3**, 277–90.

Taylor, B.L. and Gerrodette, T. (1993) The uses of statistical power in conservation biology: the vaquita and northern spotted owl. *Conservation Biology* **7**, 489–500.

Taylor, B.L. and Dizon, A.E. (1999) First policy then science: why a management unit based solely on genetic criteria cannot work. *Molecular Ecology* **8**, S11–16.

Taylor, B.L., Wade, P.R., Stehn, R.A., and Cochrane, J.F. (1996) A Bayesian approach to classification criteria for spectacled eiders. *Ecological Applications* **6**, 1077–89.

Taylor, B.L., Wade, P.R., Ramakrishnan, U., Gilpin, M., and Akçakaya, H.R. (2002) Incorporating uncertainty in population viability analyses for the purpose of classifying species by risk. In: *Population Viability Analysis* (eds. S.R. Beissinger and D.R. McCullough), pp. 239–56. University of Chicago Press, Chicago.

Taylor, C.M. and Hastings, A. (2005) Alee effects in biological invasions. *Ecology Letters* **8**, 895–908.

Taylor, P.D., Fahrig, L., Henein, K., and Merriam, G. (1993) Connectivity is a vital element of landscape structure. *Oikos* **68**, 571–3.

Taylor, R.J. (1984) *Predation*. Chapman and Hall, New York.

Templeton, A.R., Robertson, R.J., Brisson, J., and Strasburg, J. (2001) Disrupting evolutionary processes: the effect of habitat fragmentation on collared lizards in the Missouri Ozarks. *Proceedings of the National Academy of Sciences USA* **98**, 5426–32.

Tewksbury, J.J., Levey, D.J., Haddad, N.M. et al. (2002) Corridors affect plants, animals, and their interactions in fragmented landscapes. *Proceedings of the National Academy of Sciences USA* **99**, 12923–6.

Thomas, D.W., Blondel, J., Perret, P., Lambrechts, M.M., and Speakman, J.R. (2001) Energetic and fitness costs of mismatching resource supply and demand in seasonally breeding birds. *Science* **291**, 2598–600.

Thomas, J.W. (1986) Effectiveness- the hallmark of the natural resource management professional. *Transactions of the 51st North American Wildlife and Natural Resources Conference* 27–38.

Thomas, J.W. (1994) Forest Ecosystem Management Assessment Team: objectives, process and options. *Journal of Forestry*. **92**, 12–14.

Thomas, J.W. and Pletscher, D.H. (2000) The convergence of ecology, conservation biology, and wildlife biology: necessary or redundant? *Wildlife Society Bulletin* **28**, 546–9.

Thomas, J.W. and Pletscher, D.H. (2002) The "lynx affair" – professional credibility on the line. *Wildlife Society Bulletin* **30**, 1281–6.

Thomas, L. and Krebs, C.J. (1997) A review of statistical power analysis software. *Bulletin of the Ecological Society of America* **78**, 126–39.

Thomas, J.W., Raphael, M.G., Anthony, R.G. et al. (1993) *Viability assessments and Management Considerations for Species Associated with Late-Successional and Old-Growth Forests of the Pacific Northwest*. USDA Forest Service, US Government Printing Office, Washington DC.

Thompson, S.K. (2002). *Sampling*, 2nd edn. J. Wiley and Sons, New York.

Thompson, W.L., White, G.C., and C. Gowan. (1998) *Monitoring Vertebrate Populations*. Academic Press, San Diego, CA.

Todd, C.R. and Burgman, M.A. (1998) Assessment of threat and conservation priorities under realistic levels of uncertainty and reliability. *Conservation Biology* **12**, 966–74.

Tompkins, D.M., Draycott, R.A.H., and Hudson, P.J. (2000) Field evidence for apparent competition mediated via the shared parasites of two gamebird species. *Ecology Letters* **3**, 10–14.

Torchin, M.E., Lafferty, K.D., Dobson, A.P., McKenzie, V.J., and Kuris, A.M. (2003) Introduced species and their missing parasites. *Nature* **421**, 628–30.

Towns, D.R. and Broome, K.G. (2003) From small Maria to massive Campbell: forty years of rat eradications from New Zealand islands. *New Zealand Jounal of Zoology* **30**, 377–98.

Towns, D.R., Atkinson, I.A.E., and Daugherty, C.H. (2006) Have the harmful effects of introduced rats on islands been exaggerated? *Biological Invasions* **18**, 863–91.

Tracy, C.R. and George, T.L. (1992) On the determinants of extinction. *American Naturalist* **139**, 102–22.

Trivers, R.L. and Willard, D.E. (1973) Natural selection of parental ability to vary the sex ratio of offspring. *Science* **179**, 90–2.

Turchin, P. (1999) Population regulation: A synthetic view. *Oikos* **84**, 153–9.

Turchin, P. (2003) *Complex Population Dynamics: a Theoretical/Empirical Synthesis*. Princeton University Press, Princeton, NJ.

Turchin, P. and Hanski, I. (1997) An empirically based model for latitudinal gradient in vole population dynamics. *American Naturalist* **149**, 842–74.

Unsworth, J.W., Leban, F.A., Leptich, D.J., Garton, E.O., and Zager, P. (1994) *Aerial Survey: User's Manual*, 2nd edn. Idaho Fish and Game, Boise, ID.

US Department of the Interior, Department of Commerce (1996) Proposed policy on the treatment of intercrosses and intercross progeny (the issue of "hybridization"); proposed rule. *Federal Register* **61**, 4710–13.

US Fish and Wildlife Service, Nez Perce Tribe, National Park Service, and USDA Wildlife Services (2001) *Rocky Mountain Wolf Recovery 2000 Annual Report*. US Fish and Wildlife Service, Helena, MT.

US Fish and Wildlife Service (2003) Recovery plan for the Red-cockaded woodpecker (*Picoides borealis*), 2nd revision, US Fish and Wildlife Service, Atlanta, GA.

Våge, D.I., Lu, D., Klungland, H., Lien, S., Adalsteinsson, S., and Cone, R.D. (1997) A non-epistatic interaction of *agouti* and *extension* in the fox, *Vulpes vulpes*. *Nature Genetics* **15**, 311–15.

Van Horne, B. (1983) Density as a misleading indicator of habitat quality. *Journal of Wildlife Management* **47**, 893–901.

Van Noordwijk, A.J. and Scharloo, W. (1981) Inbreeding in an island population of the great tit. *Evolution* **35**, 674–88.

Van Vuren, D. (1998) Mammalian dispersal and reserve design. In: *Behavioral Ecology and Conservation Biology* (eds. T. Caro), pp. 369–93. Oxford University Press, Oxford.

Van Vuren, D. and Armitage, K.B. (1994) Survival of dispersing and philopatric yellow-bellied marmots: what is the cost of dispersal? *Oikos* **69**, 179–81.

Varley, J.D. (1993) Saving the parks: why Yellowstone and the research it fosters matter so much. *Yellowstone Science* **1**, 13–16.

Veit, R.R. and Lewis, M.A. (1996) Dispersal, population growth, and the allee effect: dynamics of the house finch invasion of eastern North America. *American Naturalist* **148**, 255–74.

Vilà, C., Sundqvist, A.-K., Flagstad, Ø. et al. (2003) Rescue of a severely bottlenecked wolf (*Canis lupus*) population by a single immigrant. *Proceedings of the Royal Society London Series B* **270**, 91–7.

Vucetich, J.A. and Creel, S. (1999) Ecological interactions, social organization, and extinction risk in African wild dogs. *Conservation Biology* **13**, 1172–82.

Vucetich, J.A. and Waite, T.A. (2000) Is one migrant per generation sufficient for the genetic management of fluctuating populations? *Animal Conservation* **3**, 261–6.

Vucetich, J.A., Peterson, R.O., and Schaefer, C.L. (2002) The effect of prey and predator densities on wolf predation. *Ecology* **83**, 3003–13.

Vucetich, J.A., Smith, D.W., and Stahler, D.R. (2005) Influence of harvest, climate and wolf predation on Yellowstone elk, 1961–2004. *Oikos* **111**, 259–70.

Vyas, N.B., Kuenzel, W.J., Hill, E.F., and Sauer, J.R. (1995) Acephate affects migratory orientation of the white-throated sparrow (*Zonotrichia albicollis*). *Environmental Toxicology and Chemistry* **14**, 1961–5.

Wade, P.R. (2000) Bayesian methods in conservation biology. *Conservation Biology* **14**, 1308–16.

Wade, P.R. (2002a). A Bayesian stock assessment of the eastern Pacific gray whale using abundance and harvest data from 1967–1996. *Journal of Cetacean Research and Management* **4**, 85–98.

Wade, P.R. (2002b) Bayesian population viability analysis. In: *Population Viability Analysis* (eds. S.R. Beissinger and D.R. McCullough). University of Chicago Press, Chicago.

Waits, J.L. and Leberg, P.L. (1999) Advances in the use of molecular markers for studies of population size and movement. In: *Transactions of the Sixty-fourth North American Wildlife and Natural Resources Conference* (eds. R.E. McCabe and S.E. Loos), pp. 191–201. Wildlife Management Institute, Washington DC.

Walsh, P.D., Abernathy, K.A., Bermejo, M. et al. (2003) Catastrophic ape decline in western equatorial Africa. *Nature* **422**, 611–14.

Walters, C.J. (1986) *Adaptive Management of Renewable Resources*. Macmillan, New York.

Walters, C.J. and Holling, C.S. (1990) Large-scale management experiments and learning by doing. *Ecology* **71**, 2060–8.

Walters, J.R. (1991) Application of ecological principles to the management of endangered species: the case of the red-cockaded woodpecker. *Annual Review of Ecology and Systematics* **22**, 505–23.

Walters, J.R., Crowder, L.B., and Priddy, J.A. (2002) Population viability analysis for red-cockaded woodpeckers using an individual-based model. *Ecological Applications* **12**, 249–60.

Waples, R. (1995) Evolutionarily significant units and the conservation of biological diversity under the Endangered Species Act. *American Fisheries Society Symposium* **17**, 8–27.

Waples, R.S. (2002) Definition and estimation of effective population size in the conservation of endangered species. In: *Population Viability Analysis* (eds. S.R. Beissinger and D.R. McCullough), pp. 147–68. University of Chicago Press, Chicago.

Warren, C.D., Peek, J.M., Servheen, G.L., and Zager, P. (1996) Habitat use and movement of two ecotypes of translocated caribou in Idaho and British Columbia. *Conservation Biology* **10**, 547–53.

Wasser, S.K., Shedlock, A.M., Comstock, K., Ostrander, E.A., Mutayoba, B., and Stephens, M. (2004) Assigning African elephant DNA to geographic region of origin: applications to the ivory trade. *Proceedings of the National Academy of Sciences USA* **101**, 14847–52.

Wedekind, C. (2002) Manipulating sex ratios for conservation: short-term risks and long-term benefits. *Animal Conservation* **5**, 13–20.

Weimerskirch, H., Brothers, N., and Jouventin, P. (1997) Population dynamics of Wandering albatross *Diomedea exulans* and Amsterdam albatross *D. amsterdamensis* in the Indian Ocean and their relationships with long-line fisheries: conservation implications. *Biological Conservation* **79**, 257–70.

Weldon, A.J. and Haddad, N.M. (2005) The effects of patch shape on indigo buntings: evidence for an ecological trap. *Ecology* **86**, 1422–31.

Welsh, Jr., H.H. and Droege, S. (2001) A case for using plethodontid salamanders for monitoring biodiversity and ecosystem integrity of North American forests. *Conservation Biology* **15**, 558–69.

Westemeier, R.L., Brawn, J.D., Simpson, S.A. et al. (1998) Tracking the long-term decline and recovery of an isolated population. *Science* **282**, 1695–8.

White, C. (2005) Hunters ring dinner bell for ravens: experimental evidence of a unique foraging strategy. *Ecology* **86**, 1057–60.

White, G.C. (2000) Population viability analysis: data requirements and essential analyses. In: *Research Techniques in Animal Ecology: Controversies and Consequences* (eds. L. Boitani and T.K. Fuller), pp. 288–331. Columbia University Press, New York.

White, G.C. and Garrott, R.A. (1990) *Analysis of Wildlife Radio-Tracking Data*. Academic Press, San Diego, CA.

White, G.C., Anderson, D.R., Burnham, K.P., and Otis, D.L. (1982) *Capture-Recapture and Removal Methods for Sampling Closed Populations*. LA-8787-NERP. Los Alamos National Laboratory, Los Alamos, NM.

White, G.C., Freddy, D.J., Gill, R.B., and Ellenberger, J.H. (2001) Effect of adult sex ratio on mule deer and elk productivity in Colorado. *Journal of Wildlife Management* **65**, 543–51.

White, G.C., Franklin, A.B., and Shenk, T.M. (2002) Estimating parameters of PVA models from data on marked animals. In: *Population Viability Analysis* (eds. S.R. Beissinger and D.R. McCullough), pp. 169–290. University of Chicago Press, Chicago.

White, P.J. and Garrott, R.A. (1999) Population dynamics of kit foxes. *Canadian Journal of Zoology* **77**, 486–93.

Whitman, K., Starfield, A.M., Quadling, H.S., and Packer, C. (2004) Sustainable trophy hunting of African lions. *Nature* **428**, 175–8.

Wiens, J.A. (1996) Wildlife in patchy environments: metapopulations, mosaics, and management. In: *Metapopulations and Wildlife Conservation* (ed. D.R. McCullough), pp. 53–84. Island Press, Washington DC.

Wilbur, H.M. (1997) Experimental ecology of food webs: complex systems in temporary ponds. *Ecology* **78**, 2279–302.

Wilcove, D.S., Rothstein, D., Dubow, J., Phillips, A., and Losos, E. (1998) Quantifying threats to imperiled species in the United States: assessing the relative importance of habitat destruction, alien species, pollution, overexploitation, and disease. *BioScience* **48**, 607–15.

Wiles, G.J., Bart, J., Beck, Jr., R.E., and Aguon, C.F. (2003) Impacts of the brown tree snake: patterns of decline and species persistence in Guam's avifauna. *Conservation Biology* **17**, 1350–60.

Williams, B.K., Johnson, F.A., and Wilkins, K. (1996) Uncertainty and the adaptive management of waterfowl harvests. *Journal of Wildlife Management* **60**, 223–32.

Williams, B.K., Nichols, J.D., and Conroy, M.J. (2002) *Analysis and Management of Animal Populations*. Academic Press, San Diego, CA.

Williams, D.R. (2003) The health of men: structured inequalities and opportunities. *Public Health Matters* **93**, 724–31.

Williams, S.E., Bolitho, E.E., and Fox, S. (2003) Climate change in Australian tropical rainforests: an impending environmental catastrophe. *Proceedings of the Royal Society of London Series B* **270**, 1887–92.

Wilson, E.O. (1987) The little things that run the world: the importance and conservation of invertebrates. *Conservation Biology* **1**, 344–6.

Wilson, E.O. (1992) *The Diversity of Life*. The Belknap Press of Harvard University Press, Cambridge, MA.

Wilson, E.O. (1995) *Naturalist*. Warner Books, New York.

Wilson, K.R. and Anderson, D.R. (1985) Evaluation of two density estimators of small mammal population size. *Journal of Mammalogy* **66**, 13–21.

Wilson, G.A. and Rannala, B. (2003) Bayesian inference of recent migration rates using multilocus genotypes. *Genetics* **163**, 1177–91.

Winterstein, S.R., Pollock, K.H., and Bunck, C.M. (2001) Analysis of survival data from radiotelemetry studies. In: *Radio Tracking and Animal Populations* (eds. J.J. Millspaugh and J.M. Marzluff), pp. 351–83. Academic Press, San Diego, CA.

Wisdom, M.J. and Cook, J.G. (2000) North American elk. In: *Ecology and Management of Large Mammals in North America* (eds. S. Demarais and P.R. Krausman), pp. 694–735. Prentice Hall, Upper Saddle River, NJ.

Wisdom, M.J. and Mills, L.S. (1997) Sensitivity analysis to guide population recovery: prairie-chickens as an example. *Journal of Wildlife Management* **61**, 302–12.

Wisdom, M.J., Mills, L.S., and Doak, D.F. (2000) Life stage simulation analysis: estimating vital-rate effects on population growth for conservation. *Ecology* **81**, 628–41.

Wisdom, M.J., Wales, B.C., Rowland, M.M. et al. (2002) Performance of greater sage-grouse models for conservation assessment in the Interior Columbia Basin. *Conservation Biology* **16**, 1232–42.

Wittmer, H.U., Sinclair, A.R.E., and McLellan, B.N. (2005) The role of predation in the decline and extirpation of woodland caribou. *Oecologia* **144**, 257–67.

Wolf, C.M., Griffith, B., Reed, C., and Temple, S.A. (1996) Avian and mammalian translocations: update and reanalysis of 1987 survey data. *Conservation Biology* **10**, 1142–54.

Wolfe, T. (1942) *You Can't Go Home Again*. Sun Dial Press, New York.

Wolff, J.O. (2003) An evolutionary and behavioral perspective on dispersal and colonization of mammals in fragmented landscapes. In: *Mammal Community Dynamics: Management and Conservation in the Coniferous Forests of Western North America* (eds. C.J. Zabel and R.G. Anthony), pp. 614–30. Cambridge University Press, Cambridge.

Woodroffe, R. and Ginsberg, J.R. (1998) Edge effects and the extinction of populations inside protected areas. *Science* **280**, 2126–8.

Woods, J.G., Paetkau, D., Lewis, D., McLellan, B.N., Proctor, M., and Strobeck, C. (1999) Genetic tagging of free-ranging black and brown bears. *Wildlife Society Bulletin* **27**, 616–27.

Woods, M., McDonald, R.A., and Harris, S. (2003) Predation of wildlife by domestic cats *Felis catus* in Great Britain. *Mammal Review* **33**, 174–88.

Wright, G.J., Peterson, R.O., Smith, D.W., and Lemke, T.O. (2006) Selection of northern Yellowstone elk by gray wolves and hunters. *Journal of Wildlife Management* **70** (in press).

Wright, S. (1931) Evolution in Mendelian populations. *Genetics* **16**, 97–159.

Wright, S. (1969) *The Theory of Gene Frequencies*. Evolution and the Genetics of Populations, vol. 2. University of Chicago Press, Chicago.

Wright, T. (1998) Sampling and census 2000: the concepts. *American Scientist* **86**, 245–53.

Yoccoz, N.G. (1991) Use, overuse, and misuse of significance tests in evolutionary biology and ecology. *The Bulletin of the Ecological Society of America* **72**, 106–11.

Zager, P., Mills, L.S., Wakkinen, W., and Tallmon, D. (1995) Woodland caribou: a conservation dilemma. *Endangered Species Update* **12**, 1–4.

Zegers, G. (2000) Genetic variability and resistance to infectious disease with particular emphasis on the major histocompatibility complex in the valley pocket gopher. PhD thesis, University of California, Santa Cruz, CA.

Species lists

Common Names

adder	*Vipera berus*
African elephant	*Loxodonia africana*
African lion	*Panthera leo*
African wild dog	*Lycaon pictus*
Allegheny woodrat	*Neotoma magister*
alpine ibex	*Capra ibex*
American beaver	*Castor canadensis*
American black bear	*Ursus americanus*
American ginseng	*Panax quinquefolius*
American peregrine falcon	*Falco peregrinus anatum*
American pronghorn	*Antilocapra americana*
arctic fox	*Alopex lagopus*
arctic ground squirrel	*Spermophilus parryii plesius*
Australian black duck	*Anas superciliosa rogersi*
Australian possum	*Trichosurus vulpecula*
bactrian camel	*Camelus bactrianus*
bald eagle	*Haliaeetus leucocephalus*
bank vole	*Clethrionomys glareolus*
barn owl	*Tyto alba*
barred owl	*Strix varia*
bawbaw frog	*Philoria frosti*
Bay checkerspot butterfly	*Euphydryas editha bayensis*
bighorn sheep	*Ovis canadensis*
bison	*Bison bison*
black brant	*Branta bernicla nigricans*
Black Death	*Yersinia pestis*
black rat	*Rattus rattus*
black rhinoceros	*Diceros bicornis*
black-footed ferret	*Mustela nigripes*
black-footed rock-wallaby	*Petrogale lateralis*
black-headed gull	*Larus ridibundus*
black-tailed prairie dog	*Cynomys ludovicianus*
black-throated blue warbler	*Dendroica caerulescens*
blind mole-rat	*Cryptomys damarensis*

blue-breasted fairy-wren	*Malurus pulcherrimus*
blue-footed booby	*Sula nebouxii*
bobcat	*Lynx rufus*
bobwhite quail	*Colinus virginianus*
brown bear	*Ursus arctos*
brown tree snake	*Boiga irregularis*
brown treecreeper	*Climacteris picumnus*
brown-headed cowbird	*Molothrus ater*
bubonic plague	*Yersinia pestis*
buffalo	*Syncerus caffer*
buffalo wolf	*Canis lupus*
bufflehead	*Bucephala albeola*
bush-tailed phascogale	*Phascogale tapoatafa*
cactus finch	*Geospiza scandens*
California condor	*Gymnogyps californianus*
California sea lion	*Zalophus californianus*
California spotted owl	*Strix occidentalis occidentalis*
Canada lynx	*Lynx canadensis*
cane toad	*Bufo marinus*
canine distemper	*Morbillivirus* spp.
canine parvovirus	*Parvovirus* spp.
canvasback	*Aythya valisineria*
caribou	*Rangifer tarandus*
carnivorous starfish	*Pisaster* spp.
Carolina chickadee	*Poecile carolinensis*
Channel Island foxes	*Urocyon littoralis*
Chatham Island black robin	*Petroica traversi*
cheetah	*Acinonyx jubatus*
collared dove	*Streptopelia decaocto*
Columbia spotted frog	*Rana luteiventris*
common chimpanzee	*Pan troglodytes*
common frog	*Rana temporaria*
common periwinkle	*Littorina littorea*
common raven	*Corvus corax*
common tuatara	*Sphenodon punctatus*
Concho water snake	*Nerodia paucimaculata*
cordgrass	*Spartina* spp.
cotton fungus	*Saprolegnia ferax*
cougar	*Felis concolor*
	Puma concolor
coyote	*Canis latrans*
crab-eating fox	*Cerdocyon thous*
Cryan's buckmoths	*Hemileuca* spp.
Cunningham's skink	*Egernia cunninghami*
deer mouse	*Peromyscus maniculatus*
dingo	*Canis lupus dingo*
dogs	*Canis* spp.
domestic cat	*Felis silvestris catus*
Douglas fir	*Pseudotsuga menziesii*
dusky seaside sparrow	*Ammodramus maritimus nigrescens*
	Ammospiza maritima nigrescens

dwarf huckleberry	*Vaccinium caespitosum*
dwarf mongoose	*Helogale parvula*
eastern barred bandicoot	*Perameles gunni*
eastern collared lizard	*Crotaphytus collaris collaris*
Edith's checkerspot butterfly	*Euphydryas editha*
elephant seal	*Mirounga* spp.
elk	*Cervus elaphus*
emus	*Dromaius* spp.
European blue tit	*Parus caeruleus*
European mink	*Mustela lutreola*
European rabbit	*Oryctolagus cuniculus*
fawn-footed Mosaic-tailed rat	*Melomys cervinipes*
feral pig	*Sus scrofa*
fire ants	*Solenopsis* spp.
fisher	*Martes pennanti*
flightless ground beetle	*Carabus violaceus*
Florida manatee	*Trichechus manatus latiostris*
Florida mottled duck	*Anas fulvigula fulvigula*
Florida panther	*Felis concolor coryi*
flying foxes	*Pteropus* spp.
flying squirrel	*Glaucomys sabrinus*
	Glaucomys volans
fruit bats	*Pteropus* spp.
fruit fly	*Drosophila* spp.
gecko	*Oedura reticulata*
giant African snail	*Achatina fulica*
glanville fritillary butterfly	*Melitaea cinxia*
goats	*Capra* spp.
golden eagle	*Aquila chrysaetos*
Golden lion tamarin	*Leontopithecus rosalia*
grand skink	*Oligosoma grande*
grassy stunt virus	*Tenuivirus* spp.
gray fox	*Urocyon cinereoargenteus*
gray partridge	*Perdix perdix*
gray squirrel	*Sciurus carolinensis*
gray whale	*Eschrichtius robustus*
gray wolf	*Canis lupus*
greater prairie-chicken	*Tympanuchus cupido pinnatus*
greater sage grouse	*Centrocercus urophasianus*
greater white-toothed shrew	*Crocidura russula*
grizzly bear	*Ursus arctos horribilis*
ground squirrel	*Spermophilus parryii*
Guácimo Colorado tree	*Luehea seemannii*
Gulf Coast beach mouse	*Peromyscus polionotus*
Hawaiian duck	*Anas wyvilliana*
Hawaiian honeycreepers	Drepanididae
headwater salamander	*Gyrinophilus porphyriticus*
horse	*Equus caballus*
house finch	*Carpodacus mexicanus*
house mouse	*Mus musculus*
humpback whale	*Megaptera novaeangliae*

Hutton's shearwater	*Puffinus huttoni*
Iberian lynx	*Lynx pardinus*
indigo bunting	*Passerina cyanea*
Indri	*Indri indri*
island fox	*Urocyon littoralis*
jaguar	*Panthera onca*
kakapo	*Strigops habroptilus*
Kirtland's warbler	*Dendroica kirtlandii*
Koala	*Phascolarctos cinereus*
largemouth bass	*Micropterus salmoides*
Laysan finch	*Telespiza cantans*
Laysan teal	*Anas laysanensis*
Leadbeater's possum	*Gymnobelideus leadbeateri*
leaf-mining moth	*Cameraria hamadryadella*
Least weasel	*Mustela nivalis nivalis*
leopards	*Panthera pardus* sspp.
loggerhead sea turtle	*Caretta caretta*
Macquarie parakeet	*Cyanoramphus novaezelandiae erythrotis*
mallard	*Anas platyrhynchos*
marbled murrelet	*Brachyramphus marmoratus*
Mariana crow	*Corvus kubaryi*
martens	*Martes* spp.
meadow vole	*Microtus pennsylvanicus*
merganser	*Mergus merganser*
Mexican jay	*Aphelocoma ultramarina*
midwife toad	*Alytes muletensis*
monk parakeet	*Myiopsitta monachus*
moose	*Alces alces*
mormon cricket	*Anabrus simplex*
mountain goat	*Oreamnos americanus*
mountain gorilla	*Gorilla gorilla*
mountain lion	*Felis concolor*
mourning dove	*Zenaida macroura*
Mule deer	*Odocoileus hemionus*
muriquis	*Brachyteles* spp.
muskrat	*Ondatra zibethicus*
New Zealand grey duck	*Anas superciliosa superciliosa*
North American mink	*Mustela vison*
North American red squirrel	*Tamiasciurus hudsonicus*
North Sea cod	*Gadus morhua*
northern bobwhite quail	*Colinus virginianus*
northern flying squirrel	*Glaucomys volans*
northern hairy-nosed wombat	*Lasiorhinus krefftii*
northern pintail	*Anas acuta*
northern right whale	*Eubalaena glacialis*
northern spotted owl	*Strix occidentalis caurina*
Norway rat	*Rattus norvegicus*
ocelot	*Leopardus pardalus*
oriental white-backed vulture	*Gyps bengalensis*
oropendolas	*Psarcolius* spp.
Otago skink	*Oligosoma otagense*

o'u	*Psittirostra psittacea*
Pacific rat	*Rattus exulans*
Pacific salmon	*Onchorynchus* spp.
painted turtle	*Chrysemys picta*
passenger pigeon	*Ectopistes migratorium*
pelican	*Pelecanus* spp.
plethodontid salamanders	*Plethodon* spp.
pocket gopher	*Thomomys bottae*
Polar bear	*Ursus maritimus*
prairie dogs	*Cynomys* spp.
prairie-chickens	*Tympanuchus* spp.
predatory whelks	*Concholepas* spp.
prickly forest skink	*Gnypetoscincus queenslandiae*
puma	*Felis concolor*
rabies	*Lyssavirus* spp.
raccoon	*Procyon lotor*
raccoon roundworm	*Baylisascaris procyonis*
ravens	*Corvus* spp.
red deer	*Cervus elaphus*
red fox	*Vulpes vulpes*
red grouse	*Lagopus lagopus scoticus*
red kangaroo	*Macropus rufus*
red squirrel	*Sciurus vulgaris*
red wolf	*Canis rufus*
red-cockaded woodpecker	*Picoides borealis*
red-legged frog	*Rana aurora*
reindeer	*Rangifer tarandus*
rinderpest virus	*Morbillivirus* spp.
ruddy duck	*Oxyura jamaicensis*
ruffed grouse	*Bonasa umbellus*
rufous hare-wallaby	*Lagorchestes hirsutus*
saber-toothed cat	*Smilodon fatalis*
sachem skipper butterfly	*Atalopedes campestris*
Sage grouse	*Centrocercus urophasianus*
Saiga antelope	*Saiga tatarica tatarica*
San Joaquin kit fox	*Vulpes macrotis mutica*
San Nicholas Island fox	*Urocyon littoralis*
scarlet tanager	*Piranga olivacea*
sea otter	*Enhydra lutris*
sea urchins	*Strongylocentrotus* spp.
seaside sparrow	*Ammodramus maritimus*
sharp-tailed grouse	*Tympanuchus phasianellus*
sheep	*Ovis* spp.
ship rat	*Rattus rattus*
Short-eared owl	*Asio flammeus*
Short-necked turtle	*Emydura macquarii*
Short-tailed albatross	*Phoebastria albatrus*
Siberian polecat	*Mustela eversmanni*
silvery minnow	*Hybopgnathus nuchalis*
Snail kite	*Rostrhamus sociabilis*
snowshoe hare	*Lepus americanus*

Soay sheep	*Ovis aries*
Song sparrow	*Melospiza melodia*
Sooty shearwater	*Puffinus griseus*
southern blue fin tuna	*Thunnus maccoyii*
southern flying squirrel	*Glaucomys sabrinus*
sparrowhawk	*Accipiter nisus*
spotted hyena	*Crocuta crocuta*
spotted knapweed	*Centaurea maculosa*
spotted owl	*Strix occidentalis*
starling	*Sturnus vulgaris*
Stoat	*Mustela erminea*
Stock dove	*Columba oenas*
striped skunk	*Mephitis mephitis*
Texas panther	*Felis concolor stanleyana*
Thomson's gazelle	*Gazella thomsonii*
Tiger	*Panthera tigris*
tree swallow	*Tachycineta bicolor*
tuataras	*Sphenodon* spp.
viperine snake	*Natrix maura*
weasels	*Mustela* spp.
western grey kangaroo	*Macropus fuliginosus*
western red-backed vole	*Clethrionomys californicus*
	Myodes californicus
western toad	*Bufo boreas*
white-crowned sparrow	*Zonotrichia leucophrys*
white-fronted bee eater	*Merops bullockoides*
white-tailed deer	*Odocoileus virginianus*
white-throated sparrow	*Zonotrichia albicollis*
wild maize	*Zea diploperennis*
wild turkey	*Meleagris gallopavo*
wildebeest	*Connochaetes taurinus*
willow	*Salix* spp.
willow ptarmigan	*Lagopus lagopus*
wolverine	*Gulo gulo*
Wood duck	*Aix sponsa*
Wood thrush	*Hylocichla mustelina*
Wood warblers	Parulidae
woodland caribou	*Rangifer tarandus caribou*
yellow-bellied marmot	*Marmota flaviventris*

Scientific Names

Accipiter nisus	sparrowhawk
Achatina fulica	giant African snail
Acinonyx jubatus	cheetah
Aix sponsa	wood duck
Alces alces	moose
Alopex lagopus	arctic fox
Alytes muletensis	midwife toad
Ammodramus maritimus	seaside sparrow

Ammodramus maritimus nigrescens	dusky seaside sparrow
Ammospiza maritima nigrescens	dusky seaside sparrow
Anabrus simplex	mormon cricket
Anas acuta	northern pintail
Anas fulvigula fulvigula	Florida mottled duck
Anas laysanensis	Laysan teal
Anas platyrhynchos	mallard
Anas superciliosa rogersi	Australian black duck
Anas superciliosa superciliosa	New Zealand grey duck
Anas wyvilliana	Hawaiian duck
Anatidae	ducks, geese, swans
Anolis spp.	lizards
Antilocapra americana	American pronghorn
Aphelocoma ultramarina	Mexican jay
Aquila chrysaetos	golden eagle
Asio flammeus	short-eared owl
Atalopedes campestris	sachem skipper butterfly
Aythya valisineria	canvasback
Baylisascaris procyonis	raccoon roundworm
Bison bison	bison
Boiga irregularis	brown tree snake
Bonasa umbellus	ruffed grouse
Brachyramphus marmoratus	marbled murrelet
Brachyteles spp.	muriquis
Branta bernicla nigricans	black brant
Bucephala albeola	bufflehead
Bufo boreas	western toad
Bufo marinus	cane toad
Camelus bactrianus	bactrian camel
Cameraria hamadryadella	leaf-mining moth
Canis spp.	dogs
Canis latrans	coyote
Canis lupus	gray wolf
Canis lupus dingo	dingo
Canis rufus	red wolf
Capra spp.	goats
Capra ibex	alpine ibex
Carabus violaceus	flightless ground beetle
Caretta caretta	loggerhead sea turtle
Carpodacus mexicanus	house finch
Castor canadensis	American beaver
Centaurea maculosa	spotted knapweed
Centrocercus urophasianus	greater sage grouse
Cerdocyon thous	crab-eating fox
Cervus elaphus	elk, red deer
Chrysemys picta	painted turtle
Clethrionomys spp.	voles
Clethrionomys californicus	western red-backed vole
Clethrionomys glareolus	bank vole
Climacteris picumnus	brown treecreeper
Colinus virginianus	northern bobwhite quail

Columba oenas	stock dove
Columbidae	doves and pigeons
Concholepas spp.	predatory whelks
Connochaetes taurinus	wildebeest
Corvus spp.	ravens
Corvus corax	common raven
Corvus kubaryi	Mariana crow
Crocidura russula	greater white-toothed shrew
Crocuta crocuta	spotted hyena
Crotaphytus collaris collaris	eastern collared lizard
Cryptomys damarensis	blind mole-rat
Cyanoramphus novaezelandiae erythrotis	Macquarie parakeet
Cynomys spp.	prairie dogs
Cynomys ludovicianus	black-tailed prairie dog
Dendroica caerulescens	black-throated blue warbler
Dendroica kirtlandii	Kirtland's warbler
Diceros bicornis	black rhinoceros
Diomedea spp.	great albatrosses
Drepanididae	Hawaiian honeycreepers
Dromaius spp.	emus
Drosophila spp.	fruit flies
Ectopistes migratorium	passenger pigeon
Egernia cunninghami	Cunningham's skink
Emydura macquarii	short-necked turtle
Enhydra lutris	sea otter
Equus caballus	horse
Eschrichtius robustus	gray whale
Eubalaena glacialis	northern right whale
Eumeces spp.	skinks
Euphydryas editha	Edith's checkerspot butterfly
Euphydryas editha bayensis	Bay checkerspot butterfly
Falco peregrinus anatum	American peregrine falcon
Felis concolor	cougar
Felis concolor coryi	Florida panther
Felis concolor stanleyana	Texas panther
Felis silvestris catus	domestic cat
Gadus morhua	North Sea cod
Gazella thomsonii	Thomson's gazelle
Geospiza scandens	cactus finch
Glaucomys sabrinus	southern flying squirrel
Glaucomys volans	northern flying squirrel
Gnypetoscincus queenslandiae	prickly forest skink
Gorilla gorilla	mountain gorilla
Gulo gulo	wolverine
Gymnobelideus leadbeateri	Leadbeater's possum
Gymnogyps californianus	California condor
Gyps bengalensis	oriental white-backed vulture
Gyrinophilus porphyriticus	headwater salamander
Haliaeetus leucocephalus	bald eagle
Helogale parvula	dwarf mongoose
Hemileuca spp.	Cryan's buckmoths

Hybopgnathus nuchalis	silvery minnow
Hylocichla mustelina	wood thrush
Indri indri	entrina
Lagopus lagopus	willow ptarmigan
Lagopus lagopus scoticus	red grouse
Lagorchestes hirsutus	rufous hare-wallaby
Larus ridibundus	black-headed gull
Lasiorhinus krefftii	northern hairy-nosed wombat
Leontopithecus rosalia	Golden lion tamarin
Leopardus pardalus	ocelot
Lepus americanus	snowshoe hare
Littorina littorea	common periwinkle
Loxodonia africana	African elephant
Luehea seemannii	Guácimo Colorado tree
Lycaon pictus	African wild dog
Lynx canadensis	Canada lynx
Lynx pardinus	Iberian lynx
Lynx rufus	bobcat
Lyssavirus spp.	rabies
Macropus spp.	kangaroos
Macropus fuliginosus	western grey kangaroo
Macropus rufus	red kangaroo
Malurus pulcherrimus	blue-breasted fairy-wren
Marmota flaviventris	yellow-bellied marmot
Martes spp.	martens
Martes pennanti	fisher
Megaptera novaeangliae	humpback whale
Meleagris gallopavo	wild turkey
Melitaea cinxia	glanville fritillary butterfly
Melomys cervinipes	fawn-footed melomys
Melospiza melodia	song sparrow
Mephitis mephitis	striped skunk
Mergus merganser	merganser
Merops bullockoides	white-fronted bee eater
Micropterus salmoides	largemouth bass
Microtus spp.	voles
Microtus pennsylvanicus	meadow vole
Mirounga spp.	elephant seal
Molothrus ater	brown-headed cowbird
Morbillivirus spp.	canine distemper, rinderpest virus
Mus domesticus	house mouse
Mus musculus	house mouse
Mustela spp.	weasels
Mustela erminea	stoat
Mustela eversmanni	Siberian polecat
Mustela lutreola	European mink
Mustela nigripes	black-footed ferret
Mustela nivalis nivalis	least weasel
Mustela vison	North American mink
Myiopsitta monachus	monk parakeet
Natrix maura	viperine snake

Neotoma magister	Allegheny woodrat
Nerodia paucimaculata	Concho water snake
Odocoileus hemionus	mule deer
Odocoileus virginianus	white-tailed deer
Oedura reticulata	gecko
Oligosoma grande	grand skink
Oligosoma otagense	Otago skink
Onchorynchus spp.	Pacific salmon
Ondatra zibethicus	muskrat
Oreamnos americanus	mountain goat
Oryctolagus cuniculus	European rabbit
Ovis spp.	sheep
Ovis aries	domestic sheep, Soay sheep
Ovis canadensis	bighorn sheep
Oxyura jamaicensis	ruddy duck
Pan troglodytes	common chimpanzee
Panax quinquefolius	American ginseng
Panthera leo	African lion
Panthera onca	jaguar
Panthera pardus sspp.	leopards
Panthera tigris	tiger
Parulidae	wood warblers
Parus caeruleus	European blue tit
Parvovirus spp.	canine parvovirus
Passerina cyanea	indigo bunting
Pelecanus spp.	pelican
Perameles gunni	eastern barred bandicoot
Perdix perdix	grey partridge
Peromyscus maniculatus	deer mouse
Peromyscus polionotus	Gulf Coast beach mouse
Petrogale lateralis	black-footed rock-wallaby
Petroica traversi	Chatham Island black robin
Phascogale tapoatafa	bush-tailed phascogale
Phascolarctos cinereus	koala
Philoria frosti	bawbaw frog
Phoebastria albatrus	short-tailed albatross
Picoides borealis	red-cockaded woodpecker
Piranga olivacea	scarlet tanager
Pisaster spp.	carnivorous starfish
Plethodon spp.	plethodontid salamanders
Plethodontidae	plethodontid salamanders
Poecile carolinensis	Carolina chickadee
Procyon lotor	raccoon
Psarcolius spp.	oropendolas
Pseudotsuga menziesii	Douglas fir
Psittirostra psittacea	o'u
Pteropodidae	flying foxes
Pteropus spp.	flying foxes
Puffinus griseus	sooty shearwater
Puffinus huttoni	Hutton's shearwater
Puma concolor	cougar

Rana aurora	red-legged frog
Rana luteiventris	Columbia spotted frog
Rana temporaria	common frog
Rangifer tarandus	caribou, reindeer
Rangifer tarandus caribou	woodland caribou
Rattus exulans	Pacific rat
Rattus norvegicus	Norway rat
Rattus rattus	black rat
Rostrhamus sociabilis	snail kite
Saiga tatarica tatarica	saiga antelope
Salix spp.	willow
Salmo salar	salmon
Saprolegnia ferax	cotton fungus
Sciurus carolinensis	gray squirrel
Sciurus vulgaris	red squirrel
Smilodon fatalis	saber-toothed cat
Solenopsis spp.	fire ants
Solenopsis invicta	fire ant
Solenopsis richteri	fire ant
Spartina spp.	cordgrass
Spermophilus parryii	ground squirrel
Spermophilus parryii plesius	arctic ground squirrel
Sphenodon spp.	tuataras
Sphenodon punctatus	common tuatara
Streptopelia decaocto	collared dove
Strigops habroptilus	kakapo
Strix occidentalis	spotted owl
Strix occidentalis caurina	northern spotted owl
Strix occidentalis occidentalis	California spotted owl
Strix varia	barred owl
Strongylocentrotus spp.	sea urchins
Sturnus vulgaris	starling
Sula nebouxii	blue-footed booby
Sus scrofa	feral pig
Syncerus caffer	buffalo
Tachycineta bicolor	tree swallow
Tamiasciurus hudsonicus	North American red squirrel
Telespiza cantans	Laysan finch
Tenuivirus spp.	grassy stunt virus
Thermus aquaticus	bacterium
Thomomys bottae	pocket gopher
Thunnus maccoyii	southern blue fin tuna
Trichechus manatus latiostris	Florida manatee
Trichosurus vulpecula	Australian possum
Tympanuchus spp.	prairie-chickens
Tympanuchus cupido pinnatus	greater prairie-chicken
Tympanuchus phasianellus	sharp-tailed grouse
Tyto alba	barn owl
Urocyon cinereoargenteus	grey fox
Urocyon littoralis	island fox
Urophora spp.	gall flies

Ursus americanus	American black bear
Ursus arctos	brown bear
Ursus arctos horribilis	grizzly bear
Ursus maritimus	polar bear
Vaccinium caespitosum	dwarf huckleberry
Vipera berus	adder
Vulpes macrotis mutica	San Joaquin kit fox
Vulpes vulpes	red fox
Yersinia pestis	Black Death, bubonic plague
Zalophus californianus	California sea lion
Zea diploperennis	wild maize
Zenaida macroura	mourning dove
Zonotrichia albicollis	white-throated sparrow
Zonotrichia leucophrys	white-crowned sparrow

Subject index

Page numbers in *italics* indicate figures, those in **bold** indicate tables.

50–500 rule, 250

A (adenine), 45, 47
Aanes, S., 297
absolute changes, 94
abundance
  absolute changes, 94
  determination, 59, 93
  and dispersal, 202
  factors affecting, 92, 214
  and habitats, 214
  and harvesting, 297
  prey, 160
  relative, 215
  sampling, 99–100
  species, 4
  threshold, 297
  see also effective population size
abundance change *see* population growth
abundance estimation, 19, **48**, 56, 59–74, 257
  canonical approach, 60–1, 65
  capture–mark–recapture models, 66–73
  vs. censuses, 60–3
  distance-sampling methods, 63–6
  equations, 60–1
  vs. indices, 60–3
  Jolly–Seber models, 72–3
  mark–recapture, 211
  natural logarithm plots, 109
  noninvasive genetic sampling, 57
  sample variance, 296–7
  and stochasticity, 296–7
  time series, *100, 258*
  transect methods, 63, *64*, 87
  waterfowl, *100*
abundance estimators, removal, 66n
abundance indices, 62–3

Accipiter nisus (sparrowhawk), population
    recovery, 240
accuracy, 21–4, 271
Achatina fulica (giant African snail),
    infestations, 237
Acinonyx jubatus (cheetah), 199, 263
  predators, 173–4
actual population size, 250
adaptation, 245
  and genetic variation, 176–7
  introduced species, 234
  native species, 234
adaptive harvest management (AHM),
    307
  principles, 304–5
  stages, 305
  waterfowl, 303–5
adder *see Vipera berus* (adder)
additive mortality, 172, 287
  and hunting, 288–91
  and survival rates, 288–90
additive predation, 172
adenine (A), 45, 47
aerial surveys, 66, 67, *100*, 303
Africa
  cheetahs, 199
  chimpanzees, 276
  doves, 233
  gorillas, 276
  ivory trade, 288
  rhinoceroses, 278
  rinderpest virus, 236
  umbrella species, 278–9
  wild meat, 241
African elephant *see Loxodonia africana*
    (African elephant)
African lion *see Panthera leo* (African lion)